# Design, Modeling and Control of Aerial Robots for Physical Interaction and Manipulation

Von der Fakultät Konstruktions-, Produktions- und Fahrzeugtechnik der Universität Stuttgart zur Erlangung der Würde eines Doktor-Ingenieurs (Dr.-Ing.) genehmigte Abhandlung

Vorgelegt von

Burak Yüksel

aus Giresun

| | |
|---|---|
| Hauptberichter: | Prof. Dr.-Ing. Frank Allgöwer |
| Mitberichter: | Prof. Dr. Heinrich H. Bülthoff |
| | HDR. Dr. Antonio Franchi |
| Vorsitzender: | Prof. Dr.-Ing. Prof. Dr. E.h. Peter Eberhard |

Tag der mündlichen Prüfung: 17. Februar 2017

Institut für Systemtheorie und Regelungstechnik

Universität Stuttgart

2017

Bibliografische Information der Deutschen Nationalbibliothek

Die Deutsche Nationalbibliothek verzeichnet diese Publikation in der
Deutschen Nationalbibliografie; detaillierte bibliografische Daten sind
im Internet über http://dnb.d-nb.de abrufbar.

ISBN 978-3-8325-4492-8

Logos Verlag Berlin GmbH
Comeniushof, Gubener Str. 47,
10243 Berlin
Tel.: +49 (0)30 42 85 10 90
Fax: +49 (0)30 42 85 10 92
INTERNET: http://www.logos-verlag.de

# Acknowledgements

Essence of life. This is, like many of us, what I have been trying to identify in the course of my time until this very moment, in a train to Tübingen on planet Earth, that is in its own trajectory in our solar system, somewhere in the Milky Way in our universe, presumably in one of many.

During this time I got the impression that the life can have more than just one *ethos*, as long as it collects some meaning from them and becomes more cheerful while doing so. I believe the joy of wisdom is one of them, and the people to share this joy with is another. I was lucky to have these two in my life, especially during the time of my doctorate thesis work, which finally has forged this book/manuscript/thesis.

There have been so many people during this time with whom I have met, collaborated, shared great ideas and moments, and of course drinks in most of these occasions. I would like to start expressing my finest gratitudes to Dr. Antonio Franchi, who has been the major mentor not only of my PhD, but also my personal evolution from the day-one of my thesis work. Besides benefiting from his vast knowledge in the field of robotics and control engineering, I have learned a lot from him by observing his way of approaching different real life problems.

I would like to thank both Prof. Heinrich Bülthoff and Prof. Frank Allgöwer for their sincere support for my doctorate work. I owe the quality of this work to the facilities provided by Prof. Bülthoff and the Max Planck Society; and I would never have the chance to be able to finalize and defend this PhD thesis if I wouldn't have Prof. Allgöwer's supervision.

I like to express my sincere gratitudes to another great mentor, Prof. Cristian Secchi, because of his guidance on the nonlinear control theory and his friendship whenever it was needed. He is the person who walked me through the beauty of the energetic functions as a *lingua franca* for understanding different physical systems and their characteristics. He changed the way I see the world and our universe that we know of.

I would like to thank the Eiffel Excellence Scholarship for funding part of my PhD work in Toulouse, France, where I have got the chance to work with excellent scientists of CNRS-LAAS.

Furthermore, I was blessed with the most friendly and intelligent colleagues: Nicholas Staub, Marco Tognon, Gabriele Buondonno, Dr. Carlo Masone, Dr. Paolo Stegagno, Dr. Markus Ryll, Marcin Odelga, Sujit Rajappapa, Dr. Aamir Ahmad, Christian Schenk and everyone else in the labs of *Max Plank Institut für biologisches Kybernetik/Tübingen, CNRS-LAAS/Toulouse, DIAG-Sapienza Università/Roma*. Thank you all for the good moments we had and your competence in your fields.

My dearest friends; Stefano, Carlo, Mario, Nicole, Kresh, Christiane, Katha, Lotta, Frank, Merve, Eva, Aurelié, ... and all the others who have permanent places in my heart. Over the time you have become a great part of my life. Thank you for your (hopefully everlasting) friendship.

Finally, and above all, I would like to cherish here my family: my mother Ünzile, my father Halil and my sister Büşra. A beautiful set of people who are full of love and caring for all others. I consider myself very lucky for having them, their support and endless love in my life. I am also blessed with another family I have found in Germany; Evi, Frieder and Madita. You always make me feel home.

And of course, selbstverständlich, *Menja*. The light of my life. Who would I be without you? Thank you, millions of times, for always being on my side.

Regionalbahn from Stuttgart to Tübingen, May 2017
Burak Yüksel

*Hep ilerleyeceğini umduğum insanlık için. . .*

*Yesterday I was clever and I wanted to change the world. Today I am wise, so I am changing myself.* — Mevlana Celaleddin Rumi (1207 - 1273, Konya.)

# Table of Contents

# Notation

## Frames, Scalars, Vectors and Matrices

In this thesis, we used in general the following conventions;

- scalars are presented using normal weighted characters, e.g. $a \in \mathbb{R}$,

- vectors are denoted with boldfaced symbols, e.g. $\mathbf{v} \in \mathbb{R}^n$,

- matrices are indicated with capital boldfaced symbols, e.g. $\mathbf{M} \in \mathbb{R}^{n \times n}$.

For a vector $\mathbf{v} = [v_x \ v_y \ v_z]^T \in \mathbb{R}^3$, the operator $[\mathbf{v}]_\wedge : \mathbb{R}^3 \to \mathrm{so}(3)$ performs the skew-symmetric operation as

$$[\mathbf{v}]_\wedge = \begin{bmatrix} 0 & -v_z & v_y \\ v_z & 0 & -v_x \\ -v_y & v_x & 0 \end{bmatrix} \in \mathrm{so}(3).$$

The frames are shown with $\mathcal{F}_\star$, which are always described with the point placed at the center of the frame and three principle Cartesian axes. For example, the inertial (world) frame is denoted with $\mathcal{F}_W : \{\mathrm{P_W}, \mathbf{x}_W, \mathbf{y}_W, \mathbf{z}_W\}$. For the body frame of an aerial robot, it is $\mathcal{F}_B : \{\mathrm{P_B}, \mathbf{x}_B, \mathbf{y}_B, \mathbf{z}_B\}$.

The interconnection matrix, $\mathcal{J}$, and the dissipation matrix $\mathcal{R}$ in Chapter 4 are shown differently from the usual matrix convention used in the overall thesis.

# Nomenclature

Here we list not all, but some important Greek/Latin characters used in this thesis.

| | |
|---|---|
| $\mathbb{R}^n$ | Set of real numbers in dimension of $n$ |
| $\mathbb{N}$ | Set of all natural numbers (in this thesis, including 0) |
| $\mathbb{Z}$ | Set of all integers |
| $\mathbf{P}_{\mathrm{B}}, \mathbf{P}_{\mathrm{C}_0}$ | Point placed at the Center of Mass (CoM) of the aerial robot |
| $\mathbf{P}_{\mathrm{G}}$ | Point placed at the Center of Actuation (CoA) of the aerial robot |
| $\mathbf{x}_*, \mathbf{y}_*, \mathbf{z}_*$ | Principal axes placed at point $\mathbf{P}_*$ |
| $\mathbf{q}_q \in \mathbb{R}^6$ | Generalized coordinates of $\mathbf{P}_{\mathrm{B}}$ |
| $\mathbf{p}_q \in \mathbb{R}^3$ | Position of $\mathbf{P}_{\mathrm{B}}$ in the world frame |
| $\mathbf{R} \in \mathbb{R}^{3 \times 3}$ | Orientation of the body frame ($\mathcal{F}_B$) w.r.t. the world frame ($\mathcal{F}_W$) |
| $\boldsymbol{\eta} \in \mathbb{R}^3$ | Minimal representation of $\mathbf{R}$ using *roll-pitch-yaw* convention |
| $\boldsymbol{\omega} \in \mathbb{R}^3$ | Angular velocity of the aerial robot rigid body in $\mathcal{F}_B$ |
| $\Omega_i \in \mathbb{R}$ | Spinning velocity of the $i$-th propeller |
| $\mathbf{T} \in \mathbb{R}^{3 \times 3}$ | Transformation matrix from $\dot{\boldsymbol{\eta}}$ to $\boldsymbol{\omega}$ |
| $\mathbf{M}_{qr} \in \mathbb{R}^{3 \times 3}$ | Quadrotor inertia matrix (of $\mathbf{P}_{\mathrm{B}}$ in $\mathcal{F}_B$) |
| $\mathbf{C}$ | Coriolis matrix for a considered system dynamics |
| $\mathbf{g}$ | Gravitational forces for a considered system dynamics |
| $\mathbf{G}$ | Control input matrix for a considered system dynamics |
| $\mathbf{u}$ | Control input vector for a considered system dynamics |
| $\mathbf{u}_r$ | Torque vector acting to the CoM of the quadrotor |
| $\mathbf{d}$ | Distance between two points |
| $\mathbf{w}_{ext}$ | External wrench acting on the quadrotor CoM |
| $u_t$ | Thrust intensity acting at the CoM of the aerial robot |
| $\phi, \theta, \psi$ | Roll,Pitch,Yaw angles, respectively. Also, $\phi$ : elastic deflection in Sec. 6.2.8. |
| $m_q$ | Mass of the aerial robot |
| $k_e$ | Linear elastic spring constant |
| $\omega_n$ | Natural frequency |
| $H$ | Total amount of energy (Hamiltonian) stored in a system |
| $V$ | Lyapunov candidate |
| $\lVert * \rVert, \lVert * \rVert_2$ | *2-norm* of $*$ |
| $\lvert * \rvert$ | Determinant of $*$ |

# Abbreviations

| | |
|---|---|
| **APhI** | Aerial Physical Interaction |
| **BL** | BrushLess |
| **BL-CTRL** | BrushLess-ConTROLler |
| **CAD** | Computer Aided Design |
| **CAN** | Controller Area Network |
| **CoA** | Center of Actuation |
| **CoM** | Center of Mass |
| **DAQ** | Data Acquisition Box |
| **DFL** | Dynamic Feedback Linearization |
| **DoF** | Degrees of Freedom |
| **F/T** | Force/Torque |
| **IDA-PBC** | Interconnection and Damping Assignment - Passivity Based Control |
| **IMU** | Inertial Measurement Unit |
| **MAV** | Micro Aerial Vehicle |
| **MoCap** | Motion Capture system |
| **NED** | North-East-Down |
| **ODE** | Ordinary Differential Equation |
| **PC** | Personal Computer |
| **PDE** | Partial Differential Equation |
| **PH** | Port Hamiltonian |
| **PWM** | Pulse Width Modulation |
| **PVTOL** | Planar Vertical Take-Off and Landing |
| **ROS** | Robot Operating System |
| **UAV** | Unmanned Aerial Vehicle |
| **USB** | Universal Serial Bus |
| **UKF** | Unscented Kalman Filter |
| **VTOL** | Vertical Take-Off and Landing |

# Abstract

Robots with flying capabilities, so called *aerial robots*, are essentially robotic platforms, which are autonomously controlled via some sophisticated control engineering tools. Similar to many *aerial vehicles* (e.g. fixed-wing planes or the helicopters which are already an important part of our lives since almost a century), they can overcome the gravitational forces thanks to their design and/or actuation type. What makes them different from the conventional aerial vehicles, is the level of their *autonomy*. Reducing the complexity for piloting of such robots/vehicles provide the human operator more freedom and comfort. Particularly the small size (or miniature) aerial robots (such as *quadrotors*) are becoming a bigger part of our lives, while they are rapidly advanced in the robotics society for improving our life quality. With their increasing autonomy, they can perform many complicated tasks by their own (such as surveillance, monitoring, or inspection), leaving the human operator the most high-level decisions to be made, if necessary. In this way they can be operated in hazardous and challenging environments, which might posses high risks to the human health. Thanks to their wide range of usage, the ongoing researches on these robots will have an increasing impact on the human life.

Over the past two decades, aerial robots have been extremely put in use for tasks e.g. surveillance, monitoring, filming, obstacle avoidance, etc. All these tasks had at least one thing in common: avoiding the flying robot any physical interaction with its environment. The obvious reason is because such interaction could lead the robot to a crash or to an unstable/uncontrolled scenario. This would be of course undesired, because it might finish the mission earlier than planned and even damage the fragile electronics onboard of the robot. In the time this thesis work had started, novel methods and technologies were becoming emerging needs on how to perform meaningful physical interaction tasks with aerial robots, while maintaining their stable flight.

Today, using the aerial robots for physical interaction and manipulation is a popular topic, with a great interest of many researchers. Including this thesis work, there have been various studies addressing the design, modeling and control problem of *Aerial Physical Interaction* (APhI) and *Aerial Manipulation*. A clear motivation of using aerial robots for physical interaction, is to benefit their great workspace and agility. Moreover, developing robots that can perform not only APhI but also aerial manipulation can bring the great workspace of the flying robots together with the vast dexterity of the manipulating arms. However achieving this is not only challenged by the limited technology, but also by the lack of sophisticated methods for handling the control of the system in a desired and stable way during physical interaction. It is important to note, that the APhI is still an open topic in many senses, and many studies are addressing it using different perspectives. This thesis work is one of those, which humbly tries to provide rigorous solutions to that problem using System/Control, Mechanics, Electronics and Computer Engineering tools.

In this doctoral thesis, the APhI and the *Aerial Manipulation* is studied in terms of *design*, *modeling* and *control* of aerial robots and manipulating arms; when they come together and become one system. Although fixed-wing planes, or helicopters can be (and are) considered as aerial robots, during the course of this thesis we focus on the *quadrotors*, mostly because of their accessibility and *Vertical Take-off and Landing* (VTOL) ability. Using the *nonlinear* mathematical models of the robots at hand, here we propose several different control methods for APhI and aerial manipulation tasks. Furthermore, we present novel design tools (e.g. new manipulating arms) to be used together with miniature aerial robots, and contribute to the robotics society not only in terms of theory but also practical implementation and experimental robotics.

# Deutsche Kurzfassung

# Entwurf, Modellierung und Regelung eines fliegenden Roboters für physikalische Interaktion und Manipulation

Roboter mit Flugfähigkeit, sogenannte *fliegende Roboter*, sind Plattformen, die mittels Regelungsalgorithmen autonom gesteuert werden können. Analog zu zahllosen anderen *Fluggeräten* (z.b. die Starrflügelflugzeuge oder Hubschrauber) , können fliegende Roboter die Gravitationskraft, aufgrund ihres Designs und ihres Antriebs, überwinden. Was die fliegenden Roboter jedoch von konventionellen Flugzeugen unterscheidet, ist ihre *Autonomie*. Diese Autonomie macht die Steuerung solcher Roboter einfacher und komfortabler. Insbesondere kleine fliegende Roboter, wie die Quadrotoren, werden durch die stetige Weiterentwicklung ein immer wichtigerer Teil unseres täglichen Lebens. Durch ihre steigende Autonomie, können solche Roboter eine Vielzahl komplizierter Aufgaben, wie zum Beispiel die Überwachung und Inspektion, eigenständig ausführen. Da diese Roboter für eine Vielzahl von Aufgaben eingesetzt werden können, werden sie einen immer wichtigeren Einfluss auf unser Leben haben.

In den letzten 20 Jahren wurden fliegende Roboter für eine wachsende Anzahl von Aufgaben eingesetzt. Bei all diesen Aufgaben wurde jedoch eines immer vermieden: Die direkte physische Interaktion des fliegenden Roboters mit seiner Umgebung. Der offensichtliche Grund dafür ist, dass eine solche Interaktion dazu führen kann, dass der Roboter in einen instabilen Zustand gerät. Dies kann zum Absturz und damit zur Zerstörung der empfindlichen Elektronik des Roboters führen. Zu dem Zeitpunkt als die vorliegende Doktorarbeit ihren Anfang nahm, wurden deshalb dringend neue Methoden und Technologien benötigt, die eine sinnvolle physische Interaktion, bei stabilem Flug, ermöglichten. Heute ist die physische Interaktion von fliegenden Robotern ein beliebtes Thema, das von vielen Wissenschaftlern bearbeitet wird. Neben der vorliegenden Arbeit, wurden zahlreiche Studien durchgeführt, die das Design, die Modellierung und die Regelung der *Aerial Physical Interaction* (APhI) und der *Aerial Manipulation* zum Thema hatten. Der enorme Vorteil den fliegende Roboter bei der physischen Interaktion, gegenüber traditionellen Methoden haben, ist ihr große Reichweite und ihre Beweglichkeit. Wenn zudem Roboter entwickelt werden, die nicht nur mit der Umgebung interagieren können, sondern auch in der Lage sind, diese zu manipulieren, wird die große Reichweite der Roboter mit der enormen Geschicklichkeit der Manipulator-Arme ergänzt. Die Verbindung der Roboter mit einem solchen Manipulator-Arm ist jedoch aufgrund technischer Limitierungen und dem Fehlen ausgereifter Methoden, die eine stabiles Verhalten des Gesamtsystems während der Interaktion ermöglichen, eine

Herausforderung. Und auch heute noch sind viele Fragen, die die Interaktion von fliegenden Robotern mit ihrer Umgebung betreffen, ungeklärt und das Thema verschiedenster Studien. Die Arbeit, die in der vorliegenden Doktorarbeit beschrieben wird, ist ein Teil davon und versucht einen Beitrag zur Lösung dieser offenen Fragen beizutragen.

In der vorliegenden Doktorarbeit wird sowohl die *Aerial Physical Interaction* als auch die *Aerial Manipulation* in Bezug auf das Design, die Modellierung und die Regelung von fliegenden Robotern, Manipulator-Armen und dem Gesamtsystem aus beiden untersucht. Obwohl auch Starrflügelflugzeuge und Hubschrauber als fliegende Roboter betrachtet werden können und sollten, liegt der Fokus der vorliegenden Arbeit auf den *Quadrotoren*, unter anderem wegen ihrer Zugänglichkeit und weil diese in der Lage sind vertikal zu starten und zu landen. Mittels *nonlinearer* mathematischer Modelle dieser Roboter, werden in der vorliegenden Arbeit unterschiedliche Methoden zur Regelung und Steuerung der physischen Interaktion und Manipulation vorgeschlagen. Des Weiteren werden neue Werkzeuge (z.B. neue Manipulator-Arme) vorgestellt, die mit den fliegenden Robotern kombiniert werden können. Auf diese Weise kann die vorliegende Arbeit, neben theoretischen Methoden, auch mittels praktischer Anwendungen und Experimenten zum Fortschritt der Robotik beitragen.

# Chapter 1

# Introduction

## 1.1 Aerial Robots

Today, in daily life, we are encountered with various types of robots. From transportation to manufacture, surgery, automation and in many other fields, the robots are actively involved in our lives (Siciliano and Khatib (2008)). In most of the scenarios, these robots are fixed to their base, making them so called *ground robots*. Example to such robots could be the fixed-base manipulators as deeply studied in Murray et al. (1994) and in Siciliano et al. (2009). Clearly, mobilization of the robots comes with the great advantage of their increased workspace. However, the vast number of the mobile robots are still constrained to the ground they are in contact with, e.g. the wheeled robots (see Campion and Chung (2008)) or the legged ones, e.g. bipeds, humanoids, etc (see Kajita and Espiau (2008)).

It is obvious that the robots which can freely fly have greater workspace than those which cannot. Such robots, called *aerial robots*, can overcome the pulling gravity force using mechanisms e.g. fixed-wings (see Fig. 1.1) or propellers (see Fig. 1.2). The flight of the aerial robots is achieved by the aerodynamics of the body or the actuated parts of the system. The *stable* flight of such systems is achieved using sophisticated control methods. Although the aerodynamics of such systems is crucial for a stable flight, in this thesis we study the autonomy of the flying robots and appropriate control methods for it.

In the scope of this thesis, by *aerial robots*, we will mean the autonomous systems, which can perform robotic tasks while performing a stable flight. A robotic task can be; trajectory tracking (see Mistler et al. (2001) and Lee et al. (2010)), search-and-rescue/surveillance (see ICARUS (2011- 2015) and SHERPA (2013- 2017)), manipulation (see ARCAS (2011- 2015) and AeRoArms (2015-2019)), etc. Compared to the other mobile robots, e.g. automobiles, humanoids and *Autonomous Underwater Vehicles* (AUV), the aerial robots, e.g. *Unmanned Aerial Vehicles* (UAV) or also called *Unmanned Aerial Systems* (UAS), face different physical challenges, since their floating bases need to counter-balance the gravity forces at all time. In a non-vacuum environment, e.g. in the earth atmosphere, this is mostly accomplished thanks to the difference in the air pressure. This can be achieved with a fixed-wing design, as for the airplanes, such as in Fig. 1.1. However such designs require high cruise velocities of the robot for a stable flight, which usually limits the robot for performing tasks that needs *hovering* condition, e.g. in the case of aerial manipulation. Hovering and stable flight can be achieved, if the aerial robot is flying thanks to the rotating propellers, like for the helicopters as given in Fig. 1.2.

The control of conventional helicopters is not a trivial task, even when this control is shared with an experienced pilot (see myCopter Project (Jan 2011- Dec 2014) for the research results on this matter). Sophisticated methods for full autonomous control of

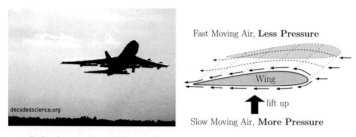

Figure 1.1: -*Left:* An airplane taking off.
-*Right:* Schematic figure of a wing of an airplane. Due to the aerodynamic design, the air below the wing moves slower than the one above, creating more pressure and hence a lifting force.

large size helicopters have been studied using simplified nonlinear dynamic models of the system, e.g. in Isidori et al. (2003). However their implementation in real robots/vehicles is still an open problem, mostly due to the large sizes of the system, which relate to their highly detailed and nonlinear dynamics (see Padfield (2007) and Ren et al. (2012)). It is important to note that the identification of the system parameters including aerodynamics effects is nontrivial as well. This makes the nonlinear mathematical models of such vehicles less reliable, hindering the implementation of sophisticated nonlinear control methods to the real systems (see Remple and Tischler (2012)).

On the other hand, it is much easier to develop advanced autonomous controllers for smaller flying robots, e.g. *Micro Aerial Vehicles* (MAV); because their mathematical models are much more reliable for developing advanced nonlinear controllers implementable into the real systems (see Cai et al. (2014) for a detailed survey on the existing MAVs). Especially the robots with stationary flight capability, e.g. *Vertical Take-off and Landing* (VTOL) vehicles are the one of the most accessible aerial robots for such implementation among the others. They are particularly interesting for us, because of their hovering capability, which upper hands them for the APhI and Aerial Manipulation tasks.

### 1.1.1 VTOLs for Aerial Robotics

There has been different types of VTOLs in the literature used for the purpose of aerial robotics. A very detailed survey is done in Cai et al. (2014) including 132 different models of small-scaled aerial vehicles used worldwide. We are particularly interested in the *Vertical Take-off and Landing* (VTOL) vehicles, and their implementation for aerial robotics. Helicopters (medium or small size) are one of those that was considered by the researchers for aerial robotics. In Naldi (2008) a detailed modeling of a miniature helicopter is explained; from its rigid body to the engine dynamics. Furthermore, different from the conventional helicopters, other types of aerial robots with hovering capabilities have also been designed and controlled. One example is the aerial robots with *ducted-fan*, which are effective designs for the applications where, the size is limited and static thrusts are important. A ducted-fan design has been developed and tested in Naldi et al. (2010), which can also use its stationary flight capacity for interacting with the environment, as described in Marconi et al. (2011). This design is consisting of one propeller and flapping mechanisms

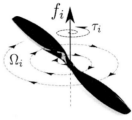

Figure 1.2: *-Left:* A Bell 206 Helicopter used by the US police.
*-Right:* A propeller produced by *HiSystems GmbH* (Mikrokopter.de) for small-sized VTOLs, e.g. quadrotors. Due to the particular shape of this propeller, when it rotates clockwise (CW) with the velocity of $\Omega_i$, it creates a thrust force $f_i$ upwards and a drag force opposite to the direction of rotation, $\tau_i$. The same notations are used in Fig. 2.1 as well.

controlled by three more inputs; making it an underactuated system with four control inputs. Another ducted-fan design is used in Hofer et al. (2016). In Papachristos et al. (2015) a *Tri–TiltRotor* design is presented, where a VTOL actuated with three propellers is realized with two front propellers can tilt together using another control input; making the system underactuated with four inputs. This design allows achieving direct actuation along the translational directions, although the standard body-pitching underactuation still remains. Implementing tilting mechanisms to the propellers of the standard VTOLs can overcome the underactuation of the system in the design phase, which is exactly what it is done in Ryll et al. (2015). There, authors developed a custom aerial VTOL robot based on a standard quadrotor (see Sec. 1.1.2) by adding an independent tilting mechanism to each propeller, which turned it into an *overactuated* aerial robot with eight control inputs (see also the middle picture of Fig. 1.3). Full actuation or even redundancy has been studied further for VTOLs by different researchers, and the examples can be found e.g. in Rajappa et al. (2015) and in Brescianini and D'Andrea (2016). Although increasing number of actuation can boost the capabilities of the VTOL system, it is intriguing to study the robots with fewer number of actuations. Studying underactuation of the mechanisms, besides an attractive challenge for the Control Engineers, can be beneficial depending on the task of the robot (e.g. the ones in Sec. 1.1.2), can allow reduced energy consumption and lighter designs, and as well as let on developing controllers accounting for some system failure, e.g. the ones studied in Mueller and D'Andrea (2016).

Over and above, the quadrotors, with four symmetrically aligned propellers all facing upwards, are one of the most used VTOL designs in the field of robotics. It is worth mentioning again, that one (but not only) important reason for this is; they are broadly available to the public, and this makes them a suited setup for generating high impact on our daily life. We dedicate the following section for its state of art.

Figure 1.3: Three different multirotor VTOL examples from the literature, used for aerial robotics. On the left, a quadrotor VTOL used also as the experimental setup in this thesis. In the middle an overactuated VTOL (see Ryll et al. (2015)). On the right, an omni-directional VTOL (see Brescianini and D'Andea (2016)).

## 1.1.2 Quadrotors

In this thesis, we chose the *Quadrotor* Vertical Take-off and Landing (VTOL) vehicles as the aerial platform to be studied and finally used in our experiments (see the first picture from left of Fig. 1.3). The main reason is that they are easily accessible to the broad range of people, they are affordable, and their underactuated dynamics create exclusive control challenges for the researchers while allowing these robots quickly accelerating along the translational directions (see Mahony et al. (2016)). A quadrotor is an aerial robot/vehicle which is lifted by four symmetrically aligned propellers, all facing upwards; opposite direction of the gravity vector (Bouabdallah et al. (2007)). Thanks to this symmetry one can formalize a very realistic mathematical model of the robot (See Mahony et al. (2012) and Chapter 2 of this thesis for the details), which finally allows developing advanced control techniques for this system. Notice that a quadrotor has *four* control inputs (the velocities of the four propellers), but it moves in the 6 Dimensional Cartesian space (3D: 3 translation and 3 rotation). Hence it is underactuated and also considering its nonlinear dynamics, its control as well as the generation of feasible trajectories for this kind of system in general is not a trivial task (Mellinger et al. (2010)). This challenge was taken by the scientists since more than a decade. In Mistler et al. (2001) the exact linearization of a quadrotor VTOL has been shown, with the flat outputs of the system. Later, a geometric tracking control for quadrotors was proposed in Lee et al. (2010). This control is developed on the special Euclidean group SE(3), allowing the quadrotor tracking complex rotational maneuvers which can exhibit singularities when they would be represented using Euler angles. In Liu et al. (2015) these technique has been improved including robustness analysis of the system (based on the work done in Schmidt et al. (2014)) and using Model Predictive Controller for trajectory planning. Turning the underactuation of the quadrotors to their advantage is nicely put in example by Mueller et al. (2011), where quadrotor is performing aggressive maneuvers for juggling a ball.

Besides control of the quadrotors, their state estimation is another challenge. Most of the works mentioned above are performed indoor and benefiting from the *Motion Capture* (MoCap) systems, allowing researchers to acquire precise (or relatively good) measurements of the robot's state. Due to the relatively small sizes of most of the quadrotors, relying only on board measurements (e.g. using cameras, Inertial Measurement Unit (IMU), etc.) means adding more hardware to the flying system, which increases its weight and reduces its load range and the flight capacity[1]. However very recent promising developments are giving the sign of overcoming this practical/technological challenge in the near future. For example, in Loianno et al. (submitted) and in Falanga et al. (submitted) authors show two different methods for performing aggressive trajectory tracking of the quadrotors using only on board sensors, where the former one can pass through gaps by tilting 90 [deg] using an off-line planning algorithm, while the latter can do the same up to 45 [deg] but performing trajectory planning in real time.

Until recently, the main field of application for the quadrotors was focused on surveillance and patrolling (Mahony and Kumar (2012)), search and rescue (SHERPA (2013- 2017)), civil monitoring and agriculture (Duggal et al. (2016)); with full autonomy or with human-in-the-loop using e.g. haptic devices (Franchi et al. (2012b)). All these examples have one thing in common; the flying robots are kind of *passive observers*, meaning that they do not actively and physically interact with their environment. Recently, a new field of application emerged in the literature, which requires the physical interaction of the aerial robots. We dedicated the next two sections for the research made in this particular area.

# 1.2 Aerial Physical Interaction (APhI)

Aerial Physical Interaction (APhI) is a case, in which the aerial robot exerts meaningful forces and torques (wrench) to its environment while preserving its stable flight. In this case, the robot does not try avoiding every obstacle in its environment, but prepare itself for embracing the effect of a physical interaction, furthermore turn this interaction into some meaningful robotic tasks. Some examples to such tasks are; surface inspection, tool operation, object transformation and manipulation. This is a relatively new research topic, promising novel theoretical and practical challenges for the researchers. In fact, by the time this thesis work has started, there were few examples of APhI in the literature, and in couple of years this has dramatically changed. An admittance control framework presented in Augugliaro and D'Andrea (2013), allowing quadrotors interact with humans, physically. The controller presented there is proposed for the partially linearized translational dynamics in near-hovering configuration of the robot, which provides a local solution in terms of physical interaction. A hybrid position and wrench control for quadrotors is presented in Bellens et al. (2012), where for dealing with poorly structured environment, an impedance control has been exploited. Authors of Gioioso et al. (2014) turned a standard near-hovering controller into a 3D force controller, and implemented it on a quadrotor for effectively exerting desired forces to its environment via a rigid tool. Using quadrotors equipped with rigid tools for APhI is further studied in Ha et al. (2015) and Nguyen and Lee (2013), where the nonlinear quadrotor dynamics is exploited for performing tool operations e.g. screw-driving. In Fumagalli et al. (2012b), researchers presented a design of a quadrotor VTOL for contact inspection purposes. The controller presented there is a *passivity-based* controller; shaping only the potential energy of the quadrotor for setting a desired stiffness behavior[2](see also Mersha et al. (2011)).

---

[1]Also with today's technology, on board state estimation is still not as good as when using external measuring methods, e.g. MoCap systems

[2]Potential energy is only one of the factors affecting the way a mechanical system interacts with the environment. Inertial properties and damping also play a major role for determining the interactive behavior. Furthermore, since the direction of the thrust of a quadrotor depends on the orientation of

In order to achieve APhI, knowledge of the interaction forces/torques (wrench) is crucial. This knowledge can be acquired either by using some (indirect) estimation methods, or directly measuring them with Force/Torque (F/T) transducers. Using estimators has advantages of avoiding additional sensors on the flying robot, reducing the computation power and also spending less money for the hardware. Examples are various; in Augugliaro and D'Andrea (2013) *Kalman filters* are used to estimate the external forces. A more general method is proposed in Ruggiero et al. (2014), where a residual *momentum-based* wrench estimator for quadrotors is presented. This method is further analyzed in Tomic and Haddadin (2014). In McKinnon and Schoellig (2016) an algorithm based on *unscented quaternion estimator* is used for estimating the external wrenches acting on a quadrotor body. During the course of this thesis, we have developed a *nonlinear Lyapunov-based* disturbance observer for estimating the external wrenches acting on the quadrotors (Yüksel et al. (2014b)). These results are summarized in Sec. 3.3 of this thesis. Notice that all the estimation methods presented so far are dependent on other sensor measurements, and they require some kind of system model. On the other hand, transducers has its own advantages, e.g. it provides reliable measurements independent of other sensors or any system model. Implementation of the F/T transducers can be done in two ways; by placing it on board of the flying robot, e.g, in Fig. 3.2, or on the surface of the interaction, e.g. in Gioioso et al. (2014). Although the former is clearly more preferable, it is not always straight-forward to implement it, due to the limited load capacity of the flying robots. In Chapter 3 we discuss this in detail, and in Sec. 3.2 we describe a setup, built as a part of this thesis, allowing us using F/T transducers on board of a flying quadrotor for measuring the interaction wrenches.

## 1.3  Aerial Manipulation

Aerial Manipulation can be considered as a subset of APhI, where the flying robot is designed and controlled in purpose of manipulating its environment. Motivating examples of aerial manipulation are; inspection and maintenance, construction and precision requiring implementations in hazardous or high-rise environments (see Fig. 1.4). Although the aerial robot itself can be used as a flying manipulator (e.g. the designs presented in Ryll et al. (2015), Ryll et al. (2016) would be suitable choices), in the most common scenarios the overall flying robot is a *binomial* of a flying base, and one or multiple mechanisms used for manipulation, e.g. manipulating arms (see Fig. 1.5 for some examples). Such designs have been adopted by the researchers so far, because it combines the flying capability of an aerial vehicle with its vast workspace and the dexterity of a manipulating arm (or even multiple arms). In Thomas et al. (2013) authors presented a quadrotor equipped with a 1 DoF rigid arm, performing an aerial grasping task. This is an agile aerial manipulation task, performed thanks to the differential flatness property of the system (we deeply studied this as well and extended this work to many directions in Chapter 6). In Yang and Lee (2014) a generic model of a quadrotor equipped with a manipulating arm is studied, where a passive-decomposition method is implemented for controlling the Center of Mass (CoM) positions of the overall system, and then a back-stepping like controller is proposed for regulating

---

the system, it is not sufficient to shape the Cartesian impedance for achieving an effective control of interaction. In light of this, we improved this controller in Yüksel et al. (2014a), which is also explained in Chapter 4 of this thesis.

Figure 1.4: A motivating real-life challenge for enabling aerial robots to perform APhI and manipulation tasks in hazardous environments. The most left figure shows pictures from such environments as the motivating challenges. In the middle today's practice is depicted. The most right figure shows the visions of three different European Projects: ARCAS (2011-2015), AeRoArms (2015-2019) and AEROworks.

the end-effector motions. In Ruggiero et al. (2016) a 6 DoF servo manipulating arm is used for aerial manipulation, on board of a multirotor VTOL, which is an eight-rotor aircraft in coaxial configuration[3] (see also the middle picture of Fig. 1.5). Such designs clearly increase the dexterity of the aerial manipulator, while increasing the complexity of the system. With increasing capacity of the flying platform, one can improve the effectiveness of the aerial manipulation. In Kondak et al. (2014) a double-rotor helicopter equipped with a KUKA LWR 7 DoF manipulating arm is presented, where the overall system can perform aerial grasping in outdoor (see also third picture of Fig. 1.5). A valve turning application is performed in Korpela et al. (2014), where a quadrotor is equipped with a dual-arm mechanism. In Garimella and Kobilarov (2015) a model-predictive controller is adopted for an aerial manipulation task.

The most of the manipulating arm designs studied in the literature are rigidly actuated, i.e. the actuator of the individual joints are rigidly attached to their links. To our best knowledge, a compliant (or elastic) manipulating arm on a quadrotor has been first time considered in Yüksel et al. (2015), where design, modeling and identification of a light-weight elastic-joint arm has been used for link velocity amplification and aerial physical interaction (also see Chapter 5). This study has been improved in different directions, which is a part of this thesis work and given in Chapter 6.

Alternatively, in Nguyen et al. (2015) a framework for multiple quadrotors connected to a tool is shown, where the quadrotors are used as rotating thrust generators (flying actuators) and the overall setup is an aerial tool. Aerial robots with suspended cables are also studied for aerial manipulation, e.g. in Sreenath and Kumar (2013). Besides all

---

[3]In that work the arm is not dynamically controlled, but using its kinematics only. The reason is that its actuators cannot be used in torque-control mode. The method we proposed in Section 6.4.2 is elegantly overcoming this issue for such actuators, while allowing dynamics-aware control of the overall system.

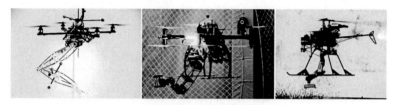

Figure 1.5: Three different aerial manipulator examples from the literature. On the left, a quadrotor equipped with a 2 DoF manipulating arm (AeRoArms (2015-2019)) and its details are given in Section 6.4.3 of this thesis. In the middle a multirotor VTOL equipped with a 6 DoF servo manipulating arm as part of ARCAS (2011-2015) and used in Ruggiero et al. (2016). On the right a double-rotor helicopter from *SwissDrones* is equipped with a KUKA LWR 7 DoF manipulating arm and a gripper (ARCAS (2011-2015),AeRoArms (2015-2019), Kondak et al. (2014)).

these works, we note that there are at least two European Union projects focusing on aerial manipulation; ARCAS (2011-2015) (finished) and AeRoArms (2015-2019) (ongoing).

## 1.4 On the Control of Underactuated Aerial Robots for APhI and Manipulation

The robot designs we consider in this thesis are decidedly including an underactuated flying base, mainly for the reasons explained in Section 1.1.2. By underactuation we mean that the system cannot be commended to follow arbitrary trajectories in its configuration space. Particularly in this thesis, we will consider the trivial underactuation, i.e. that the number of the control inputs is less than the number of the independent DoF of the system. In practice, underactuation can be thought of controlling a system using lesser number of actuators (e.g. motors), which might help reducing the overall operation costs. Moreover it can be even useful for specific tasks, as for the quadrotors it helps them swiftly accelerate along the translational directions, so that they can perform agile and dynamic trajectory tracking tasks. However underactuation of a system is a challenging problem from the control theory point of view, and it needs to be handled properly (see Spong (1998) and Fantoni and Lozano (2002)). Especially in the case of physical interaction, or control of a flying manipulator; the underactuation might become a greater problem, due to the reasons mentioned in Section 1.2 and Section 1.3.

### 1.4.1 Control of APhI

In this thesis, for APhI of an underactuated aerial robot, quadrotor, we used *passivity-based* controllers. It is well known that the term *passivity* has an important role in system analysis and it leads to powerful methods for designing feedback laws for nonlinear systems (also see Isidori (2013)). Furthermore, passivity as a property of a nonlinear system, can be used as a stabilization tool. In fact, in Sepulchre et al. (1997) it is exampled as the Nyquist-Bode 180 degree phase lag criterion for the nonlinear systems. Without giving a concrete and detailed definition of passivity, let us here try to conceptualize it in brief words, by starting with the definitions of two important concepts: *storage function*- a positive semidefinite energetic function of the state, and the *supply rate*- a term in units of power computed as a bilinear function of the input and the output of the system. Then, in short words, passivity is a property, which satisfies that the rate of increase of the storage function of a system is not higher than its supply rate, setting an *upper bound* to the system energy. This also means that the increase in any storage function of the system can only be done exogenously. This making sure that the external inputs, i.e. control inputs or the external (disturbing or meaningful) forces and torques, are the only reasons of an increase in the storage function, allows us developing control methods for achieving input/output stability in case of e.g. physical interaction (or APhI) using the concepts developed based on Theorem 1. The precise definition of passivity can be found in Willems (1972) and in Khalil (2001). The aforementioned *upper bound* on the system energy is fundamentally linked to the stability of the system. In fact by employing the storage function as the Lyapunov function of the system, one can show that passivity also implies Lyapunov stability (see Appendix C.2 of this thesis and also Sepulchre et al. (1997) with Khalil (2001)). Similarly, one can use the total energy of the system as the storage function, i.e. *Hamiltonian*, for analyzing the passivity. In fact, this is exactly what we do in Section 4.1.1 of this thesis, by taking the system underactuation into account. There, using the Hamiltonian as the storage function, we first bring the quadrotor dynamics into its *Port-Hamiltonian* (PH) form (so that in a later stage we can *reshape* its physical properties), and finally analyze the stability of the controlled system using passivity.

We would like to note that the passivity is certainly not the ultimate way of analyzing/monitoring/achieving stability, since the term *stability* is a very general concept. Surely many other criteria exist for providing a stable behavior to a nonlinear system, and from energetic point of view they can be linked to each other (see Khalil (2001) and the Appendix C of this thesis). Passivity is one of these methods, which can be used for rendering a nonlinear system to a stable behavior. It is also noticeable, that in Zames (1966) it is shown how the passivity is a more *conservative* way for providing stability, when it is compared to other methods developed based on *input-output stability*, e.g. *conic sector stability theorem* (more detail can be found given in Forbes (2012) and Bridgeman and Forbes (2015).

## 1.4.2 Control of Aerial Manipulators

In this thesis, besides APhI, we also study the control of aerial manipulation. To do so, we focus mainly on the motion tracking control of the aerial manipulators. For this, we greatly benefit from two important properties of the nonlinear systems; the *differential flatness* and *exact linearizability*. A differentially flat system has certain outputs (which can be detected using a proper coordinate transform of the state), that can represent all the states and the control inputs of the nonlinear system using only these outputs and their finite number of derivatives (see Murray et al. (1995) and also Section 6.2.2 of this thesis). These system outputs are in general called as the *flat outputs*, as also within this thesis. Obviously, being able to represent all the states and the inputs as functions of the flat outputs is practical and important property, especially in the phase of control and trajectory planning. In this thesis (Chapter 6), we present the flat outputs of various aerial manipulator models, together with Dynamic Feedback Linearization (DFL) controllers which brings the overall nonlinear system dynamics to its linear controllable form using its flat (or exactly linearizing) outputs[4]. DFL can use the flatness property of the system, and allow implementing widely used linear control methods for nonlinear systems (see e.g. Section 6.2.2). Notice that wiht DFL, we perform in this thesis an exact linearization, which is completely different from linearization of a nonlinear system around a point at the state space. On the contrary it is the exact linearization via dynamic feedback in an open and dense set of the state space (see Fact 1 of Section 6.2.2 and De Luca and Oriolo (2002)). Note that due to the matching of the total relative degree of the system with the dimension of the augmented (total) state after DFL (see Definition 7), the nonlinear system is rendered to a *minimum phase* system, i.e. all destabilizing effects in the zero dynamics are canceled (or they become trivial as elegantly stated in Isidori (1995) and Isidori (2013)). Furthermore, with DFL, exact tracking of the flat outputs can be achieved. More emphasis on this matter is given in Sections 6.1.1 and 6.2.2. Chapter 6 of this thesis is dedicated for exposing what are the flat outputs of different types of aerial manipulators, which of them are interesting to control, and how their tracking control can be achieved using DFL.

It is noticeable that despite the useful properties of DFL and its convenience on turning a highly nonlinear system to a linear controllable one; it has drawbacks as well. First of all, as mentioned in Sepulchre et al. (1997), DFL cancels all the nonlinearities, including the useful ones. This might mean an additional effort on the control without any use, or it might even be harmful in terms of system stability. Second, DFL is sensitive to uncertainties; i.e. not robust. The first drawback might be inevitable, or be avoided using more adaptive linear controllers, although more sophisticated methods are available in the literature, e.g. *back-stepping/forwarding controllers* (see Sepulchre et al. (1997), Khalil (2001) and Krstic et al. (1995)). A notable characteristic of this method is, that the control is achieved through *feedback passivation*, i.e. in every step of the back-stepping/forwarding the considered output is constructed in a way that the entire system is minimum phase, and at each step the considered system is relative degree one (see Sepulchre et al. (1997)). The second drawback of DFL can be avoided improving the robustness of the linear controller (e.g. the integrator term we added in the linear controllers described in Chapter 6), which might be a conservative way of controlling a system. Hence it might be more preferable considering other robust nonlinear control techniques, e.g. *sliding-mode controllers* (Khalil (2001)) or the methods described in Krstic et al. (1995). On top of these, *flatness-based controllers*, e.g. the one presented in Section 6.4, are another way of controlling the nonlinear system considering its flatness, without doing an exact linearization. In fact, as shown in Sec. 6.4.1 they are more implementable than e.g. DFL for the practical reasons.

We are certainly interested in considering different nonlinear control techniques for APhI and Aerial Manipulation in the future, e.g. back-stepping/forwarding and sliding-mode like controllers, which are not outlined within this thesis.

# 1.5 Outline of this Thesis

Let us give the outline of this thesis in the following;

- In Chapter 2, we describe the mathematical model of a quadrotor VTOL and its experimental setup. We made the choice of using quadrotor as the aerial platform, due

---

[4]If the zero dynamics of the system has an asymptotically stable equilibrium, then it can be stabilized using a dynamic output feedback. With this in mind, we study the aerial manipulators in Chapter 6. More on this matter can be found in Isidori (1995).

to the reasons explained in Chapter 1, and highlighted in Section 1.1.2. In Section 2.1 we describe the well-known kinematics of a quadrotor, and two dynamics model used within this thesis. The controllers studied in this thesis are numerically validated using different simulation environments, and they are described in Section 2.2. The experimental setup of a quadrotor, including all its fundamental mechanics, electronics, sensors (hardware) and their managing softwares are described in detail in Section 2.3.

- In Chapter 3 we study the external wrench measurement and estimation for the quadrotors. For achieving APhI, the knowledge of the external wrench is crucial. Here, we first present the possible measurement technologies for acquiring the wrench information using transducers, as put in Section 3.2. Then in Section 3.3 we present a *Lyapunov-based nonlinear disturbance observer* for estimating the external wrenches acting on a quadrotor. These two ways of collecting the wrench information have their own advantages and disadvantages, which are compared and discussed in Section 3.4.

- Chapter 4 is presenting a novel controller for the quadrotors, when they are performing APhI. This controller is developed based on the Interconnection and Damping Assignment - Passivity Based Control (IDA-PBC) method, allowing controlling the APhI by reshaping the physical properties of the flying robot. After giving some preliminaries on passivity theorem in Sec. 4.1.1, we introduce our controller in Section 4.2. We present both numerical and experimental results of the proposed controller in Sec. 4.3, where the flying robot is exposed to certain external forces/torques; e.g. hitting, or it is performing meaningful interaction tasks, e.g. sliding on a ceiling surface.

- In Chapter 5, we introduce the design, modeling and identification of a novel elastic-joint arm, to be used for APhI and aerial manipulation. In this chapter we approach to these two problems from design point of view, and provide experimental results showing the advantages of using compliant actuators.

- Aerial manipulation is studied deeply in Chapter 6. Here by an aerial manipulator, we consider a binomial of a flying platform and one or more manipulating arms. There, we study the tracking control problem of such systems by analyzing their *differential flatness property* (some preliminaries are given in Sec. 6.1). In Section 6.2 we consider a Planar-VTOL (PVTOL) aerial robot, equipped with a joint arm, which can be actuated rigidly or via some compliant elements. This entire section is dedicated to the modeling, and control of such systems, as well as exposing their differential flatness property and providing Dynamic Feedback Linearizing (DFL) controllers. We validate our theoretical results in Section 6.2.9 with extensive and realistic simulations, and in Sec. 6.2.10 with preliminary experiments for a quadrotor equipped with a Variable Stiffness Actuator (VSA). In Section 6.3 we extend our theory for the aerial manipulators, which having a PVTOL again as a flying platform, but this time equipped with a generic number of manipulating arms, each having any numbers of DoF, can be actuated rigidly or with some elastic elements. Their differential flatness property is shown here together with a proper DFL controller. The numerical simulation results are provided in Sec. 6.3.4 for an aerial grasping scenario. In Section 6.4 we extend our previous findings into 3D, where a decentralized flatness-based controller is presented. Its experimental validation is resulted in 6.4.3.

We conclude this thesis in Chapter 7, where the summary of the discussions together with idea of future perspectives are shared.

Note that we have provided the technical proofs of the propositions made in this thesis within Appendix A, and necessary technical computations in Appendix B. They are part of this thesis as much as any other chapter, but for brevity reasons they are added as appendices, ready to be reached at reader's will. Furthermore, a summarizing knowledge on the nonlinear system stability is shared in Appendix C, where we highlight the relationships between different stability concepts for the nonlinear systems.

## 1.6 Open Problems and Contributions

By the time of this thesis work has started, there were few works on the APhI and aerial manipulation. Both theoretical and practical novelties were required, and in parallel to many other researchers, we have contributed to the robotics society with several publications, which eventually brought out this thesis.

In the first year of this thesis work, we have focused on the control theory for controlling the APhI of the quadrotor VTOLs. This has been achieved by IDA-PBC, and described in Chapter 4 of this thesis, which is written based on our published papers in Yüksel et al. (2014a) and Yüksel et al. (2014b). For enabling this kind of controls, knowledge of the external wrench is required. We proposed an estimation method for external forces and torques acting on a quadrotor in Chapter 3, which is partially written based on what we have published in Yüksel et al. (2014b).

Second year of the work was mainly dedicated on the design of the mechanisms, and improvement of the electronics and the software of our experimental setup. In this time period, together with two master level students; Saber Mahboubi (funded by Max Planck Society) and Nicolas A. Rongione (funded by DAAD-RISE) we have developed an in-home light-weight elastic joint-arm for APhI. This work is published in Yüksel et al. (2015), and finally became the Chapter 5 of this thesis. Later, the majority of the workload was spared for the hardware/software improvement of the experimental setup. With great helps of Anthony Mallet from LAAS-CNRS, and Dr. Paolo Stegagno from MPI-Tübingen, we have developed and implemented new interfaces that enabled the experimental validations of the proposed theories. This part is detailed in Section 2.3 of the thesis, where the mentioned softwares are made available to the public.

In the final part of this thesis work, we dedicated our focus on aerial manipulation. Especially after being experienced with a quadrotor+elastic-joint arm setup (as mentioned in Chapter 5), we further studied such systems. Overall Chapter 6 is dedicated to the control of aerial manipulators, and their differential flatness property. In Yüksel et al. (2016b) and in Yüksel and Franchi (2016) we have published our first studies on the control of aerial robots equipped with rigid or elastic-joint arms, and they shaped the Section 6.2 of this thesis. We further studied such systems considering a case, where the aerial robot can be equipped with a generic number of manipulating arm, each having any numbers of DoF, rigidly or elastically actuated. Together with their differential flatness property, we studied their control and published our results in Yüksel et al. (2016a), which essentially made Section 6.3. Finally based on this, we proposed a decentralized flatness based controller for aerial manipulators, and this work appeared as Tognon et al. (2017) created Section 6.4 of this thesis.

Although we have approached the open problems of APhI and aerial manipulation in terms of both theory and application, they are still not completely closed. Many improvements need to be done, and instead of listing them here, we prefer to discuss them at the end of each relevant chapter or section of this thesis. Also we note that a list of future possible developments are listed in Chapter 7.

# Chapter 2

# Quadrotor: Mathematical Model and Experimental Setup

Quadrotors are in general light-weight aerial robots with high agility and great workspace. They are also off-the-shelf products today which makes them easy to acquire, and preferable platforms for the researchers (see Sec. 1.1 for an extensive literature review). In this chapter, we provide the base model of a quadrotor, together with its kinematics and dynamics. Our goal is to provide a clear mathematical model of this kind of aerial robot, which is going to be used for developing the control methods in the following chapters. In addition, we present here the real quadrotor setup in detail, which is used in our experiments. From electronics to software, we give the details of the tools we developed and used in the frame of this thesis.

## 2.1 Quadrotor Model

In this section we describe the frames of references, kinematics and two different dynamic models of a quadrotor VTOL, which will be used through this thesis.

Consider the model of a quadrotor as sketched in Fig. 2.1. The North-East-Down (NED) convention is used for every frame. We denote with $\mathcal{F}_W : \{P_W, \mathbf{x}_W, \mathbf{y}_W, \mathbf{z}_W\}$ the world (inertial) frame. The body (base) frame of the quadrotor is placed at its Center of Mass (CoM) and denoted with $\mathcal{F}_B : \{P_B, \mathbf{x}_B, \mathbf{y}_B, \mathbf{z}_B\}$. The rotation matrix representing the orientation of $\mathcal{F}_B$ in $\mathcal{F}_W$ follows the *roll-pitch-yaw* convention (Siciliano and Khatib (2008)) and it is

$$\mathbf{R}(\boldsymbol{\eta}) = \begin{bmatrix} c_\theta c_\psi & c_\psi s_\theta s_\phi - c_\phi s_\psi & s_\phi s_\psi + c_\phi c_\psi s_\theta \\ c_\theta s_\psi & c_\phi c_\psi + s_\theta s_\phi s_\psi & c_\phi s_\theta s_\psi - c_\psi s_\phi \\ -s_\theta & c_\theta s_\phi & c_\theta c_\phi \end{bmatrix} \in SO(3), \tag{2.1}$$

where $\boldsymbol{\eta} = [\phi \ \theta \ \psi]^T \in \mathbb{R}^3$ are the roll, pitch, and yaw angles in the minimalistic representation of the orientation, and clearly $\mathbf{R}(\boldsymbol{\eta}) \in SO(3)$ is the associated orthogonal rotational matrix with $|\mathbf{R}| = 1$. Notice that $s_*$ and $c_*$ stands for $\sin(*)$ and $\cos(*)$, respectively. Furthermore, the transformation between the body frame angular velocity $\boldsymbol{\omega} = [\omega_x \ \omega_y \ \omega_z]^T \in \mathbb{R}^3$ to the *euler rates* $\dot{\boldsymbol{\eta}} = [\dot{\phi} \ \dot{\theta} \ \dot{\psi}]^T \in \mathbb{R}^3$ can be done using

$$\boldsymbol{\omega} = \mathbf{T}(\boldsymbol{\eta})\dot{\boldsymbol{\eta}}, \quad \dot{\boldsymbol{\eta}} = \mathbf{T}(\boldsymbol{\eta})^{-1}\boldsymbol{\omega}$$

$$\mathbf{T}(\boldsymbol{\eta}) = \begin{bmatrix} 1 & 0 & -s_\theta \\ 0 & c_\phi & s_\phi c_\theta \\ 0 & -s_\phi & c_\phi c_\theta \end{bmatrix}, \quad \mathbf{T}(\boldsymbol{\eta})^{-1} = \begin{bmatrix} 1 & s_\phi t_\theta & c_\phi t_\theta \\ 0 & c_\phi & -s_\phi \\ 0 & s_\phi/c_\theta & c_\phi/c_\theta \end{bmatrix} \tag{2.2}$$

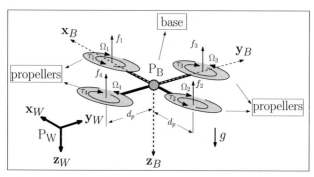

Figure 2.1: Schematic figure of a quadrotor and notation used in the thesis. For the $i$-th propeller, $\Omega_i$ is the spinning velocity, $f_i$ the thruster forces, $\tau_i$ is the reaction torque of the $i$-th motor due to the shaft acceleration and the propeller blade drag. Also see Fig. 1.2. Gravity is facing the positive $\mathbf{z}_W$.

with $t_*$ is $\tan(*)$. Notice that $\boldsymbol{\omega} \in \mathbb{R}^3$ is the angular velocity of $\mathcal{F}_B$ w.r.t. $\mathcal{F}_W$ expressed in $\mathcal{F}_B$. Each propeller of the quadrotor is assumed to be placed with the same distance from the CoM, and it is $d_p \in \mathbb{R}_{>0}$. In our configuration the first and second propellers are counter-rotating to those of third and fourth (see Fig. 2.1). Hence using the following constant transformation, we can map the spinning velocities of the propellers to the well known body-fixed forces of a quadrotor

$$\mathbf{u} = \begin{bmatrix} u_t \\ \hdashline \mathbf{u}_r \end{bmatrix} = \begin{bmatrix} u_t \\ u_\phi \\ u_\theta \\ u_\psi \end{bmatrix} = \begin{bmatrix} c_f & c_f & c_f & c_f \\ 0 & 0 & -d_p c_f & d_p c_f \\ d_p c_f & -d_p c_f & 0 & 0 \\ -c_t & -c_t & c_t & c_t \end{bmatrix} \begin{bmatrix} \Omega_1^2 \\ \Omega_2^2 \\ \Omega_3^2 \\ \Omega_4^2 \end{bmatrix} \in \mathbb{R}^{4\times1}, \qquad (2.3)$$

where $u_t$ is the total thrust force, $u_\phi$ is the roll, $u_\theta$ is the pitch and $u_\psi$ is the yaw torque. Note that positive $u_t$ is facing the $-\mathbf{z}_B$ direction, which is upwards. The spinning velocity of the $i$-th propeller is depicted with $\Omega_i$ and its direction is as in Fig. 2.1. The constants $c_f$ and $c_t$ are to be identified parameters (see Sec. 2.3.1), which satisfy $f_i = c_f \Omega_i^2$ and $\tau_i = c_t \Omega_i^2$.

Now with these in mind, we can derive the dynamic model for the CoM of a quadrotor.

**Newton-Euler Dynamics**

Consider the CoM of the quadrotor placed at $P_B$. Similar to Lee et al. (2013), the Newton-Euler dynamics of this point will be given in two parts; rotational

$$\Sigma_r : \begin{cases} \mathbf{M}_{qr}\dot{\boldsymbol{\omega}} &= [\boldsymbol{\omega}]_\wedge \mathbf{M}_{qr}\boldsymbol{\omega} + \mathbf{u}_r + \boldsymbol{\tau}_{ext} \\ \dot{\boldsymbol{\eta}} &= \mathbf{T}(\boldsymbol{\eta})\boldsymbol{\omega}, \end{cases} \qquad (2.4)$$

and the translational

$$\Sigma_t : \left\{ m_q \ddot{\mathbf{p}}_q = -u_t \mathbf{R}(\boldsymbol{\eta})\mathbf{e}_3 + m_q g \mathbf{e}_3 + \mathbf{f}_{ext}, \right. \qquad (2.5)$$

where $m_q \in \mathbb{R}_{>0}$ is the mass of the quadrotor; $\mathbf{e}_3 = [0\ 0\ 1]^T \in \mathbb{R}^3$; $\mathbf{R}(\boldsymbol{\eta})$ is the rotation matrix as in (2.1); $\mathbf{p}_q = [x_q\ y_q\ z_q]^T \in \mathbb{R}^3$ is the position of $\mathrm{P_B}$ in $\mathcal{F}_W$; $u_t \in \mathbb{R}$ is the thrust[1] control input along $-\mathbf{z}_B$; $\boldsymbol{\omega} \in \mathbb{R}^3$ is the body frame angular velocity as mentioned above; $\mathbf{M}_{qr} = \mathrm{diag}([\mathrm{J_{xx}}, \mathrm{J_{yy}}, \mathrm{J_{zz}}]) \in \mathbb{R}^{3\times3}$ is the inertia matrix w.r.t. the body frame; $g \in \mathbb{R}^+$ is the gravity acceleration directed along $\mathbf{z}_W$; $\mathbf{u}_r \in \mathbb{R}^3$ and $\boldsymbol{\tau}_{ext} \in \mathbb{R}^3$ represent the torque inputs and the external (environment) torques, respectively, both expressed in $\mathcal{F}_B$; $\mathbf{f}_{ext} \in \mathbb{R}^3$ is the external (environment) force expressed in $\mathcal{F}_W$; $\mathbf{T}(\boldsymbol{\eta}) \in \mathbb{R}^3$ is as in (2.2); and $[\star]_\wedge : \mathbb{R}^3 \to \mathrm{so}(3)$ is the skew-symmetric operator.

**Lagrange Dynamics**

In this thesis we will also use the Lagrange formalism of the quadrotor CoM (placed at $\mathrm{P_B}$) depicted in Fig. 2.1. The translational dynamics of the system remains the same as in (2.5). The rotational dynamics however is

$$\Sigma_r : \left\{ {}^W\mathbf{M}_{qr}(\boldsymbol{\eta})\ddot{\boldsymbol{\eta}} \right. = -\mathbf{C}_r(\boldsymbol{\eta}, \dot{\boldsymbol{\eta}})\dot{\boldsymbol{\eta}} + \mathbf{u}_r + \boldsymbol{\tau}_{ext}, \tag{2.6}$$

where ${}^W\mathbf{M}_{qr}(\boldsymbol{\eta}) = \mathbf{T}(\boldsymbol{\eta})^T \mathbf{M}_{qr}\mathbf{T}(\boldsymbol{\eta}) \in \mathbb{R}^{3\times3}$ is the rotational inertia matrix w.r.t $\mathcal{F}_W$ and $\mathbf{C}_r(\boldsymbol{\eta}, \dot{\boldsymbol{\eta}}) \in \mathbb{R}^{3\times3}$ is the matrix representing the Coriolis terms for the rotational dynamics, whose components $ij$-th are $(C_{ij})$;

$$
\begin{aligned}
C_{11} &= 0 \\
C_{12} &= (J_{yy} - J_{zz})(\dot{\theta}c_\phi s_\phi + \dot{\psi}s_\phi^2 c_\theta) + (J_{zz} - J_{yy})\dot{\psi}c_\phi^2 c_\theta - J_{xx}\dot{\psi}c_\theta \\
C_{13} &= (J_{zz} - J_{yy})\dot{\psi}c_\phi s_\phi c_\theta^2 \\
C_{21} &= (J_{zz} - J_{yy})(\dot{\theta}c_\phi s_\phi + \dot{\psi}s_\phi^2 c_\theta) + (J_{yy} - J_{zz})\dot{\psi}c_\phi^2 c_\theta - J_{xx}\dot{\psi}c_\theta \\
C_{22} &= (J_{zz} - J_{yy})\dot{\phi}c_\phi s_\phi \\
C_{23} &= -J_{xx}\dot{\psi}s_\theta c_\theta + J_{yy}\dot{\psi}s_\phi^2 c_\theta s_\theta + J_{zz}\dot{\psi}c_\phi^2 s_\theta c_\theta \\
C_{31} &= (J_{yy} - J_{zz})\dot{\psi}c_\theta^2 s_\phi c_\phi - J_{xx}\dot{\theta}c_\theta \\
C_{32} &= (J_{zz} - J_{yy})(\dot{\theta}c_\phi s_\phi s_\theta + \dot{\phi}s_\phi^2 c_\theta) + (J_{yy} - J_{zz})\dot{\phi}c_\phi^2 c_\theta + \\
& \quad + J_{xx}\dot{\psi}s_\theta c_\theta - J_{yy}\dot{\psi}s_\phi^2 s_\theta c_\theta - J_{zz}\dot{\psi}c_\phi^2 s_\theta c_\theta \\
C_{33} &= (J_{yy} - J_{zz})\dot{\phi}c_\phi s_\phi c_\theta^2 - J_{yy}\dot{\theta}s_\phi^2 c_\theta s_\theta - J_{zz}\dot{\theta}c_\phi^2 c_\theta s_\theta + J_{xx}\dot{\theta}c_\theta s_\theta.
\end{aligned}
\tag{2.7}
$$

## 2.2 Simulation Setups of the Quadrotor

In the course of this thesis, we present several control methods for achieving aerial physical interaction and manipulation. Before validating these methods experimentally using the real robot, we first test them *numerically* using computer simulations.

The most frequently used simulation environment for this thesis is Matlab/Simulink with its ODE solvers. In fact, the numerical results in Sec. 4.3.1, Sec. 6.2.9 and in Sec. 6.3.4 are acquired using Matlab/Simulink. For simulations, we put additional effort to make them realistic, by adding noises to the measured quantities, implement low-sampled discretization on some of them replicating the transmission frequency of the real sensors, and even adding

---

[1]In normal situations it is actually $u_t > 0$. However, if needed by the task, negative thrust can always be achieved in the implementation, as, e.g., in Cutler et al. (2011).

| Parameters | Notation | Value | Unit |
|---|---|---|---|
| Mass | $m_q$ | 1.2 | [kg] |
| Moment of Inertia | $\mathbf{M}_{qr}$ | diag([0.015, 0.015, 0.026]) | [kgm$^2$] |
| Link distance | $d_p$ | 0.2 | [m] |
| Max. Propeller Velocity | $\Omega_i^{max}$ | 95 | [Hz] |
| Max. Propeller Thrust | $f_i^{max}$ | 7 | [N] |
| Identified Propeller Parameters | $c_f, c_t$ | 0.00065, 0.00001 | See (2.3) and Sec. 2.3.1 |

Table 2.1: Approximate physical parameters of the quadrotor setup shown in Fig. 2.2. Notice that depending on the design or experiment, these parameters can (and will) change.

parametric uncertainties making the controllers less confident about the real model of the system.

We also implemented some of the controllers presented in this thesis using physics-based simulators, e.g. Sim Mechanics of Matlab. There, a CAD model of the robot is designed, which is more realistic than using the mathematical model of the system. Then the controllers developed based on the mathematical model of the system are tested for controlling this CAD model. For example, when testing the controller presented in Sec. 6.2.7 for a quadrotor + VSA setup, right after the Matlab/Simulink simulations and before the experimental validation, we implemented it together with the CAD model of the system, as depicted in Fig. 6.18.

For solving the optimal control problem, described in (6.74) of Sec. 6.2.8, we used ACADO numerical optimizer. The details of ACADO can be found in Houska et al. (2011).

Finally, mostly for visualization purposes of some of the Matlab/Simulink simulations, we used UNITY Game Engine[2], which is used to simulate the motion of the CAD model of the quadrotor.

## 2.3 Experimental Setup of the Quadrotor

In this section we will briefly present the quadrotor we used for our experiments, as well as the hardware and the software for the experimental setup. The main body of the quadrotor setup is manufactured by *HiSystems GmbH*, and named as *Mikrokopter* Quadrotor[3]. A picture of the quadrotor setup is given in Fig. 2.2, and its physical parameters are summarized in Table 2.1. This is the flying robot base, and depending on the experiment, we mount additional parts on it (e.g. *Force/Torque* (F/T) sensor onbard as shown in Fig. 3.2, or any additional manipulator arm, e.g. in Fig. 5.7, Fig. 6.18 or in Fig. 6.26).

### 2.3.1 Hardware

The flying robot, quadrotor, is consisting of various mechanical parts and electronics. The quadrotor has four rigid bars, connecting four burshless motors (will be referred as BL-Motors or BLDC) and their propellers to the main body of the robot (see Fig. 2.2). Notice that propellers are rigidly attached to their motors, as well as the motors to the

---

[2]https://unity3d.com/
[3]http://www.mikrokopter.de/en/home

► Mo-Cap Marker
► BL-Motor and Propeller

► Flight-Ctrl + IMU
► Brushless-Ctrl ▲
▲
► Battery

Figure 2.2: Experimental setup of the quadrotor VTOL. From top to bottom, there is first
the MoCap markers to be detected by the MoCap system. Later there is the
flight controller, which has the IMU on it. Below there are four brushless (BL)
motor controllers (symmetrically placed around the flight controller), connected
to the flight controller via an I²C Bus. These brushless motor controllers are also
connected to a square-shaped power board, which is powering all the electronics
on board and connected to a 16 [V] DC battery. Between battery and the
powerboard, four rigid bars are rigidly attached to the whole system, each
carrying one brushless motor. Each motor is rigidly attached to its propeller
via screws.

bars, and bars to the body. On top of them there are four brushless motor controllers, a
flight controller with an *Inertial Measurement Unit* (IMU) on it and markers for a *Motion
Capture* (MoCap) system, in this order. Below the rigid bars there is the battery as the
energy source.

**Micro-controllers and the Brushless Motors**

On the platform, there are four control circuits for the brushless motors, each having
one ATMEGA168 $\mu$-controller[4] (see Fig. 2.2). These controllers are connected to a *flight
controller* via I²C bus, which has an *Inertial Measurement Unit* (IMU) on board. Through
a serial channel we communicate with the flight controller, allowing us to read/write data
from/to both the flight controller and the brushless motor controllers.

A Motion Capture System (MoCap) is placed in the room of the experiments[5], which is
acquiring the poses of specific markers using 6 different fixed near-infrared cameras (see

---

[4]http://wiki.mikrokopter.de/en/BL-Ctrl_2.0
[5]https://www.vicon.com/

19

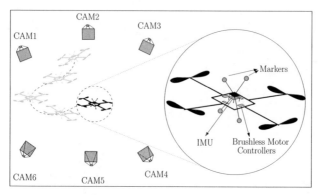

Figure 2.3: Schematic figure of an indoor experimental setup with a quadrotor. In the room there are 6 different near-infrared cameras tracking the *pose* of the *markers* in the global frame, and this information is given by a *Motion Capture System* at 120 [Hz]. The *Inertial Measurement Unit* (IMU) is providing the linear acceleration and the rotational velocity of the quadrotor body, former in the global and the latter in the body frame. This information is provided by the flight controller, through the serial communication at 1 [kHz]. The brushless motor controllers are for controlling the propeller velocities.

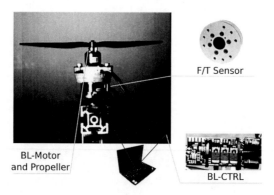

Figure 2.4: Test setup for identification of the propeller parameters $c_f$ and $c_t$ to be used in (2.3). The force/torque sensor used in this setup is an *ATI-Mini45* with calibration according to Si-145-5 and a readily available software. The BL-CTRL can set the desired velocity (so it is an *Electronic Speed Control - ESC*) of the propellers and measure it using the software explained in Sec. 2.3.2.

Fig. 2.3). These special markers are rigidly fixed on the quadrotor body.

**Sensors for the State Estimation**

For state estimation of the robot, we use an external MoCap system and the IMU on board of the quadrotor. MoCap system is providing the pose of the quadrotor, $\mathbf{q}_q = [\mathbf{p}_q^T, \boldsymbol{\eta}^T]^T \in \mathbb{R}^6$ in the world frame at 120 [Hz]; while the IMU is giving the linear acceleration, $\ddot{\mathbf{p}}_q \in \mathbb{R}^3$, in the world frame and the angular velocity of the body (in the body frame), $\boldsymbol{\omega} \in \mathbb{R}^3$, both at 1 [kHz].

**F/T Sensor Hardware**

In our experiments we used two different Force/Torque (F/T) sensors. The first one is ATI Mini45 F/T sensor, calibrated according to SI-145-5, and located at LAAS-CNRS, Toulouse, which is a heavy sensor considering its electronics for data acquisition (all together weights about 4 [kg]). Since the total weight is way more than the load limit of our quadrotor, this sensor is not suitable to be placed on board of it. However thanks to its readily working hardware/software (of the ATI sensor with its drivers[6] and the brushless controllers allowing the velocity control of the propellers), we used it to identify the propeller parameters, $c_f$ and $c_t$, which were required for calculating the propeller velocities, $\Omega_i$, which can be mapped directly to the quadrotor control input vector $\mathbf{u}$, using the conversion in (2.3). So for implementing any given control method providing $\mathbf{u}$, we need the parameters $c_f$ and $c_t$ for computing the necessary propeller velocities $\Omega_i$, which will be sent as the commands of the four BL-CTRLs on board. The setup used for identification is shown in Fig. 2.4, where the propeller+brushless motor (BL-motor) is rigidly attached on the ATI-Mini45 sensor, and the sensor is rigidly connected to a platform below. The BL-motor is connected to a brushless controller (BL-CTRL), which can measure and control the propeller velocity (using the software explained in Sec. 2.3.2). The sensor is connected to its data acquisition (DAQ) box (not given in the figure). Both the DAQ box and the BL-CTR are connected to a personal computer (PC), where an authentic software runs to control the identification process. There, we repeatedly set the desired BL-motor velocity for various values, and measure both its velocity (via BL-CTRL) and the forces/torques (via the F/T sensor). At the end, the propeller values are identified as $c_f = 0.00065$ and $c_t = 0.00001$, which are also depicted in Table 2.1.

The second F/T Sensor is the *FTSens*, produced by IIT originally for the ICub humanoid robots (see Fumagalli et al. (2012a)). There are two reasons why we chose this sensor; first it was relatively cheaper than its peers in the market, and second it weights 122gr together with its electronics. This is definitely in the load range of our quadrotor. A challenge of using this sensor was implementing the software for acquiring the meaningful force/torque measurements, from scratch. The FTSens communicates through the *Controller Area Network* (CAN) bus channel, in which it receives the commands and sends the sensor data based on the CAN protocol. An additional hardware setting is required for acquiring the F/T data through CAN using the conventional computers which have serial channels, e.g. the *Universal Serial Bus* (USB). Such conversion can be done using the hardware setup shown and explained in detail in Fig. 2.5. For this setup to work, both the computer and the sensor need to be programmed properly. The details of the software packages developed in this thesis are explained in Sec. 2.3.2.

---

[6]http://robotpkg.openrobots.org/robotpkg/hardware/daqflex-libs/index.html

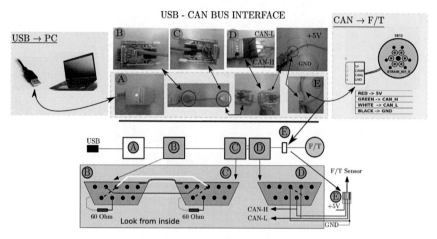

Figure 2.5: Hardware setup and detailed connections of the USB-CAN bus communication for the FTSens F/T sensor. From USB to the F/T sensor, the following connections are used: (A): PEAK USB-CAN converter, (B-C-D): CAN-BUS connectors properly adjusted, (E): CAN-BUS input of the F/T sensor. Above in the figure, these connections are shown using the real pictures of the setup. Below, a sketch of the proper connections are given for the convenience of the reader. Here, we respected the color code of the real connections.

## 2.3.2 Software

During the course of this thesis, we have improved the tools available for the APhI not only theoretically, but also practically. For implementation of the controllers we have presented in this thesis, it was necessary to set the hardwares (as explained in Sec. 2.3.1) and implement/create proper softwares for them to work.

For our experiments, we benefited greatly from the *Robot Operation System* (ROS): a collection of different robotic software frameworks providing standard operating system services, e.g. hardware abstraction, low-level control, high level operation, etc. (see also Quigley et al. (2009)). Furthermore, we have implemented all of our software packages compatible with ROS, and *TeleKyb* framework: an open-source end-to-end ROS-based software for general purpose of mobile robot control developed at Max Planck Institute for Biological Cybernetics, Tübingen (for details, see Grabe et al. (2013). Hence all the estimators (state and force/torque) or controllers presented in this thesis, and the softwares for e.g. flight controllers, brushless motor controllers, sensors, are implemented in a way that they can work with ROS and TeleKyb.

We would like to thank Dr. Paolo Stegagno and Dr. Anthony Mallet for their involvement during the development of these software packages. Especially we note that Dr. Anthony Mallet has developed personally most of the low-level software (LAAS-CNRS side) as explained in the following.

**Telekyb and the Low-Level Software**

The quadrotor setup we use in our experiments has one flight-controller (with IMU on it) and four brushless-motor controllers regulating the spinning velocities of the propellers, $\Omega_i$ (see Fig. 2.2).

Until recently, as for the most of the quadrotor setups in the market, the spinning of the propellers were controlled by setting the duty cycle of the *Pulse Width Modulated* (PWM) signal of the input voltage of the brushless motors. This *open-loop* way of controlling the propeller rotation rate assumes that the battery voltage of the quadrotor is not changing, and there is a constant one-to-one mapping between the propeller (or brushless motor) velocity and the PWM duty cycle. Clearly, in reality this is not the case, since with the dropping battery voltage during the flight of the quadrotor, a non-linear mapping between the propeller velocity ($\Omega_i$ in Fig. 2.1) and the PWM input occurs, which is hard to identify. Furthermore a static nonlinear mapping is valid only at steady state while inertial and dissipative phenomena have to be taken forcibly into account, when the rotational speed is changing, in order to achieve good tracking performances. For this reason, a low-level controller on the brushless controllers (BL-CTLR) for tracking the propeller velocity, $\Omega_i$, was needed. A *Sliding Mode Controller* is developed at LAAS-CNRS by Dr. Antonio Franchi and Dr. Anthony Mallet for this purpose and implemented on each $\mu$-controller of the four brushless motors[7]. This very low-level control loop makes sure that the propellers are rotating at the velocity, which is computed by a higher control loop using the transformation given in (2.3). This setup allows us implementing the controllers developed based on the explicit system dynamics, e.g. the ones presented in Chapter 4 and Chapter 6.

The flight controller on board (with IMU) connects the brushless controllers to a higher-level control framework. By doing so, it provides the IMU measurements to a computer, while sending the desired propeller spinning velocities to the brushless motors. The software managing this is also implemented at LAAS-CNRS[7] with extended capabilities, e.g. measurements of the motor currents, velocities, battery voltages, etc.

We have implemented an interface, connecting both these low-level softwares with a higher control loop in TeleKyb framework[8]. This interface can work with already developed packages of TeleKyb, e.g. obstacle avoidance, state estimation, etc., while allowing us testing our model-based controllers using cutting-edge low-level softwares developed in the time of this thesis[7].

**Data Fusion for State Estimation**

It is obvious that for realizing a closed-loop control for the system at hand, the state estimation is crucial. As mentioned before, for this purpose two types of sensors are used; a MoCap system implemented in the experiment room measuring the pose of the MoCap markers', hence quadrotor's, pose $\mathbf{q}_q = [\mathbf{p}_q^T, \boldsymbol{\eta}^T]^T \in \mathbb{R}^6$ in $\mathcal{F}_W$ at 120 [Hz][9], and an IMU implemented on board of the quadrotor measuring its linear acceleration $\ddot{\mathbf{p}}_q \in \mathbb{R}^3$ in $\mathcal{F}_W$ and its rotational velocity $\boldsymbol{\omega} \in \mathbb{R}^3$ in $\mathcal{F}_B$ both at 1 [kHz]. However, for our controllers what we need is the state of the quadrotor, i.e. $\mathbf{q}_q$ and $\dot{\mathbf{q}}_q$. For this reason, we use an *Unscented*

---

[7]https://git.openrobots.org/projects/tk3-mikrokopter

[8]https://svn.tuebingen.mpg.de/humus-telekyb/hydro/trunk/packages/telekyb_uavs/tk_mkomegacontrolinterface/

[9]Actually, MoCap measurements give the quaternions instead of the Euler angles; however the transition from one to another is clear considering the convention used in (2.1).

*Kalman Filter* (UKF) for fusing the data of the different sensors and estimating the quadrotor state at 1 [kHz]. The algorithm for this UKF is developed at LAAS-CNRS and made available for the public[10].

## F/T Sensor Software

In our experiments we used the IIT's F/T sensor developed for ICub robots[11], which is a 6D force/torque sensor, relatively cheap w.r.t. its peers and it does not require additional heavy data acquisition unit. Its low cost and light weight makes it a good choice to be used for APhI experiments. The sensor has been successfully tested and used before in ICub robots, as reported in Fumagalli et al. (2012a).

By the time we acquired it, we only received the hardware with no software package directly providing the sensor measurements in proper units, e.g. N or Nm. The sensor transmits its raw data through a CAN bus, hence within the time of this thesis we have implemented a CAN bus-Serial communication software (driver) providing the raw sensor data, and a ROS based software acquiring the force and torque measurements in meaningful units.

The software package we have created for this sensor is available for the public use[12]. There we provide:

- how to set up your computer (for both Intel or ARM processors) for using the CAN-USB converter,

- how to get the calibration data from the sensor,

- how to *ping* the sensor and let it send the raw data to your computer.

The details on the communication protocol of the sensor are available in the wiki-page of the ICub[11]. Using this *driver* and following the instructions explained in Fig. 2.5, it is now straightforward to receive the *raw data* from the FTSens F/T sensor.

For processing this raw data, we implemented a ROS (C++) based software within the TeleKyb framework. This software is tested with ROS-Indigo in Ubuntu 14.04 OS. It[13] receives the raw data from the serial channel the sensor is connected to (through a CAN-USB converter) and as output returns the force and torque measurements in meaningful units in a ROS topic. In this way the output can also be used by other ROS-based packages, e.g. the controller tested in Sec. 4.3.2. Notice that this code is strongly depended on both ROS and TeleKyb message types and their existing packages. Its usage for our experiments is also made available to the public[14], but for the initial access a permission from Max Planck Society would be needed.

---

[10]http://robotpkg.openrobots.org/robotpkg/localization/pom-genom3/index.html
[11]http://wiki.icub.org/wiki/FT_sensor
[12]https://redmine.laas.fr/projects/byueksel/repository/ftsens_iit
[13]https://svn.tuebingen.mpg.de/humus-telekyb/hydro/trunk/packages/telekyb_users/tk_
byueksel/src/ftsens_subpub.cpp
[14]https://svn.tuebingen.mpg.de/humus-telekyb/hydro/trunk/packages/telekyb_users/tk_
byueksel/

# Chapter 3

# External Wrench Measurement and Estimation for an Aerial Robot

To perform a meaningful physical interaction task, the knowledge of the interactive forces and torques (F/T, or *wrench*) is required. In robotics, especially for the ground robots, it is common to use an F/T sensor or transducer for measuring this information. Using an F/T sensor comes with the advantage of having accurate measurements, which are certainly independent of any system model of the robot/the environment they are used with/at. However, this measurement brings additional costs in terms of power, money, but especially the weight. Particularly for aerial robots, the additional weight of the sensor could dramatically limit the capabilities of the system. Moreover, the measurement is limited to the location of its transducer. These in mind, estimation of the external wrench is a computationally cheap way, which is adding no cost of weight or money and can be performed for any point on the system. For this reason it is reasonable, and necessary to study the estimation methods for the external wrenches, as well as improve the existing transducers for the use of aerial physical interaction.

In this chapter we will first introduce some of the external measurement methods available in the market, and describe the transducer we used which is suitable for aerial physical interaction. Recall from Sec. 2.3.2, during the time of this thesis work we have developed necessary drivers and software for this sensor. Later, we will introduce an external wrench estimator for quadrotor VTOLs, developed based on *Lyapunov-based nonlinear disturbance observers* and present its performance using numerical data. Finally we compare both wrench measurement and the estimation methods using the experimental data, and discuss the results. Based on this discussion, we decide for one of these two methods for acquiring the external wrench information to be used in the controller presented in Chapter 4.

We note that the content of this chapter, to be exact the Section 3.3, where we explain the nonlinear wrench observer, is published in Yüksel et al. (2014b).

## 3.1 Introduction

The physical interaction of flying systems is a challenging control and design problem and it became recently the interest of many researchers. Especially quadrotor VTOLs are becoming popular tools for physical interaction tasks, and for a broad literature review please refer to Section 1.2.

One of the challenges when flying machines are interacting with their environment is the stabilization and the control of this interaction in a meaningful way, i.e., they can exploit interactive forces for achieving a desired task[1]. Another crucial issue for aerial physical interaction is measuring the external wrenches (i.e., forces and torques) acting on the body of the flying system. This can be done using force/torque sensors that, on one side, give reliable independent and accurate measure. On the other side, using sensor can increase the cost of the equipment, the demand from the power supply and the weight of the robot, which consequently can decrease the flight capacity of the aerial vehicle. Another viable solution is the use of a wrench estimator, which proposes a cheaper solution. The estimator might give a sufficiently accurate estimate of the wrench, if it is properly designed and if the other sensors (for e.g. velocity, pose and, if available, acceleration) are accurate and reliable enough. Furthermore, using a wrench estimator would mean lighter and more power efficient aerial robots, which is a critical fact considering their low load capabilities.

With these insights, we set our goals for this chapter as: *i)* finding an appropriate F/T sensor hardware for APhI, and implementing it on a quadrotor, *ii)* developing an F/T estimation method for quadrotors, *iii)* comparing these two techniques to find the most suitable one to be used later for APhI.

Now, in the following we start with the F/T sensor setup we implemented to be used for APhI. Then we present a wrench estimation method in Sec. 3.3. In Sec. 3.4 we compare these two techniques and discuss their advantages over each other.

## 3.2 Measurement of the External Wrench

Acquiring the knowledge of the forces and the torques in 3D (wrench) is possible using F/T sensors, which are already in use for robotic manipulators and humanoids (see Siciliano and Khatib (2008)). Recently they have been in use also for the aerial robots. In Gioioso et al. (2014) it has been shown how to turn a quadrotor into a 3D force tool, and for the experimental setup an ATI-Gamma F/T sensor was used. In that work, like in many, the F/T sensor is either placed in the environment, e.g. mounted on a wall, or on the robot but only when it is not completely flying, i.e. when the robot is fixed to a test bench as in Yu and Ding (2012) and in Schiano et al. (2014). One of the main reason why these sensors are not yet used on board of a flying aerial vehicle, is because of their weight. Especially considering their electronics, e.g. the data acquisition box of the ATI-Gamma sensor mentioned above, most of the aerial robots used in research are not capable of flying with them on board.

This have been told, there are more light-weight 6D F/T sensors available in the market. For the part of the experiments of this thesis, we have decided to use the *FTSens* 6D F/T sensor, developed by the *Italian Institute of Technology* (IIT), as introduced in Fumagalli et al. (2012a). This sensor weights 122 [gr] with all electronics (but without the CAN-USB hardware explained in Fig. 2.5), which makes it a suitable candidate to be used on board of an aerial robot. In this way, the F/T sensors can be used not only as ground truths, but also as the direct measurements fed back to the control algorithms.

We explained the hardware and the software for the communication with the F/T sensor, in Sec. 2.3.1 and in Sec. 2.3.2, respectively. Now let us give the details on its usage on board of a quadrotor VTOL. A sketch of our quadrotor and the F/T sensor setup is

---

[1]There are various ways of approaching this challenge, as listed in Sec. 1.2, and the way we see this problem is summarized in Sec. 1.4.1. Moreover, we address the problem of APhI not here but in Chapter 4.

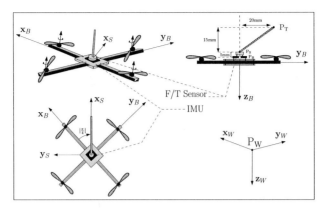

Figure 3.1: Placement of the interaction tooltip on top of the quadrotor: IMU ($P_B$), Force/Torque sensor ($P_S$), the rigid tool and its tip ($P_T$). This setup is later realized both as CAD model and in reality as shown in Fig. 3.2.

given in Fig. 3.1. On top of the F/T sensor a rigid tool is placed, intended to be used as the *interaction* tool with the environment. We intentionally placed this rigid tool in a way there is $|\pi/4|$ [rad] between the tool tip and the quadrotor frame; so that during a physical interaction the propellers will be away from the obstacles. The CAD design of this setup is shown in Fig. 3.2, where we also give its realization in detail. We note here that such design is also depicted in Fig. 4.4, where it is used for numerical simulations of the controller presented in Chapter 4. Moreover, the real setup shown in Fig. 3.2 is used for its experimental validation of the same controller in Sec. 4.3.2.

Now let us then compute the external wrenches acting on different parts of the quadrotor, using the measurements acquired from the F/T sensor. We had previously defined $\mathcal{F}_W$ : $\{P_W, \mathbf{x}_W, \mathbf{y}_W, \mathbf{z}_W\}$ as the world frame, and $\mathcal{F}_B$ : $\{P_B, \mathbf{x}_B, \mathbf{y}_B, \mathbf{z}_B\}$ is the body-fixed frame of the quadrotor. Now, let us define the $\mathcal{F}_S$ : $\{P_S, \mathbf{x}_S, \mathbf{y}_S, \mathbf{z}_S\}$ as the F/T sensor frame. Assume that the IMU frame is same as $\mathcal{F}_B$. Then define $\mathcal{F}_{Sb}$ : $\{P_S, \mathbf{x}_{Sb}, \mathbf{y}_{Sb}, \mathbf{z}_{Sb}\}$ as the frame of the F/T sensor, after its orientation is aligned with the orientation of the body-fixed frame. Then let us define the following wrench informations:

- The external wrench acting at and about the tip point of the tool ($P_T$), is defined with $\mathbf{w}_t \in \mathbb{R}^6$ in $\mathcal{F}_W$, since the external forces and torques are coming from the world frame.

- The wrench measured by the sensor, is defined with $\tilde{\mathbf{w}}_s \in \mathbb{R}^6$ in $\mathcal{F}_S$, since the measurements are done in the sensor frame,

- The wrench measured by the sensor and adapted to the body frame, is defined with $\mathbf{w}_s \in \mathbb{R}^6$ in $\mathcal{F}_{Sb}$, since the sensor is fixed on the body of the quadrotor,

- The wrench entering to the quadrotor dynamics, is defined with $\mathbf{w}_{ext} \in \mathbb{R}^6$, where the forces are defined in $\mathcal{F}_W$ and the torques are in $\mathcal{F}_B$ so equivalently in $\mathcal{F}_{Sb}$. This is because of the choice made when writing the quadrotor equations of motion in

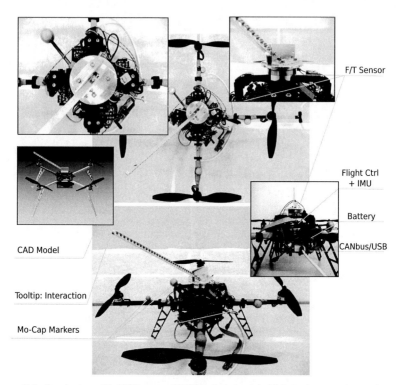

Figure 3.2: Quadrotor with F/T sensor (*FTSens*) on board. This is the experimental setup used for validating the IDA-PBC controller presented in Chapter 4. Notice that the CAD model is developed based on the description in Fig. 3.1, and so is the real robot. MoCap markers are placed different than the one in Fig. 2.2 so that both the F/T sensor and the interaction tool can be placed on top of the quadrotor. The hardware for the CAN-USB communication, described in Fig. 2.5 is placed on the bottom of the robot. For this setup, it is $m_q = 1.49$ [kg], and $\mathbf{M}_{qr} = \mathrm{diag}([0.01708,\ 0.0172,\ 0.0274]) \in \mathbb{R}^{3\times3}$ in units of [kgm$^2$].

Sec. 2.1, where the translational dynamics is written in the world frame, while the rotational one is in the body frame.

Now, it is clear that the only measurement we get is $\tilde{\mathbf{w}}_s \in \mathbb{R}^6$ in $\mathcal{F}_S$, but we need $\mathbf{w}_{ext} \in \mathbb{R}^6$ for the controller presented in Chapter 4, and maybe also $\mathbf{w}_t \in \mathbb{R}^6$ in $\mathcal{F}_W$ for visualization or for another type of controller. Then let us clarify the following relationship between the different wrench informations:

- Find $\mathbf{w}_s \in \mathbb{R}^6$ in $\mathcal{F}_{Sb}$. To do so, change the F/T sensor frame from $\mathcal{F}_S$ to $\mathcal{F}_{Sb}$. Notice that it can be done using time invariant rotations only (see Fig. 3.1 for the orientations of the frames). More specifically, remembering that $\mathcal{F}_S : \{\mathrm{P}_S, \mathbf{x}_S, \mathbf{y}_S, \mathbf{z}_S\}$

and $\mathcal{F}_{Sb} : \{\mathrm{P_S}, \mathbf{x}_{Sb}, \mathbf{y}_{Sb}, \mathbf{z}_{Sb}\}$, we have

$$\begin{bmatrix} \mathbf{x}_{Sb} \\ \mathbf{y}_{Sb} \\ \mathbf{z}_{Sb} \end{bmatrix} = \mathbf{R}_S^B \begin{bmatrix} \mathbf{x}_S \\ \mathbf{y}_S \\ \mathbf{z}_S \end{bmatrix}, \quad \mathbf{R}_S^B = \mathbf{R}_{x_S}(\pi)\mathbf{R}_{z_S}(\pi/4), \tag{3.1}$$

where $\mathbf{R}_{*_S}$ is the rotation matrix defined in $\mathcal{F}_S$ and around $*$-axis. Hence, if the $\tilde{\mathbf{w}}_s \in \mathbb{R}^6$ is the measurement of the F/T sensor defined in $\mathcal{F}_S$ (because it is fixed in the sensor frame), then

$$\mathbf{w}_s = \begin{bmatrix} \mathbf{R}_S^B & \mathbf{0}_3 \\ \mathbf{0}_3 & \mathbf{R}_S^B \end{bmatrix} \tilde{\mathbf{w}}_s \tag{3.2}$$

is the measurement but defined[2] in $\mathcal{F}_{Sb}$. Notice that $\mathbf{R}_S^B$ is a constant (time invariant) matrix, and $\mathbf{0}_i$ is a $i \times i$ zero matrix.

- Find $\mathbf{w}_t \in \mathbb{R}^6$ in $\mathcal{F}_W$. To do so, let us use the recently computed $\mathbf{w}_s$. This can be done using the following relation:

$$\mathbf{w}_s = \begin{bmatrix} \mathbf{R}_W^B(\boldsymbol{\eta}) & \mathbf{0}_3 \\ [\mathbf{d}_l]_\wedge \mathbf{R}_W^B(\boldsymbol{\eta}) & \mathbf{R}_W^B(\boldsymbol{\eta}) \end{bmatrix} \mathbf{w}_t, \tag{3.3}$$

where $\mathbf{R}_W^B(\boldsymbol{\eta})$ is the rotation matrix representing the orientation of $\mathcal{F}_W$ in $\mathcal{F}_B$, which is time variant due to the dependency of the quadrotor orientation $\boldsymbol{\eta} \in \mathbb{R}^3$. Notice that $\mathbf{d}_l \in \mathbb{R}^3$ is the distance between $\mathrm{P_T}$ and $\mathrm{P_S}$ in $\mathcal{F}_{Sb}$ frame[3], and $[\star]_\wedge : \mathbb{R}^3 \to so(3)$ is the skew-symmetric operator. Hence, using $\mathbf{w}_s$ from (3.2), we can acquire $\mathbf{w}_t$ using the relation in (3.3).

- Find $\mathbf{w}_{ext} \in \mathbb{R}^6$. To do so, use the rigid transformation from $\mathbf{w}_s$ to $\mathbf{w}_{ext}$:

$$\mathbf{w}_{ext} = \begin{bmatrix} \mathbf{R}_B^W(\boldsymbol{\eta}) & \mathbf{0}_3 \\ [\mathbf{d}_s]_\wedge \mathbf{R}_B^W(\boldsymbol{\eta}) & \mathbf{I}_3 \end{bmatrix} \mathbf{w}_s, \tag{3.4}$$

where $\mathbf{R}_B^W(\boldsymbol{\eta})$ is the rotation matrix representing the orientation of $\mathcal{F}_B$ in $\mathcal{F}_W$, $\mathbf{I}_i$ is a $i \times i$ identity matrix, and $\mathbf{d}_s$ is the distance between $\mathrm{P_S}$ and $\mathrm{P_B}$ in $\mathcal{F}_B$, which is $\mathbf{d}_s = [0\ 0\ -0.05]$ [m].

Hence, for finding the effect of $\mathbf{w}_t$ defined $\mathcal{F}_W$ and applied in the body-fixed frame of the quadrotor (this effect is defined as $\mathbf{w}_{ext}$), one can first use the F/T sensor measurements $\tilde{\mathbf{w}}_s$ in $\mathcal{F}_S$, then compute $\mathbf{w}_s$ in $\mathcal{F}_{Sb}$, and then finally use (3.3) and (3.4), respectively.

Notice that when using the NED convention, the rotation matrix from body to the world frame is $\mathbf{R}_B^W(\boldsymbol{\eta}) = \mathbf{R}$ where $\mathbf{R} \in SO(3)$ is given in (2.1) and it is true that $\mathbf{R}_W^B(\boldsymbol{\eta}) = \mathbf{R}_B^{W^T}(\boldsymbol{\eta})$, and $\boldsymbol{\eta} = [\phi\ \theta\ \psi]^T \in \mathbb{R}^3$ is the minimal representation of the rotations using Euler angles in the following order: *roll-pitch-yaw*.

We note that the F/T sensors we have used in this thesis are strain-gauge based transducers, i.e. a so called *spring element* measures the forces and torques based on

---

[2]This implies the following: from the force/torque sensor we get $\tilde{\mathbf{w}}_s \in \mathbb{R}^6$ which is naturally given in the sensors body frame, $\mathcal{F}_S$. However for our convenience we want to transform it to $\mathbf{w}_s$ defined in $\mathcal{F}_{Sb}$, because it has the same orientation as the body frame of the quadrotor. To do so, we apply (3.2).

[3]Notice that if $\bar{\mathbf{d}}_l$ is the distance between $\mathrm{P_T}$ and $\mathrm{P_S}$ in $\mathcal{F}_S$, then according to Fig. 3.1 it is true that $\mathbf{d}_l = \mathbf{R}_S^B \bar{\mathbf{d}}_l$, where $\bar{\mathbf{d}}_l = [0.2\ 0\ 0.15]^T$ [m].

the deflections of this element. Recently, new type of transducers are being produced using optical grade elastomers[4], making them more robust to the environmental change, e.g. temperature and pressure. With their increasing availability, they can be an alternative to the strain-gauge based sensors.

## 3.3 Estimation of the External Wrench

As discussed in Sec. 3.1, it is very convenient in many cases to use a wrench estimator rather than a force sensor, which is mounted on a specific point of the system. For this reason, in this section, we will present a wrench estimator that is inspired by the nonlinear disturbance observers presented for the robotic manipulators in Chen et al. (2000) and more general in Nikoobin and Haghighi (2009).

First of all, consider the Lagrange dynamics of a quadrotor VTOL as described in Sec. 2.1 (to be exact; translational from (2.5) and the rotational dynamics from (2.6)). Assigning the configuration variables of the quadrotor CoM using $\mathbf{q}_q = [\mathbf{p}_q^T \ \boldsymbol{\eta}^T]^T = [q_1 \ \cdots \ q_6]^T \in \mathbb{R}^6$, we can write its dynamics as

$$\mathbf{w}_{ext} = \mathbf{B}(\mathbf{q}_q)\ddot{\mathbf{q}}_q + \mathbf{C}(\mathbf{q}_q, \dot{\mathbf{q}}_q)\dot{\mathbf{q}}_q + \mathbf{g} - \mathbf{G}(\mathbf{q}_q)\mathbf{u}, \tag{3.5}$$

with

$$\mathbf{B}(\mathbf{q}_q) = \begin{bmatrix} m_q \mathbf{I}_3 & * \\ \mathbf{0}_3 & {}^W\mathbf{M}_{qr}(\boldsymbol{\eta}) \end{bmatrix} = \mathbf{B}^T \in \mathbb{R}^{6\times6} \qquad \mathbf{C}(\mathbf{q}_q, \dot{\mathbf{q}}_q) = \begin{bmatrix} \mathbf{0}_3 & \mathbf{0}_3 \\ \mathbf{0}_3 & \mathbf{C}_r \end{bmatrix} \in \mathbb{R}^{6\times6} \tag{3.6}$$

$$\mathbf{g} = \begin{bmatrix} -m_q g \mathbf{e}_3 \\ \mathbf{0}_{3\times1} \end{bmatrix} \in \mathbb{R}^6 \quad \mathbf{G}(\mathbf{q}_q) = \begin{bmatrix} -\mathbf{R}(\boldsymbol{\eta})\mathbf{e}_3 & \mathbf{0}_3 \\ \mathbf{0}_{3\times1} & \mathbf{I}_3 \end{bmatrix} \in \mathbb{R}^{6\times4},$$

where $\mathbf{B}$ is the generalized inertia matrix, $\mathbf{C}_r$ is the matrix representing the Coriolis terms as in (2.7), $\mathbf{g}$ is the vector for the gravitational forces, $\mathbf{G}$ is the control input matrix, $\mathbf{u} = [u_t \ \mathbf{u}_r^T]^T \in \mathbb{R}^4$ is the control input as in (2.3), and finally $\mathbf{w}_{ext} = [\mathbf{f}_{ext}^T \ \boldsymbol{\tau}_{ext}^T]^T \in \mathbb{R}^6$ represents the external wrench acting on the quadrotor.

Let us then propose the following disturbance observer based on Chen et al. (2000) and Nikoobin and Haghighi (2009):

$$\begin{aligned} \dot{\hat{\mathbf{w}}}_{ext} &= \mathbf{L}(\mathbf{q}_q, \dot{\mathbf{q}}_q)(\mathbf{w}_{ext} - \hat{\mathbf{w}}_{ext}) \\ &= -\mathbf{L}(\mathbf{q}_q, \dot{\mathbf{q}}_q)\hat{\mathbf{w}}_{ext} + \mathbf{L}(\mathbf{q}_q, \dot{\mathbf{q}}_q)\Big( \mathbf{B}(\mathbf{q}_q)\ddot{\mathbf{q}}_q + \mathbf{C}(\mathbf{q}_q, \dot{\mathbf{q}}_q)\dot{\mathbf{q}}_q + \mathbf{g} - \mathbf{G}(\mathbf{q}_q)\mathbf{u} \Big), \end{aligned} \tag{3.7}$$

where $\hat{\mathbf{w}}_{ext} = [\hat{\mathbf{f}}_{ext}^T \ \hat{\boldsymbol{\tau}}_{ext}^T]^T \in \mathbb{R}^6$ is the estimated wrench and $\mathbf{L}(\mathbf{q}_q, \dot{\mathbf{q}}_q) \in \mathbb{R}^{6\times6}$ will be designed in order to ensure the convergence of the observer. Since we do not assume any specific model for the external wrench, we have no prior information about the derivative of the external wrenches (or disturbances). Therefore it is assumed that

$$\dot{\mathbf{w}}_{ext} = \mathbf{0}, \tag{3.8}$$

which is inevitable if one does not know anything about the environment geometry and dynamics. However, one can improve the observer performance if a good model of the environment is known. Now, by defining the observer error

$$\mathbf{e}_o = \mathbf{w}_{ext} - \hat{\mathbf{w}}_{ext} \tag{3.9}$$

---

[4]http://optoforce.com/

and considering both (3.7) and (3.8), we can now calculate

$$\dot{\mathbf{e}}_o = \dot{\mathbf{w}}_{ext} - \dot{\hat{\mathbf{w}}}_{ext} = \mathbf{L}(\mathbf{q}_q, \dot{\mathbf{q}}_q)\hat{\mathbf{w}}_{ext} - \mathbf{L}(\mathbf{q}_q, \dot{\mathbf{q}}_q)\mathbf{w}_{ext}, \tag{3.10}$$

which can be expressed, by considering (3.9), as below;

$$\dot{\mathbf{e}}_o + \mathbf{L}(\mathbf{q}_q, \dot{\mathbf{q}}_q)\mathbf{e}_o = \mathbf{0}. \tag{3.11}$$

This means that the choice of $\mathbf{L}(\mathbf{q}_q, \dot{\mathbf{q}}_q)$ will directly affect the asymptotic stability of the error dynamics. In order to implement (3.7) one needs the knowledge of $\dot{\mathbf{q}}_q$, $\dot{\mathbf{q}}_q$, and $\ddot{\mathbf{q}}_q$. Measuring or estimating $\mathbf{q}_q$ and $\dot{\mathbf{q}}_q$ is quite standard in current platforms (see, e.g., Abeywardena et al. (2013), and Section 2.3 of this thesis). However, for many applications, a reliable measurement of the acceleration $\ddot{\mathbf{q}}_q$ (i.e., both the linear and angular acceleration) is not always available. For this purpose we define the auxiliary vector:

$$\boldsymbol{\Psi} = \hat{\mathbf{w}}_{ext} - \boldsymbol{\gamma}(\dot{\mathbf{q}}_q). \tag{3.12}$$

Taking the time derivative of (3.12) we have

$$\dot{\hat{\mathbf{w}}}_{ext} = \dot{\boldsymbol{\Psi}} + \frac{\partial \boldsymbol{\gamma}(\dot{\mathbf{q}}_q)}{\partial \dot{\mathbf{q}}_q} \ddot{\mathbf{q}}_q. \tag{3.13}$$

By equating (3.7) and (3.13) we get

$$\dot{\boldsymbol{\Psi}} + \frac{\partial \boldsymbol{\gamma}(\dot{\mathbf{q}}_q)}{\partial \dot{\mathbf{q}}_q} \ddot{\mathbf{q}}_q = -\mathbf{L}(\mathbf{q}_q, \dot{\mathbf{q}}_q) + \mathbf{L}(\mathbf{q}_q, \dot{\mathbf{q}}_q)\Big(\mathbf{B}(\mathbf{q}_q)\ddot{\mathbf{q}}_q + \mathbf{C}(\mathbf{q}_q, \dot{\mathbf{q}}_q)\dot{\mathbf{q}}_q + \mathbf{g} - \mathbf{G}(\mathbf{q}_q)\mathbf{u}\Big). \tag{3.14}$$

By choosing

$$\frac{\partial \boldsymbol{\gamma}(\dot{\mathbf{q}}_q)}{\partial \dot{\mathbf{q}}_q} = \mathbf{L}(\mathbf{q}_q, \dot{\mathbf{q}}_q)\mathbf{B}(\mathbf{q}_q) \tag{3.15}$$

the dynamics of the nonlinear observer can be written as following

$$\dot{\boldsymbol{\Psi}} = -\mathbf{L}(\mathbf{q}_q, \dot{\mathbf{q}}_q)\boldsymbol{\Psi} + \mathbf{L}(\mathbf{q}_q, \dot{\mathbf{q}}_q)\Big(\mathbf{C}(\mathbf{q}_q, \dot{\mathbf{q}}_q)\dot{\mathbf{q}}_q + \mathbf{g} - \mathbf{G}(\mathbf{q}_q)\mathbf{u} - \boldsymbol{\gamma}(\dot{\mathbf{q}}_q)\Big)$$
$$\hat{\mathbf{w}}_{ext} = \boldsymbol{\Psi} + \boldsymbol{\gamma}(\dot{\mathbf{q}}_q), \tag{3.16}$$

which is not depending anymore on $\ddot{\mathbf{q}}_q$. Therefore this scheme can be implemented without measuring the acceleration of the generalized coordinates $\mathbf{q}_q$ on commonly available quadrotor platforms[5]. As it is seen from (3.11), we must choose $\mathbf{L}(\mathbf{q}_q, \dot{\mathbf{q}}_q)$ such a way that the error dynamics become asymptotically stable. Moreover, the decision made in (3.15) brings a strict dependency of $\mathbf{L}(\mathbf{q}_q, \dot{\mathbf{q}}_q)$ on the choice of $\boldsymbol{\gamma}(\dot{\mathbf{q}}_q)$. Consider the following choice

$$\boldsymbol{\gamma}(\dot{\mathbf{q}}_q) = c_o\dot{\mathbf{q}}_q, \tag{3.17}$$

where $c_o > 0$ is an *observer gain*. The choice of $\boldsymbol{\gamma}(\dot{\mathbf{q}}_q)$ is different from the one made for robot manipulators as shown in Chen et al. (2000) and Nikoobin and Haghighi (2009), since we are dealing with quadrotor dynamics. We obtain that

$$\mathbf{L}(\mathbf{q}_q, \dot{\mathbf{q}}_q) = c_o\mathbf{B}(\mathbf{q}_q)^{-1}. \tag{3.18}$$

---

[5]Note again that actually $\ddot{\mathbf{p}}_q$ is available from the IMU of the quadrotor, as explained in Sec. 2.3. One can use this measurement directly if it is accurate enough. If not, or if it is quite noisy, both the translational and rotational parts of the estimator presented here provide smoothen estimations.

---

**Proposition 1.** *Consider the wrench estimator (3.16) and assume that the roll and pitch velocities are bounded, i.e. $|\dot{\phi}| < \tilde{\phi}$ and $|\dot{\theta}| < \tilde{\theta}$, where $\tilde{\phi}, \tilde{\theta} \in \mathbb{R}^+$. If (3.8) holds and if $\mathbf{L}(\mathbf{q}_q, \dot{\mathbf{q}}_q)$ is defined as in (3.18), then it is possible to have $\hat{\mathbf{w}}_{ext} \to \mathbf{w}_{ext}$.*

*Proof.* See Appendix A.1. □

---

## 3.4 Measurement vs Estimation

For the controllers developed for achieving APhI, e.g. the one presented in Sec. 4.2.2, the knowledge on the external wrench is crucial. The external wrench observer presented in Sec. 3.3 is tested numerically in Sec. 4.3.1, which will be explained better and more in detail in the next chapter. However, for now we can state that in numerical simulations, very good wrench estimation performances have been achieved, even when the noises of the other measurements are taken into account. This performance of the proposed observer strongly relies on the choice of the observer gain, $c_o$, which is introduced in (3.17).

Although tuning this gain in the numerical simulations was relatively easy, for the real experiments it was hard to find a compromise between the convergence of the estimation (see Proposition 1) and its performance. For this reason, here we compare the performances of the F/T sensor measurement (ground truth) explained in Sec. 3.2 with the wrench observer introduced in Sec. 3.3. For the experimental setup, we used the aerial robot in Fig. 3.2, where the overall quadrotor is controlled using IDA-PBC controller (explained in Sec. 4.2.2). In addition, we implemented the nonlinear wrench observer (given in Sec. 3.3), all working together within the ROS environment. Then we disturbed the hovering quadrotor by imposing external forces and torques at the tip point of the rigid link (see Fig. 3.2), which is rigidly attached to the F/T sensor, that is placed on board of the quadrotor.

The online collected external force/torque data is presented in Fig. 3.3, where the F/T sensor measurements are compared with the wrench observer values. The blue curves stand for the raw sensor measurements, while the green one is when they are low-pass filtered. Further fining is done by removing the sensor bias online, and the final sensor data is shown with black curves. The observer data is shown with red. As it is seen from Fig. 3.3, the observer follows the sensor data (which is also the ground truth), but with some oscillations and even with some bias. This is mainly due to the poor tuning of the observer gain $c_o$, and partly due to the small imprecision of the mathematical model. Especially for $f_{e_x}$ and $f_{e_y}$, the observer performs weaker w.r.t. the sensor data. However, notice that for $f_{e_z}$, observer tracks the sensor data much better, which is the direction where the aerial platform is fully actuated.

## 3.5 Discussions

Using external wrench estimators based on the nonlinear model of the system, e.g. the one presented in Sec. 3.3, is a straight-forward way since especially for the quadrotor VTOLs we have relatively reliable mathematical models of the system (see Sec. 2.1). However, as also discussed in McKinnon and Schoellig (2016), this requires fine tuning of the estimator, which in real experiments might not be always as easy as it is for the simulations (see Sec. 3.4).

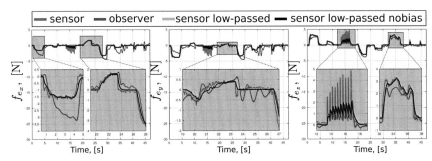

Figure 3.3: Comparison between the F/T sensor (see Sec. 3.2) and the observer (see Sec. 3.3). Only the measured/estimated forces are shown, in units of [N]. Raw sensor readings are depicted with blue curves, and the output of the observer is with red ones. Green color is used for the measured forces which are *low-pass filtered*. Dark curves are used when the bias of this low-pass filtered data is removed in real time. The magnified plots of the each grayed box is placed close by, for better comparison of the different values.

In Fig. 3.3, a comparison between the F/T sensor measurement (provided by the setup explained in Sec. 3.2) and the wrench estimation (provided by the method described in Sec. 3.3) is given. These results are from a real time experiment, where a human is interacting with the flying quadrotor, as explained in Sec. 3.4. Clearly a varying external F/T profile is imposed to the quadrotor, and for a fixed observer gain $c_o$, the estimator is sometimes doing a good job, and sometimes not. For example, for $f_{e_x}$ values (first plot, left), in the first grayed box the performance of the estimator is quite poor, while it gets better in the second grayed box. An adaptive gain tuning method for the estimator might solve this problem. On the other hand, the F/T sensor measurement (also the ground truth) is always providing reliable measurements.

We note that different estimation methods, e.g. in McKinnon and Schoellig (2016), might perform better under certain conditions. However, it is noticeable that using an F/T sensor allows acquiring the exact wrench information, independent from any system model. Moreover, in this case the wrench information would not be corrupted by any other sensor measurement, e.g. the ones provide the state of the robot (see Sec. 2.3). There could be a case, in which let's say the state of the robot might be miscalculated, which might not be crucial when the robot is in *free flight*, i.e. no APhI, but in case of APhI this might bring instability if the wrench estimation is used in the controller. Such a case can occur more frequently, especially when the robot is performing an outdoor task, where accurate state estimation of the flying robots is already a great challenge considering different weather, light and environment conditions.

Consequently, using a low-cost, light weight F/T sensor could be a beneficial choice, providing robust measurements for the indoor and future outdoor experiments; also considering the recent technological developments in transducers, e.g. usage of the optical grade elastomers[6].

---

[6]http://optoforce.com/

# Chapter 4

# Control of Aerial Physical Interaction using IDA-PBC

Using aerial robots as *passive observers* for tasks such as surveillance, monitoring, target tracking while avoiding any kind of physical interaction (e.g. obstacles) was the main objective of the major research done in control of the aerial robots (please refer to Sec. 1.1.2 for a literature review). But, what if we want them to *bump* to the objects, interact with their environment *physically*, and exert meaningful forces and torques on them?

In this chapter we address the control problem of the aerial robots when they are physically interacting with their environments. To do so, we exploit the *cyclo-passive* property of a quadrotor VTOL aerial robotic system. Using the *Interconnection and Damping Assignment - Passivity Based Control* (IDA-PBC) method; we steer the quadrotor through a desired behavior during a physical interaction, by reshaping its physical properties at will, and in a stable manner.

We note that the control method presented in this chapter and its numerical validation is published in Yüksel et al. (2014a), in Yüksel et al. (2014b) and finally in Yüksel et al. (2017).

## 4.1 Introduction

Physical interaction between objects, robots, or living organisms occurs all the time inevitably, this is due to the nature of the physical laws were are all exposed to. This actually turns the term *physical interaction* into a very fundamental question, since our understanding of the '*matter*' itself is not yet in a complete stage. Recalling from Physics lectures, there are four conventionally excepted fundamental interactions/forces[1], that clearly modifies the definition of interaction, based on from which scale one looks at the problem. In this thesis, we look at this problem in the level of so called *weak* or *gravitational* forces, where the physical interaction acts on the mass of a system and varies its energy (kinetic and/or potential). In fact, by physical interaction, we will mean exerting some real energy to some system, e.g. by establishing some level of contact with it. Here, the mechanical systems are in our interest, and the exertion of this energy will be considered through applying some forces and torques on it.

In particular, here we will be studying Aerial Physical Interaction (APhI) and show how one can shape the physical properties of a flying robot when it is interacting with

---

[1]`https://web.archive.org/web/20160304133522/https://www.pha.jhu.edu/~dfehling/particle.gif`

its environment. Energy clearly plays an important role for both modeling and control of the mechanical systems, and we will exploit this fact in this chapter for APhI of the quadrotors. In Sec. 4.1.1 we give the preliminaries on some important energetic properties of the nonlinear systems, and their connections to the concept of *stability*. We then propose the Interconnection and Damping Assignment - Passivity Based Control (IDA-PBC) method for quadrotors in Sec. 4.2, which can render the nonlinear quadrotor dynamics into a stable behavior (thanks to its passiveness), and reshape its dynamics properties (thanks to the port-Hamiltonian (PH) formalization of the system) during a physical interaction. In Sec. 4.3.1 we present the numerical results of this controller for relevant APhI tasks, e.g. sliding on a ceiling surface. We further experimentally validate IDA-PBC method in Sec. 4.3.2 for similar tasks.

## 4.1.1 Preliminaries

In this section we briefly recall the basics of *passivity theory*, and its link to stability, and finally to the port-Hamiltonian (PH) systems. Later we describe how to implement the IDA-PBC method for the PH systems.

A detailed, yet still narrowed summary of the nonlinear system stability and its connection the concept of passivity is given in Appendix C of this thesis.

### Dissipative Systems and Passivity

Consider a dynamical system satisfying Def.1.16 of Secchi et al. (2007), represented by the affine nonlinear function

$$\begin{cases} \dot{\mathbf{x}} &= \mathbf{f}(\mathbf{x}) + \mathbf{g}(\mathbf{x})\mathbf{u} \\ \mathbf{y} &= \mathbf{h}(\mathbf{x}), \end{cases} \tag{4.1}$$

where $\mathbf{x} \in \mathcal{X}$ is the state, $\mathbf{f}$, $\mathbf{g}$ are smooth vector fields and $\mathbf{h}$ is a smooth mapping. Moreover, $\mathbf{u} \in \mathcal{U}$ is the input and $\mathbf{y} \in \mathcal{Y}$ is the output of this system. Call $\rho : \mathcal{U} \times \mathcal{Y} \to \mathbb{R}$ as the *supply rate*.

---

**Definition 1** (Dissipative system-Secchi et al. (2007)). *A system of the form* (4.1) *is said to be dissipative w.r.t. the supply rate* $\rho$*; if* $\forall t \geq 0, \forall \mathbf{u} \in \mathcal{U}, \forall \mathbf{x}_o \in \mathcal{X}, \exists$ *a continuous function* $H : \mathcal{X} \to \mathbb{R}^+$*, s.t. it holds*

$$H(\mathbf{x}(t)) - H(\mathbf{x}_0) \leq \int_0^t \rho(\tau)d\tau.$$

*Moreover, this function is called as the* storage *function.*

---

A detailed description of the dissipative systems can be found in Willems (1972).

---

**Definition 2** (Passive system-Secchi et al. (2007)). *A system of the form* (4.1) *is said to be passive if it is dissipative w.r.t. the supply rate* $\rho(u, y) = \langle u, y \rangle = y^T u$*.*

---

In Definition 2 we used *brackets*, a well-known Dirac notation, for representing the input-output pairs (see details in Secchi et al. (2007)). From the time derivative of the dissipation inequality given in Definition 1, it is trivial to observe that for a passive system $\dot{H} \leq \mathbf{y}^T \mathbf{u}$. holds.

Another definition of the passivity can be done as the following:

---

**Definition 3** (Passivity-Sepulchre et al. (1997)). *A nonlinear system given in (4.1) is said to be passive if*

*(i) it is* relative degree one,

*(ii) it is* weakly minimum phase.

---

The *relative degree* of a nonlinear system is explained in Definitions 4 and 5 of this thesis. For the first condition of Definition 3 to hold, the control input matrix $\mathbf{g}$ of (4.1) must be *full rank*. For IDA-PBC, to be explained later, this means the *matching condition* given in (4.6) holds. The second condition holds when the system in (4.1) is *weakly minimum phase*, i.e. it is passive with a $C^1$ positive definite storage function $H(x)$ (see Proposition 2.46 of Sepulchre et al. (1997)).

### Passivity and Stability

Passivity, or feedback passivity, can be used as a tool for the stabilization of the nonlinear systems. For brevity, by stability of a system, we mean the stability of its equilibrium. There are different ways of analyzing the stability of a nonlinear system, and *Lyapunov stability* and the *input-output stability* are two celebrated and widely used concepts in control theory (see also Appendix C). Without giving a detailed explanation of the Lyapunov stability, it can be named as a continuity property of $\mathbf{x}(t; \mathbf{x}_0, t_0)$ of (4.1), for $\mathbf{u} = \mathbf{0}$ (so called unforced state equation), with respect to $\mathbf{x}_0$. Notice that $\mathbf{x}_0$ is the solution of (4.1) for $\mathbf{u} = \mathbf{0}$ at time $t_0$. More details on Lyapunov stability, its connection to *asymptotic stability* and its *global* properties can be found in Khalil (2001), Sepulchre et al. (1997).

Let us briefly give the relation between stability and passivity in the following theorem:

---

**Theorem 1** (Passivity and stability-Sepulchre et al. (1997)). *Let the system in (4.1) be passive with a storage function $H$, and $\mathbf{h}(\mathbf{x}, \mathbf{u})$ be $C^1$ in $\mathbf{u}$, $\forall \mathbf{x}$. Then the following properties hold:*

*(i) if $H$ is positive definite, then the equilibrium $\mathbf{x} = \mathbf{0}$ of (4.1) with $\mathbf{u} = \mathbf{0}$ is stable,*

*(ii) if $H$ is* zero-state detectable *(ZSD, see Definition C.2.2), then the equilibrium $\mathbf{x} = \mathbf{0}$ of (4.1) with $\mathbf{u} = \mathbf{0}$ is stable,*

*(iii) when there is no throughput, $\mathbf{y} = \mathbf{h}(\mathbf{x})$, then the feedback $\mathbf{u} = -\mathbf{y}$ achieves asymptotic stability of $\mathbf{x} = \mathbf{0}$, if and only if (4.1) is ZSD.*

---

The proof is available together with the Theorem 2.28 of Sepulchre et al. (1997). A detailed explanation of passivity and its connection to several stability criteria for the nonlinear systems can be found in Sepulchre et al. (1997) and Khalil (2001). Especially in the Chapter

2 of Secchi et al. (2007) and Chapter 6 of Khalil (2001) it is reported that this storage function is actually a *Lyapunov candidate*, hence if the system in (4.1) is passive, then it is also Lyapunov stable (also see Theorem C.1.1 and Appendix C.2).

**Passivity and Port-Hamiltonian (PH) Systems**

The port-Hamiltonian (PH) framework is a generalization of the standard Hamiltonian mechanics and energetic features play a primary role in the modeling process. The most common representation of a port-Hamiltonian system is the following:

$$\begin{cases} \dot{\mathbf{x}} = [\mathcal{J}(\mathbf{x}) - \mathcal{R}(\mathbf{x})] \frac{\partial H}{\partial \mathbf{x}} + \mathbf{G}(\mathbf{x})\mathbf{u} \\ \mathbf{y} = \mathbf{G}(\mathbf{x})^T \frac{\partial H}{\partial \mathbf{x}}, \end{cases} \tag{4.2}$$

where $\mathbf{x} \in \mathbb{R}^n$ is the state and $H(\mathbf{x}) : \mathbb{R}^n \to \mathbb{R}$ represents the total amount of energy (*Hamiltonian*) stored in the system and is non negative. Matrices $\mathcal{J}(\mathbf{x}) = -\mathcal{J}(\mathbf{x})^T$ and $\mathcal{R}(\mathbf{x}) \geq 0$ represent the internal energetic interconnections and the dissipation of the port-Hamiltonian system, respectively. Furthermore, $\mathbf{G}(\mathbf{x})$ is the input matrix and the input-output pair $\langle \mathbf{u}, \mathbf{y} \rangle$ represents a power port, namely a pair of variables whose product gives (generalized) power that is either stored or dissipated by the system.

In Proposition 2.18 of Secchi et al. (2007) it is formally proven that a port-Hamiltonian (PH) system in form of (4.2) is a passive system, and the storage function is its *Hamiltonian* function.

**Using IDA-PBC for PH Systems**

By using IDA-PBC from Ortega et al. (2002) together with its extension proposed in Wang et al. (2009), it is possible to control a port-Hamiltonian system in such a way that it behaves as a target dynamics, namely as a new port-Hamiltonian system with a desired interconnection matrix, damping matrix and energy function and even with a different state variable $\bar{\mathbf{x}} \in \mathbb{R}^n$. Formally, let

$$\mathbf{x} = \mathbf{\Phi}(\bar{\mathbf{x}}, t) \tag{4.3}$$

be the map relating $\bar{\mathbf{x}}$ and $\mathbf{x}$, where $\mathbf{\Phi}$ and $\frac{\partial \mathbf{\Phi}}{\partial \bar{\mathbf{x}}}$ are invertible at any time $t$. Let $\mathcal{J}_d$, $\mathcal{R}_d$ and $H_d$ be the desired interconnection matrix, dissipation matrix and energy function, respectively. The port-Hamiltonian system in (4.2) can be transformed into the target port-Hamiltonian dynamics described by

$$\dot{\bar{\mathbf{x}}} = [\mathcal{J}_d(\bar{\mathbf{x}}) - \mathcal{R}_d(\bar{\mathbf{x}})] \frac{\partial H_d}{\partial \bar{\mathbf{x}}} \tag{4.4}$$

using

$$\mathbf{u} = (\mathbf{G}^T(\mathbf{x})\mathbf{G}(\mathbf{x}))^{-1}\mathbf{G}^T(\mathbf{x}) \left[ \frac{\partial \mathbf{\Phi}}{\partial \bar{\mathbf{x}}}(\mathcal{J}_d(\bar{\mathbf{x}}) - \mathcal{R}_d(\bar{\mathbf{x}}))\frac{\partial H_d}{\partial \bar{\mathbf{x}}} - (\mathcal{J}(\mathbf{x}) - \mathcal{R}(\mathbf{x}))\frac{\partial H}{\partial \mathbf{x}} + \frac{\partial \mathbf{\Phi}}{\partial t} \right], \tag{4.5}$$

where $(\mathbf{G}^T(\mathbf{x})\mathbf{G}(\mathbf{x}))^{-1}\mathbf{G}^T(\mathbf{x})$ is the pseudoinverse of $\mathbf{G}(\mathbf{x})$, if and only if the following *matching equation* holds:

$$\mathbf{G}^{\perp}(\mathbf{x}) \left[ \frac{\partial \mathbf{\Phi}}{\partial \bar{\mathbf{x}}}(\mathcal{J}_d(\bar{\mathbf{x}}) - \mathcal{R}_d(\bar{\mathbf{x}}))\frac{\partial H_d}{\partial \bar{\mathbf{x}}} + \frac{\partial \mathbf{\Phi}}{\partial t} - (\mathcal{J}(\mathbf{x}) - \mathcal{R}(\mathbf{x}))\frac{\partial H}{\partial \mathbf{x}} \right] = \mathbf{0}, \tag{4.6}$$

where $\mathbf{G}^\perp(\mathbf{x})$ is the full rank left annihilator of $\mathbf{G}(\mathbf{x})$ .

The main drawback of IDA-PBC is the necessity of solving the nonlinear *partial differential equations* (PDE) (4.6). In general it is not possible to find a closed form solution of the matching equation and, therefore, it is not possible to find all the possible achievable target dynamics. In practice, it is necessary to test if the desired target dynamics is achievable and, if not, to modify it until (4.6) is satisfied.

More information on PH systems and IDA-PBC can be found in Secchi et al. (2007), Ortega et al. (2002) and Wang et al. (2009).

# 4.2 Reshaping the Physical Properties of a Quadrotor for APhI

In this section, we improve the IDA-PBC method for quadrotor VTOLs in a way that we can assign a desired physical behavior to the system. It is, as if we *reshape* its physics, for the purpose of APhI. To do so, in the following we first bring the quadrotor dynamics into its PH formulation, similar to (4.2), and then later control it using the IDA-PBC method.

## 4.2.1 Port-Hamiltonian Dynamics of a Quadrotor

Consider the Newton-Euler dynamics of a quadrotor, where the rotational dynamics is as in (2.4) and the translational dynamics is as in (2.5). In order to simplify the structure of the matching condition and, consequently, to enlarge the set of target dynamics that can be achieved, we consider a control input $\bar{\mathbf{u}}_r$ similar to Lee et al. (2013) but without near-hovering purposes, defined as

$$\mathbf{u}_r = \mathbf{M}_{qr}\mathbf{T}^{-1}\left[(-k_d\mathbf{I} + \mathbf{W})\dot{\boldsymbol{\eta}} + \bar{\mathbf{u}}_r + (\mathbf{I} - \mathbf{M}_{qr}^{-1})\boldsymbol{\tau}_{ext}\right], \qquad (4.7)$$

where $\mathbf{I}$ is the identity matrix of proper dimension, $k_d \in \mathbb{R}^+$, $\mathbf{u}_r = [u_\phi\ u_\theta\ u_\psi]^T \in \mathbb{R}^3$ the control torque and

$$\mathbf{W} = \mathbf{T}\dot{\mathbf{T}}^{-1} + \mathbf{T}\mathbf{M}_{qr}^{-1}[\boldsymbol{\omega}]_\wedge \mathbf{M}_{qr}\mathbf{T}^{-1}. \qquad (4.8)$$

Notice that $\mathbf{T}$ and $\mathbf{T}^{-1}$ are available from (2.2). Therefore, substituting (4.7) in (2.4), we can rewrite the rotational dynamics as follows

$$\ddot{\boldsymbol{\eta}} = -k_d\dot{\boldsymbol{\eta}} + \bar{\mathbf{u}}_r + \boldsymbol{\tau}_{ext}. \qquad (4.9)$$

The plant represented by (2.5) and (4.9) can be modeled as a mechanical port-Hamiltonian system. Let $\mathbf{M} \in \mathbb{R}^{6\times6}$ be

$$\mathbf{M} = \begin{bmatrix} m_q\mathbf{I} & \mathbf{0} \\ \mathbf{0} & \mathbf{I} \end{bmatrix} \in \mathbb{R}^{6\times6}, \qquad (4.10)$$

where $\mathbf{0}$ is the zero matrix of proper dimension. Recall that $\mathbf{q}_q = [\mathbf{p}_q^T\ \boldsymbol{\eta}^T]^T = [q_1, \cdots, q_6] \in \mathbb{R}^6$ and $\mathbf{p} = \mathbf{M}\dot{\mathbf{q}}_q \in \mathbb{R}^6$ are the configuration and momentum variables. Furthermore, let $\mathbf{u}_i = [u_t\ \bar{\mathbf{u}}_r^T]^T \in \mathbb{R}^4$ be the input vector. The quadrotor dynamics can be rewritten as:

$$\begin{bmatrix} \dot{\mathbf{q}}_q \\ \dot{\mathbf{p}} \end{bmatrix} = \left[ \begin{pmatrix} \mathbf{0} & \mathbf{I} \\ -\mathbf{I} & \mathbf{0} \end{pmatrix} - \begin{pmatrix} \mathbf{0} & \mathbf{0} \\ \mathbf{0} & \mathcal{R} \end{pmatrix} \right] \begin{bmatrix} \frac{\partial H}{\partial \mathbf{q}_q} \\ \frac{\partial H}{\partial \mathbf{p}} \end{bmatrix} + \begin{bmatrix} \mathbf{0} & \mathbf{0} \\ \mathbf{G} & \mathbf{I} \end{bmatrix} \begin{bmatrix} \mathbf{u}_i \\ \mathbf{w}_{ext} \end{bmatrix}, \qquad (4.11)$$

where $\mathcal{R} = k_d\mathbf{I}$ models the dissipation introduced by (4.7) and $\mathbf{w}_{ext} = [\mathbf{f}_{ext}^T \ \boldsymbol{\tau}_{ext}^T]^T$ is the external wrench acting on the quadrotor. Remember that this wrench can either be measured or estimated using the methods described in Chapter 3. The total energy function and the control input $\mathbf{G}$ are given by:

$$H(\mathbf{q}_q, \mathbf{p}) = \frac{1}{2}\mathbf{p}^T\mathbf{M}^{-1}\mathbf{p} + V(\mathbf{q}_q) = \frac{1}{2}\mathbf{p}^T\mathbf{M}^{-1}\mathbf{p} - m_q g q_3, \tag{4.12}$$

$$\mathbf{G} = \begin{bmatrix} \mathbf{g}_1 & \mathbf{0} \\ \mathbf{0} & \mathbf{I} \end{bmatrix} \in \mathbb{R}^{6\times4} \quad \text{with} \quad \mathbf{g}_1 = -\mathbf{R}\mathbf{e}_3 \in \mathbb{R}^3. \tag{4.13}$$

It can be shown that the quadrotor has the property of *cyclo-passivity* (see also Willems (1972)), namely it cannot create energy over closed paths in the state space. Passivity, a stronger property, cannot be proven because the gravitational potential energy $V(\mathbf{q}_q)$, and, consequently, the total energy (4.12) is not lower bounded.

---

**Proposition 2.** *The system* (4.11) *is cyclo-passive with respect to the pair*

$$\left\langle \begin{bmatrix} \mathbf{u}_i \\ \mathbf{w}_{ext} \end{bmatrix}, \begin{bmatrix} \mathbf{G}^T\frac{\partial H}{\partial \mathbf{p}} \\ \frac{\partial H}{\partial \mathbf{p}} \end{bmatrix} \right\rangle.$$

---

*Proof.* See Appendix A.2. □

---

**Remark 1.** *The cyclopassivity property can be interpreted as an extension of the more standard passivity property. It requires that the system behaves as a physical system from an energetic point of view (i.e., that the energy introduced into the system from the external world is either stored or dissipated) but it does not require that the energy function is lower bounded. Cyclopassivity, unlike passivity, prevents from proving the stability of an equilibrium point of the unforced system but, nevertheless, this is consistent with the physics of the quadrotor that has no equilibrium points in case all the inputs (both the control input and the external wrench) are null.*

---

### 4.2.2 IDA-PBC of Quadrotors for Aerial Physical Interaction

In this section we will exploit and extend the IDA-PBC formulation presented in Wang et al. (2009) in order to completely change the physical properties of a quadrotor and the way it reacts to external forces and torques. In other words, rather than controlling the position or the velocity, we aim at transforming the quadrotor into a physically different quadrotor that reacts as a new desired physical system to external solicitations.

More formally, we aim at controlling (4.11) in such a way that it behaves as a new mechanical system described by:

$$\begin{bmatrix} \dot{\mathbf{q}} \\ \dot{\mathbf{p}} \end{bmatrix} = \left[ \begin{pmatrix} \mathbf{0} & \mathbf{I} \\ -\mathbf{I} & \mathbf{0} \end{pmatrix} - \begin{pmatrix} \mathbf{0} & \mathbf{0} \\ \mathbf{0} & \mathcal{R}_d \end{pmatrix} \right] \begin{bmatrix} \frac{\partial H_d}{\partial \mathbf{q}_q} \\ \frac{\partial H_d}{\partial \bar{\mathbf{p}}} \end{bmatrix} + \begin{bmatrix} \mathbf{0} \\ \mathbf{I} \end{bmatrix} \tilde{\mathbf{w}}_{ext}, \tag{4.14}$$

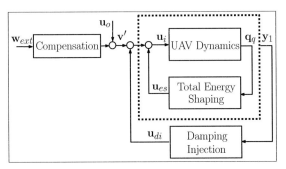

Figure 4.1: Control design using IDA-PBC and Damping Injection. Dynamic transformation and energy shaping is performed inside the dashed set of blocks. After damping injection, the external wrench is compensated.

where the new state $\bar{\mathbf{p}} = \mathbf{M}_d \dot{\mathbf{q}}_q$ is the new momentum, associated with the new inertia matrix $\mathbf{M}_d$ that is chosen to be constant and with the following structure:

$$\mathbf{M}_d = \begin{bmatrix} m_d \mathbf{I} & \mathbf{0} \\ \mathbf{0} & \mathbf{N} \end{bmatrix} \in \mathbb{R}^{6 \times 6}, \tag{4.15}$$

where $m_d \in \mathbb{R}^+$ and $\mathbf{N} \in \mathbb{R}^{3 \times 3}$ is a symmetric positive definite matrix, representing the desired mass and the desired rotational inertia respectively. The desired energy function is

$$H_d = \frac{1}{2} \bar{\mathbf{p}}^T \mathbf{M}_d^{-1} \bar{\mathbf{p}} + V_d(\mathbf{q}_q). \tag{4.16}$$

The choice of $\mathbf{M}_d$ has been made in order to mimic the structure of (4.10) such that the controlled system will have an inertia that is consistent with the mechanics of the quadrotor. Furthermore, (4.15) has the advantage of decoupling rotational and Cartesian kinetic energy, simplifying the design of the IDA-PBC control law. The desired potential function $V_d$ can be any function such that the matching equation of the IDA-PBC is satisfied. $\mathcal{R}_d$ is the desired dissipation matrix that will also be constrained by the underactuation of the quadrotor. Finally, $\tilde{\mathbf{w}}_{ext}$ is the partially compensated external wrench and it will be defined more clearly later in the following.

The control law, whose block diagram is depicted in Fig. 4.1, will be designed in two steps. In the first step (developed in Sec. 4.2.3) the non conservative wrenches will be disregarded and the internal energetic structure of the quadrotor will be shaped. In the second step (detailed in Sec. 4.2.4) dissipation and external wrench will be considered and the control input will be adjusted in such a way to achieve the target dynamics (4.14).

### 4.2.3 Total Energy Shaping

For the reasons reported in Gomez-Estern and van der Schaft (2004), when the plant contains some inherent dissipation as in (4.11), it is convenient to firstly shape the energy disregarding the inherent dissipation and then to tune the dissipation by damping injection.

Thus, in order to shape the energy of the plant, we consider the following undamped plant, where also the external wrench is disregarded

$$\begin{bmatrix} \dot{\mathbf{q}} \\ \dot{\mathbf{p}} \end{bmatrix} = \begin{bmatrix} 0 & \mathbf{I} \\ -\mathbf{I} & 0 \end{bmatrix} \begin{bmatrix} \frac{\partial H}{\partial \mathbf{q}_q} \\ \frac{\partial H}{\partial \mathbf{p}} \end{bmatrix} + \begin{bmatrix} 0 \\ \mathbf{G} \end{bmatrix} \mathbf{u}_{es}, \tag{4.17}$$

and we design the input $\mathbf{u}_{es}$ in order to obtain an undamped controlled system with the desired energy function $H_d$ and with the desired momentum $\bar{\mathbf{p}}$:

$$\begin{bmatrix} \dot{\mathbf{q}}_q \\ \dot{\bar{\mathbf{p}}} \end{bmatrix} = \begin{bmatrix} 0 & \mathbf{I} \\ -\mathbf{I} & 0 \end{bmatrix} \begin{bmatrix} \frac{\partial H_d}{\partial \mathbf{q}_q} \\ \frac{\partial H_d}{\partial \bar{\mathbf{p}}} \end{bmatrix} \tag{4.18}$$

Since $\bar{\mathbf{p}} = \mathbf{M}_d \dot{\mathbf{q}}_q = \mathbf{M}_d \mathbf{M}^{-1} \mathbf{M} \dot{\mathbf{q}}_q = \mathbf{M}_d \mathbf{M}^{-1} \mathbf{p}$, we have that the relation between the state of (4.17) and the state of the target dynamics (4.18) is given by:

$$\mathbf{x} = \begin{bmatrix} \mathbf{q}_q \\ \mathbf{p} \end{bmatrix} = \begin{bmatrix} \mathbf{I} & 0 \\ 0 & \mathbf{M}\mathbf{M}_d^{-1} \end{bmatrix} \begin{bmatrix} \mathbf{q}_q \\ \bar{\mathbf{p}} \end{bmatrix} = \mathbf{F} \begin{bmatrix} \mathbf{q}_q \\ \bar{\mathbf{p}} \end{bmatrix} = \mathbf{\Phi}(\bar{\mathbf{x}}), \tag{4.19}$$

and, consequently,

$$\frac{\partial \mathbf{\Phi}}{\partial \bar{\mathbf{x}}} = \mathbf{F}, \qquad \frac{\partial \mathbf{\Phi}}{\partial t} = \mathbf{0}. \tag{4.20}$$

Substituting (4.17), (4.18), and (4.20) in (4.6) we obtain the following matching equations:

$$\begin{cases} \frac{\partial H_d}{\partial \bar{\mathbf{p}}} - \frac{\partial H}{\partial \mathbf{p}} = 0 \\ \mathbf{G}^\perp \left\{ \frac{\partial H}{\partial \mathbf{q}_q} - \mathbf{M}\mathbf{M}_d^{-1} \frac{\partial H_d}{\partial \mathbf{q}_q} \right\} = \mathbf{0}. \end{cases} \tag{4.21}$$

It easy to check that the first equation is always satisfied. Furthermore, since both $\mathbf{M}$ and $\mathbf{M}_d$ are constant, using (4.12) and (4.16) the second condition can be rewritten as:

$$\mathbf{G}^\perp \left\{ \frac{\partial V}{\partial \mathbf{q}_q} - \mathbf{M}\mathbf{M}_d^{-1} \frac{\partial V_d}{\partial \mathbf{q}_q} \right\} = \mathbf{0}. \tag{4.22}$$

Thus, it is possible to choose $m_d$ and $\mathbf{N}$ in (4.15) arbitrarily while the desired potential energy for the controlled system must satisfy (4.22).

A possible choice for the full rank left annihilator $\mathbf{G}^\perp$ is

$$\mathbf{G}^\perp = \begin{bmatrix} 0 & -1 & \frac{\mathbf{g}_1(2)}{\mathbf{g}_1(3)} & 0 & 0 & 0 \\ -1 & 0 & \frac{\mathbf{g}_1(1)}{\mathbf{g}_1(3)} & 0 & 0 & 0 \end{bmatrix}, \tag{4.23}$$

where $\mathbf{g}_1(i)$ indicates the $i$-th component of the vector $\mathbf{g}_1$, which was defined in (4.13). Using (4.23) with (4.22) yields:

$$\begin{cases} \frac{\partial V_d}{\partial q_2} - \frac{\mathbf{g}_1(2)}{\mathbf{g}_1(3)} \left( \frac{\partial V}{\partial q_3} - \frac{m_q}{m_d} \frac{\partial V_d}{\partial q_3} \right) = 0 \\ \frac{\partial V_d}{\partial q_1} - \frac{\mathbf{g}_1(1)}{\mathbf{g}_1(3)} \left( \frac{\partial V}{\partial q_3} - \frac{m_q}{m_d} \frac{\partial V_d}{\partial q_3} \right) = \mathbf{0}. \end{cases} \tag{4.24}$$

Admissible potentials are all and only the solutions of the PDEs (4.24). A possible simple solution is:

$$V_d(\mathbf{q}_q) = -m_d g q_3 + \bar{V}_d(q_4, q_5, q_6). \tag{4.25}$$

This potential energy function is consistent with the desired mass $m_d$ since it scales the gravity force accordingly and it allows to arbitrarily shape the potential energy of the rotational part.

---

**Remark 2.** *The non-constant terms of (4.23), and consequently (4.24) have a singularity corresponding to the configurations where the pitch or the roll are at $\frac{\pi}{2} + k\pi$, where $k \in \mathbb{Z}$. In order for the controller to work properly, the quadrotor should be kept away from these configurations.*

---

**Remark 3.** *The limits in the choice of the potential are due to the underactuation of the quadrotor. Since the attitude is fully actuated, it is possible to arbitrarily choose a potential on the orientation while the underactuation in the Cartesian coordinates limits the choice of a translational potential.*

---

Thus, once an admissible potential has been chosen, using (4.5), the control input shaping the dynamics of (4.17) in (4.18) is given by:

$$\mathbf{u}_{es} = (\mathbf{G}^T \mathbf{G})^{-1} \mathbf{G}^T \left( \frac{\partial H}{\partial \mathbf{q}_q} - \mathbf{M} \mathbf{M}_d^{-1} \frac{\partial H_d}{\partial \mathbf{q}_q} \right). \tag{4.26}$$

## 4.2.4 Dissipation and External Wrench Shaping

We will now consider the full model of the plant and we will design the input $\mathbf{u} = \mathbf{u}_{es} + \mathbf{u}'$ for shaping the damping and the external wrenches.

Considering (4.19) it is possible to rewrite (4.11) as:

$$\begin{bmatrix} \dot{\mathbf{q}}_q \\ \dot{\mathbf{p}} \end{bmatrix} = \mathbf{F}^{-1} \begin{bmatrix} \mathbf{0} & \mathbf{I} \\ -\mathbf{I} & \mathbf{0} \end{bmatrix} \begin{bmatrix} \frac{\partial H}{\partial \mathbf{q}_q} \\ \frac{\partial H}{\partial \mathbf{p}} \end{bmatrix} + \mathbf{F}^{-1} \begin{bmatrix} \mathbf{0} \\ \mathbf{G} \end{bmatrix} \mathbf{u}_{es} -$$
$$- \mathbf{F}^{-1} \begin{bmatrix} \mathbf{0} & \mathbf{0} \\ \mathbf{0} & \mathcal{R} \end{bmatrix} \begin{bmatrix} \frac{\partial H}{\partial \mathbf{q}_q} \\ \frac{\partial H}{\partial \mathbf{p}} \end{bmatrix} + \mathbf{F}^{-1} \begin{bmatrix} \mathbf{0} \\ \mathbf{G} \end{bmatrix} \mathbf{u}' + \mathbf{F}^{-1} \begin{bmatrix} \mathbf{0} \\ \mathbf{I} \end{bmatrix} \mathbf{w}_{ext}. \tag{4.27}$$

Considering the results of Sec. 4.2.3 and recalling that

$$\frac{\partial H}{\partial \mathbf{p}} = \frac{\partial H_d}{\partial \bar{\mathbf{p}}},$$

we can rewrite (4.27) as:

$$\begin{bmatrix} \dot{\mathbf{q}}_q \\ \dot{\mathbf{p}} \end{bmatrix} = \begin{bmatrix} \mathbf{0} & \mathbf{I} \\ -\mathbf{I} & \mathbf{0} \end{bmatrix} \begin{bmatrix} \frac{\partial H_d}{\partial \mathbf{q}_q} \\ \frac{\partial H_d}{\partial \bar{\mathbf{p}}} \end{bmatrix} - \begin{bmatrix} \mathbf{0} & \mathbf{0} \\ \mathbf{0} & \mathbf{M}_d \mathbf{M}^{-1} \mathcal{R} \end{bmatrix} \begin{bmatrix} \frac{\partial H}{\partial \mathbf{q}_q} \\ \frac{\partial H_d}{\partial \bar{\mathbf{p}}} \end{bmatrix} + \begin{bmatrix} \mathbf{0} \\ \mathbf{M}_d \mathbf{M}^{-1} \mathbf{G} \end{bmatrix} \mathbf{u}' + \begin{bmatrix} \mathbf{0} \\ \mathbf{M}_d \mathbf{M}^{-1} \end{bmatrix} \mathbf{w}_{ext}. \tag{4.28}$$

Decompose the input as $\mathbf{u}' = \mathbf{u}_{di} + \mathbf{v}'$ and set

$$\mathbf{u}_{di} = -\mathbf{K}_v \mathbf{y}_1, \tag{4.29}$$

where

$$\mathbf{y}_1 = \mathbf{G}^T \mathbf{M}^{-T} \mathbf{M}_d^T \frac{\partial H_d}{\partial \bar{\mathbf{p}}}$$

is the natural velocity-like output of (4.28) and

$$\mathbf{K}_v = \begin{bmatrix} k_T & \mathbf{0} \\ \mathbf{0} & \mathbf{K}_R \end{bmatrix} \in \mathbb{R}^{4\times 4},$$

with $k_T \in \mathbb{R}^+$ and $\mathbb{R}^{3\times 3} \ni \mathbf{K}_R > \mathbf{0}$. The input $\mathbf{u}_{di}$ can be used for tuning the desired damping. Thus, it is possible to rewrite (4.28) as:

$$\begin{bmatrix} \dot{\mathbf{q}}_q \\ \dot{\mathbf{p}} \end{bmatrix} = \left[ \begin{pmatrix} \mathbf{0} & \mathbf{I} \\ -\mathbf{I} & \mathbf{0} \end{pmatrix} - \begin{pmatrix} \mathbf{0} & \mathbf{0} \\ \mathbf{0} & \mathcal{R}_d \end{pmatrix} \right] \begin{bmatrix} \frac{\partial H_d}{\partial \mathbf{q}_q} \\ \frac{\partial H_d}{\partial \mathbf{p}} \end{bmatrix} + \begin{bmatrix} \mathbf{0} \\ \mathbf{M}_d\mathbf{M}^{-1}\mathbf{G} \end{bmatrix} \mathbf{v}' + \begin{bmatrix} \mathbf{0} \\ \mathbf{M}_d\mathbf{M}^{-1} \end{bmatrix} \mathbf{w}_{ext}, \qquad (4.30)$$

where

$$\mathcal{R}_d = \mathbf{M}_d\mathbf{M}^{-1}\mathcal{R} + \mathbf{M}_d\mathbf{M}^{-1}\mathbf{G}\mathbf{K}_v\mathbf{G}^T\mathbf{M}^{-T}\mathbf{M}_d^T. \qquad (4.31)$$

Because of scaling due to the change of the momentum, (4.30) is not a standard damping injection and it is necessary to verify that $\mathcal{R}_d$ is always positive definite. In general the product of two positive definite matrices is not always positive definite.

---

**Proposition 3.** *The desired dissipation matrix $\mathcal{R}_d$ in (4.31) is always positive definite. Moreover, by setting*

$$\begin{cases} k_T = \left(\frac{m}{m_d}\right)^2 \bar{k}_T \\ \mathbf{K}_R = \mathbf{N}^{-1}(\bar{\mathbf{K}}_R - k_d\mathbf{N})\mathbf{N}^{-1} \end{cases} \qquad (4.32)$$

*it is possible to achieve any desired damping $\bar{k}_T \in \mathbb{R}^+$ along the actuated Cartesian direction and any rotational damping matrix $\mathbb{R}^{3\times 3} \ni \bar{\mathbf{K}}_R > \mathbf{0}$.*

---

*Proof.* See Appendix A.3. ∎

The change of momentum for the desired target dynamics introduces a scaling also on the way the external wrench $\mathbf{w}_{ext}$ influences the evolution of the system. Ideally, the external force should influence the evolution of the controlled system in the same way it does in (4.11). If the external wrench can be measured, then it is possible to exploit the control input for eliminating the scaling.

In order to obtain the ideal behavior, we can see from (4.30) that the input $\mathbf{v}'$ should be chosen in such a way that:

$$\mathbf{M}_d\mathbf{M}^{-1}\mathbf{G}\mathbf{v}' + \mathbf{M}_d\mathbf{M}^{-1}\mathbf{w}_{ext} = \mathbf{w}_{ext}. \qquad (4.33)$$

Hovewer, because of the underactuation of the quadrotor, it is possible to have only a partial compensation that can be achieved setting

$$\mathbf{v}' = \mathbf{G}^+(\mathbf{M}\mathbf{M}_d^{-1}(\mathbf{I} - \mathbf{M}_d\mathbf{M}^{-1})\mathbf{w}_{ext}) + \mathbf{u}_o, \qquad (4.34)$$

where $\mathbf{G}^+$ is the pseudo-inverse of $\mathbf{G}$ and the term $\mathbf{u}_o$ is an extra outer control input, e.g. for trajectory tracking of the quadrotor. Replacing (4.34) in (4.33) and setting $\mathbf{u}_o = \mathbf{0}$ we obtain that:

$$\mathbf{M}_d\mathbf{M}^{-1}\mathbf{G}\mathbf{G}^+(\mathbf{M}\mathbf{M}_d^{-1}(\mathbf{I} - \mathbf{M}_d\mathbf{M}^{-1})\mathbf{w}_{ext}) + \mathbf{M}_d\mathbf{M}^{-1}\mathbf{w}_{ext} = \tilde{\mathbf{w}}_{ext}, \qquad (4.35)$$

where $\tilde{\mathbf{w}}_{ext}$ is the best compensation that can be achieved.

> **Remark 4.** *By simple computations, it can be seen from (4.34) that the scaling on the external torques can be perfectly compensated and the approximation remains only on the compensation of the translational part.*

Finally, putting together (4.26), (4.29) and (4.34), we obtain that the control input $\mathbf{u}_i$ is given by (see also Fig. 4.1):

$$
\begin{aligned}
\mathbf{u}_i =& \mathbf{u}_{es} + \mathbf{u}_{di} + \mathbf{v}' = (\mathbf{G}^T\mathbf{G})^{-1}\mathbf{G}^T\left(\frac{\partial H}{\partial \mathbf{q}_q} - \mathbf{M}\mathbf{M}_d^{-1}\frac{\partial H_d}{\partial \mathbf{q}_q}\right) - \\
& - \mathbf{K}_v\mathbf{G}^T\mathbf{M}^{-T}\mathbf{M}_d^T\frac{\partial H_d}{\partial \bar{\mathbf{p}}} + \mathbf{G}^+(\mathbf{M}\mathbf{M}_d^{-1}(\mathbf{I} - \mathbf{M}_d\mathbf{M}^{-1})\mathbf{w}_{ext}) + \mathbf{u}_o,
\end{aligned}
\tag{4.36}
$$

which leads to the closed-loop system

$$
\begin{bmatrix} \dot{\mathbf{q}}_q \\ \dot{\bar{\mathbf{p}}} \end{bmatrix} = \left[ \begin{pmatrix} \mathbf{0} & \mathbf{I} \\ -\mathbf{I} & \mathbf{0} \end{pmatrix} - \begin{pmatrix} \mathbf{0} & \mathbf{0} \\ \mathbf{0} & \mathcal{R}_d \end{pmatrix} \right] \begin{bmatrix} \frac{\partial H_d}{\partial \mathbf{q}_q} \\ \frac{\partial H_d}{\partial \bar{\mathbf{p}}} \end{bmatrix} + \begin{bmatrix} \mathbf{0} \\ \mathbf{I} \end{bmatrix} \tilde{\mathbf{w}}_{ext} + \begin{bmatrix} \mathbf{0} \\ \mathbf{M}_d\mathbf{M}^{-1}\mathbf{G} \end{bmatrix} \mathbf{u}_o.
\tag{4.37}
$$

If we set $\mathbf{u}_o = \mathbf{0}$; the desired dynamics in (4.14) as a new quadrotor with a new inertia, damping and potential structure is achieved. The external input $\mathbf{u}_o$ can be used for controlling such a physically modified quadrotor. In other words, the controller in (4.36) can be used as an inner (low-level) control loop for changing the physical characteristics of the quadrotor and the input $\mathbf{u}_0$ can be exploited for building outer loops controlling this new system, taking advantage of its new desired physics.

> **Proposition 4.** *The controlled system* (4.37) *is cyclo-passive with respect to the input-output pair:*
> $$
> \left\langle \begin{bmatrix} \mathbf{u}_o \\ \tilde{\mathbf{w}}_{ext} \end{bmatrix}, \begin{bmatrix} \mathbf{G}^T\mathbf{M}^{-T}\mathbf{M}_d^T\frac{\partial H_d}{\partial \bar{\mathbf{p}}} \\ \frac{\partial H_d}{\partial \bar{\mathbf{p}}} \end{bmatrix} \right\rangle.
> $$

*Proof.* See Appendix A.4. □

> **Remark 5.** *Even if the the compensation of the external wrench is only partial, the target dynamics that is achieved is still well behaved from a physical point of view and no regenerative effects are present. Furthermore if $V_d$ can be chosen to be lower bounded in (e.g., in a desired range of operation), then the achievable target dynamics is passive.*

> **Remark 6.** *The IDA-PBC method presented here, where the control input is computed as in (4.36), aims at transforming the quadrotor into a physically different quadrotor that reacts as a new desired physical system to external solicitations, rather than controlling its position or the velocity. In this sense, it can be considered as a* low-level *controller developed purely for APhI tasks, which also admits high-level inputs, e.g. $\mathbf{u}_o$, that can be computed for position/velocity or force/torque tracking purposes. In this case, this high-level input controls the desired (or physically reshaped) quadrotor dynamics.*

## 4.2.5 Using IDA-PBC for Turning Quadrotors into 3D Force Effectors

Here we will briefly show how to use the recently presented IDA-PBC controller for turning the quadrotors into 3D force effectors. Let us first consdider the linear equation in (2.5) for the reshaped quadrotor as

$$m_d \ddot{\mathbf{p}}_q = \underbrace{-u_t \mathbf{R}(\boldsymbol{\eta})\mathbf{e}_3 + m_d g \mathbf{e}_3}_{=\mathbf{f}_{ne}} + \mathbf{f}_{ext}, \qquad (4.38)$$

where we denoted with $\mathbf{f}_{ne}$ the sum of the total thrust force and the gravity force, i.e., all the forces contributing to the linear motion of the quadrotor except for the external force $\mathbf{f}_{ext}$. If $\ddot{\mathbf{p}}_q = \mathbf{0}$, i.e., the quadrotor moves at constant (possibly zero) velocity, then we have that $\mathbf{f}_{ne} = -\mathbf{f}_{ext}$, i.e., $\mathbf{f}_{ne}$ represents, in this case, the force that the quadrotor is exerting on the external world. In this section we present a method that allows using the IDA-PBC framework presented before, in order to regulate $\mathbf{f}_{ne}$ to a certain desired value. This property can be used, e.g., to counterbalance an external disturbance like a constant wind or to press against a wall or an object with a certain given force (see numerical results in Sec. 4.3.1).

Denote with $\mathbf{f}^* = [f_x^*\ f_y^*\ f_z^*]^T \in \mathbb{R}^3$ the value of the desired $\mathbf{f}_{ne}$. Imposing $\mathbf{f}_{ne} = \mathbf{f}^*$ we obtain the following nonlinear system of equations

$$u_t \begin{bmatrix} s_\phi s_\psi + c_\phi c_\psi s_\theta \\ c_\phi s_\theta s_\psi - c_\psi s_\phi \\ c_\theta c_\phi \end{bmatrix} = \begin{bmatrix} -f_x^* \\ -f_y^* \\ -f_z^* + m_d g \end{bmatrix}, \qquad (4.39)$$

with $c_* = \cos(*)$ and $s_* = \sin(*)$. After some straightforward algebra we obtain

$$u_t \begin{bmatrix} c_\phi s_\theta \\ -s_\phi \\ c_\theta c_\phi \end{bmatrix} = \begin{bmatrix} -f_x^* c_\psi - f_y^* s_\psi \\ f_x^* s_\psi - f_y^* c_\psi \\ -f_z^* + m_d g \end{bmatrix}, \qquad (4.40)$$

which, assuming that also $\psi$ is known and denoting its value with $\psi_{ss}$, it can be solved in the unknown $\rho$, $\phi$, and $\theta$ resulting in

$$u_t^* = \sqrt{f_x^{*2} + f_y^{*2} + (f_z^* - m_d g)^2} \qquad (4.41)$$

$$\phi^* = \arcsin\left(\frac{-f_x^* c_{\psi_{ss}} - f_y^* s_{\psi_{ss}}}{u_t^*}\right) \qquad (4.42)$$

$$\theta^* = -\arcsin\left(\frac{f_x^* s_{\psi_{ss}} - f_y^* c_{\psi_{ss}}}{u_t^* c_{\phi^*}}\right). \qquad (4.43)$$

Consider now the rotational dynamics of the reshaped quadrotor. We choose $\bar{V}_d(q_4, q_5, q_6)$ in (4.25) as

$$\bar{V}_d(\mathbf{q}) = \frac{1}{2}\boldsymbol{\eta}_e^T \mathbf{K}_p \boldsymbol{\eta}_e, \qquad (4.44)$$

where $\mathbf{K}_p = \mathrm{diag}([\mathrm{k}_p^\phi, \mathrm{k}_p^\theta, \mathrm{k}_p^\psi]) \in \mathbb{R}^{3 \times 3}$ and

$$\boldsymbol{\eta}_e = \boldsymbol{\eta} - \bar{\boldsymbol{\eta}} = \begin{bmatrix} \phi - \bar{\phi} \\ \theta - \bar{\theta} \\ \psi \end{bmatrix}, \qquad (4.45)$$

with $\bar{\phi}$, $\bar{\theta}$ are parameters to be designed in order to obtain $\phi \to \phi^*$ and $\theta \to \theta^*$. By virtue of (4.44) the rotational dynamics of the reshaped quadrotor can be expressed as

$$\ddot{\boldsymbol{\eta}} = -k_d \dot{\boldsymbol{\eta}} + \mathbf{K}_p(\bar{\boldsymbol{\eta}} - \boldsymbol{\eta}) + \tilde{\boldsymbol{\tau}}_{ext}, \tag{4.46}$$

where $\tilde{\boldsymbol{\tau}}_{ext} = [\tilde{\tau}_{e_x} \ \tilde{\tau}_{e_y} \ \tilde{\tau}_{e_z}]^T \in \mathbb{R}^3$ denotes the external torque acting on the reshaped quadrotor (that, we recall, can be, measured or estimated using the methods presented in Chapter 3). Assuming that $\tilde{\boldsymbol{\tau}}_{ext}$ is constant we obtain the following equilibrium at steady state,

$$\mathbf{K}_p(\boldsymbol{\eta}_{ss} - \bar{\boldsymbol{\eta}}) = \tilde{\boldsymbol{\tau}}_{ext}, \tag{4.47}$$

where $\boldsymbol{\eta}_{ss} = [\phi_{ss} \ \theta_{ss} \ \psi_{ss}]^T \in \mathbb{R}^3$ represent the steady state attitude. It is straightforward to see that

$$\psi_{ss} = \frac{\tilde{\tau}_{e_z}}{k_p^\psi},$$

which can be used in (4.42) and (4.43) in order to find the exact values of $\phi^*$ and $\theta^*$ that are needed to achieve the desired $\mathbf{f}^*$. Given those values of $\phi^*$ and $\theta^*$ we choose $\bar{\phi}$ and $\bar{\theta}$ such that $\phi_{ss} = \phi^*$ and $\theta_{ss} = \theta^*$, i.e.,

$$\bar{\phi} = \phi^* - \frac{\tilde{\tau}_{e_x}}{k_p^\phi} \tag{4.48}$$

$$\bar{\theta} = \theta^* - \frac{\tilde{\tau}_{e_y}}{k_p^\theta}. \tag{4.49}$$

In summary, by choosing the thrust as in (4.41) and the desired potential as in (4.44), with $\bar{\phi}$ and $\bar{\theta}$ given by (4.48) and (4.49), respectively; we can let $\mathbf{f}_{ne}$ converge to $\mathbf{f}^*$ even in the presence of a disturbing (but constant) external torque $\tilde{\boldsymbol{\tau}}_{ext}$.

This technique can be applied for example to balance an external constant force $\mathbf{f}_{ext}$ produced by a wind or any other external agent. To this aim one has to measure or estimate $\mathbf{f}_{ext}$ using the sensor or the observer presented in Chapter 3 and then select $\mathbf{f}^* = -\mathbf{f}_{ext}$, which results in a compensation of the external force acting on the quadrotor. Another possible application of this technique the exertion of a constant force to a wall or to load for the purpose, e.g., of pushing, lifting, and so on.

## 4.3 Numerical and Experimental Results

This section is reserved for the numerical and experimental validations of the IDA-PBC for a quadrotor VTOL, when it is physically interacting with its environment while achieving a stable flight. We first show the numerical results in Sec. 4.3.1, for reshaping the physics of the quadrotor under the effect of disturbing forces and torques. After observing its behavior for different desired physical properties, we use this to perform an interesting physical interaction task: sliding on a ceiling surface. For this, a rigid tool on the quadrotor is thought, whose tip is in contact with the environment, while its base is rigidly attached to the quadrotor. A sketch of this can be seen Fig. 4.4, and its realization in Fig. 3.2. Such task can be imagined as surface inspection, cleaning or painting. Then we test our controller when the external forces and torques are provided via the wrench estimator

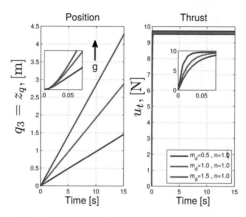

Figure 4.2: Free fall for different desired dynamics. The figure on left is the position of the quadrotor along $\mathbf{z}_W$ (height) and the one on the right is the thrust applied by the controller. The small windows show the first 0.08 [s] for both position and thrust, respectively.

proposed in Sec. 3.3, including the noises of the measurements. Notice that for the numerical simulations, no high-level control input is used besides the IDA-PBC, except the terms necessary for the gravity compensation. In this way, the effect of the IDA-PBC framework on the APhI of a quadrotor is more apparent.

Later in Sec. 4.3.2, we present the experimental results, validating the performance of the IDA-PBC for a quadrotor interacting with its environment. There the real system is tested for similar tasks as for the numerical validation, however with a outer-loop position tracking controller ensuring that the real system does not *float around* uncontrolled. The reason for this practical implementation is that, IDA-PBC alone is not a *tracking control*, but it is a *low-level* controller making sure the physical interaction is stable, by also reshaping the physics of the system. Unlike the numerical simulations, for experimental tests a high-level tracking controller was required so that the real quadrotor setup can be protected from dangerous crashes.

## 4.3.1 Numerical Validation

In this section we present some simulation results to support the theory of the control method proposed in Sec. 4.2. The parameters for the original system dynamics are chosen as follows; mass is $m_q = 1$ [kg], the gravity constant is $g = 9.81$ [m/s²], and the rotational inertia of the platform is $\mathbf{M}_{qr} = \text{diag}([0.013, \ 0.013, \ 0.022])$ in units of [kgm²]. The dissipation for the rotational dynamics, as presented in (4.9), is set to $k_d = 1$ based on our experiences Franchi et al. (2012a,b); Lee et al. (2013).

### Shaping the Weight

We have assumed the aerodynamic drag as $k_{drag} = 0.5$ in every direction. For more detail, one can check Brandt and Selig (2011). Let us first investigate the behavior of different

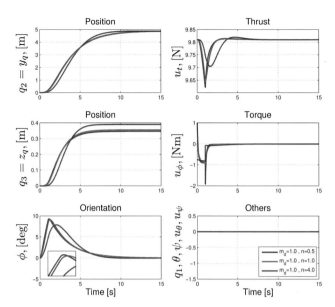

Figure 4.3: Impulse response of the rotational dynamics. The external torque is applied around $\mathbf{x}_B$ axis. The small window in orientation is highlighting the change of $\phi$ for different desired inertias. The states and control inputs, which are not effected by the impulse, remain zero.

desired masses in *free fall* case, where there are no external forces or high-level control inputs, i.e., $\mathbf{w}_{ext} = \mathbf{0}, \mathbf{u}_0 = \mathbf{0}$. Recall that $\mathbf{w}_{ext} = [\mathbf{f}_{ext}^T \; \boldsymbol{\tau}_{ext}^T]^T$ with $\mathbf{f}_{ext} = [f_{e_x} \; f_{e_y} \; f_{e_z}]^T$ and $\boldsymbol{\tau}_{ext} = [\tau_{e_x} \; \tau_{e_y} \; \tau_{e_z}]^T$. The desired system parameters are chosen as follows: $\bar{k}_T = 50$, $\bar{\mathbf{K}}_R = \mathrm{diag}([\bar{k}_R^\phi, \; \bar{k}_R^\theta, \; \bar{k}_R^\psi])$ with $\bar{k}_R^\phi = \bar{k}_R^\theta = \bar{k}_R^\psi = 5$, $\bar{V}_d = -m_d g e_3 + \frac{1}{2}\boldsymbol{\eta}_e^T \mathbf{K}_p \boldsymbol{\eta}_e$ where $\mathbf{K}_p = \mathrm{diag}([k_p^\phi, \; k_p^\theta, \; k_p^\psi])$ and $k_p^\phi = k_p^\theta = k_p^\psi = 2$, and $\boldsymbol{\eta}_e = \boldsymbol{\eta}$ for keeping the orientation of the quadrotor always at zero. Fig.4.2 shows the results for three different desired masses; $m_d = 0.5$ [kg], $m_d = 1$ [kg] and $m_d = 1.5$ [kg], while around all rotations the desired inertia value is set to $\mathbf{N} = \mathrm{diag}([n, \; n, \; n]) \in \mathbb{R}^{3\times3}$ for $n = 1$ [kgm$^2$]. In the figure, the position $q_3$ is shown on the left, and the thrust applied by the controller on the right. The direction of gravity is shown in the plot. It is seen that under the desired viscosity, which is tuned by $k_T$, the bigger mass falls faster than the smaller mass. The controller adjusts the quadrotor in a way that the system behaves as a desired mass.

**Shaping the Inertia**

Now, let us show how one can change the rotational dynamics of the system by shaping its desired energy. The rotational dynamics of the quadrotor system is fully actuated, hence we have full control on rotational properties. For this, we investigate the impulse response of the rotational dynamics, where the system is in hovering. For hovering, we used the high-level control input $\mathbf{u}_o = [m_d g \; 0 \; 0 \; 0]^T$ to balance the gravity effect for the desired

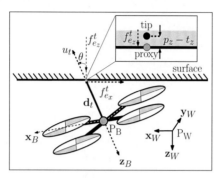

Figure 4.4: The tip of a rigid link driven by a quadrotor is in contact with a surface and sliding along one direction. The gray point represents the tooltip penetrating the surface (such as ceiling), which is the proxy and depicted with a black colored point. Notice that the system this sketch is actually realized as shown in Fig. 3.2, which is later used for the experiments in Fig. 4.13.

system. The impulse 1 [Nm] around $\mathbf{x}_B$ is applied for 1 s. The results are presented in Fig.4.3, where the system behavior with three different desired inertias are compared (for $\mathbf{N} = \mathrm{diag}([n, \; n, \; n]) \in \mathbb{R}^{3 \times 3}$, where $n = 0.5$ or $n = 1.0$ or $n = 4.0$). For all three cases the desired mass is kept at $m_d = 1$ [kg]. There it is clear that the system with smaller inertia behaves more compliant to the external torques. The change in orientation $\phi$ reveals that the second order system response, where smaller inertia has bigger magnitude and it requires higher torques to stabilize the system. When we assign a bigger inertia, the system behavior becomes stiffer and rejects the external torques. System reacts instantly and stabilizes itself with less change in orientation. The small inertia might come in handy when for example in safe human-robot-interaction. The big inertia on the other hand might be useful for tasks where the quadrotor needs to reject disturbances quickly, such as maintaining stable contact with a flat surface.

**Sliding on a Ceiling Surface**

Reshaping the physics, especially for the rotational dynamics of an underactuated quadrotor system might provide huge advantage for physical interaction of such systems. In order to show this fact, here we simulate a sliding on a surface task, where a tool in shape of a rigid stick is connected to the Center of Mass (CoM) of the quadrotor system, and its tip (tooltip) is in contact with a surface. An illustration is shown in Fig.4.4. This can be interpreted as ceiling painting, cleaning, surface inspection, e.t.c.

Let us consider two different sliding scenarios; first the tooltip is sliding on a flat surface, and second it is sliding on a rough surface, where there are dents and bulges. The surface is placed above (considering $+\mathbf{z}_W$ shows the below) the CoM of the quadrotor. We choose to slide along the positive $\mathbf{x}_W$ (See Fig.4.4). For this, quadrotor needs to be tilted with a certain tilting angle, in this case with $\theta^* < 0$. A desired attitude can be achieved by shaping the desired potential in a way that it goes to minimum in a desired configuration.

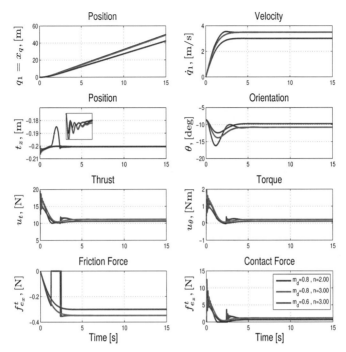

Figure 4.5: Stable contact and sliding of tooltip along a flat surface. Note that gravity vector is aligned with the $+\mathbf{z}_B$. Small inertia is more vulnerable to the external forces, and smaller mass establishes the contact faster. The hybrid contact modeling is not a problem for passivity based controller in sense of stability.

Consider the desired rotational potential as

$$\bar{V}_d(\mathbf{q}) = \frac{1}{2}\boldsymbol{\eta}_e^T \mathbf{K}_p \boldsymbol{\eta}_e, \tag{4.50}$$

where

$$\boldsymbol{\eta}_e = \boldsymbol{\eta} - \boldsymbol{\eta}^* = \begin{bmatrix} \phi - \phi^* \\ \theta - \theta^* \\ \psi - \psi^* \end{bmatrix}. \tag{4.51}$$

The desired attitude $\boldsymbol{\eta}^*$ is the equilibrium in orientation where the rotational potential goes to minimum. Once the desired attitude is achieved, we need to apply a constant thrust to the system, to maintain the contact with the surface and to win against the friction forces, so the tooltip can slide along the $+\mathbf{x}_W$ axis. The external forces acting on the tooltip can be considered as: the (contact) reaction force from surface along the $+\mathbf{z}_W$ direction, and friction force against the direction of the sliding motion on the surface. For modeling the reaction of the surface, we used proxy model conceptually introduced in Zilles and Salisbury

(1995), only along the $+\mathbf{z}_W$ direction. In our case, the position of the tooltip is the real position, and the height of the contact surface represents the proxy (See Fig.4.4). The reaction force from the surface is calculated as

$$f_{e_z}^t = k_{\text{wall}}(p_z - t_z),\tag{4.52}$$

where $k_{\text{wall}}$ is the *spring gain* depending on the characteristics of the surface, $p_z$ is the proxy (or surface) position, and $t_z$ is the tooltip position along the $\mathbf{z}_B$ axis. For the surface friction, we used a simple viscous friction model, e.g. from Olsson et al. (1998), such as

$$f_{e_x}^t = -\mu \dot{q}_1,\tag{4.53}$$

where $\mu$ is the coefficient of friction, depending on the tooltip and surface characteristics, and $\dot{q}_1$ is the velocity of the tooltip (and quadrotor) along $+\mathbf{x}_W$. In our simulation, we consider a hybrid contact model, where if the tooltip penetrates to the surface, then both reaction and friction forces are acting, otherwise there are no external forces, i.e.,

$$\begin{cases} \mathbf{f}_{ext}^t = \mathbf{0} \in \mathbb{R}^3, & \text{if } t_z > p_z \\ \mathbf{f}_{ext}^t = [f_{e_x}^t \ 0 \ f_{e_z}^t]^T, & \text{if } t_z \leq p_z. \end{cases}\tag{4.54}$$

To calculate the external wrench acting on the CoM of the quadrotor, we use the following transformation

$$\mathbf{w}_{ext} = \begin{bmatrix} \mathbf{I} & \mathbf{0} \\ [\mathbf{d}_t]_\wedge \mathbf{R}(\boldsymbol{\eta})^T & \mathbf{R}(\boldsymbol{\eta})^T \end{bmatrix} \mathbf{w}_{ext}^t,\tag{4.55}$$

where $\mathbf{d}_t$ is the distance between CoM of the quadrotor and the tooltip, $\mathbf{R}(\boldsymbol{\eta})$ is the rotation matrix as in (2.1), $\mathbf{w}_e^t = [\mathbf{f}_{ext}^t{}^T \ \boldsymbol{\tau}_{ext}^t{}^T]^T \in \mathbb{R}^6$ is the external wrench acting on the tooltip. In our case, $\boldsymbol{\tau}_{ext}^t = \mathbf{0}$. For the wall characteristics, we assigned $k_{wall} = 2000$, and $\mu = 0.1$. To win the friction force and start sliding, it is necessary that the angle between normal of the surface and the applied force must satisfy

$$|\theta^*| > \tan^{-1}(\mu).\tag{4.56}$$

Hence, we choose $\theta^* = -0.15$ [rad] $\simeq -8.6$ [deg]. A constant thrust of $2m_dg$ [N] is applied using $\mathbf{u}_o$ to maintain the contact and to slide along $\mathbf{x}_W$, as $\mathbf{u}_o = [2m_dg \ \mathbf{0}_{1\times3}]^T$. The distance between tooltip and CoM of the quadrotor is chosen as $\mathbf{d}_t = [0.2 \ 0 \ -0.2]^T$, in units of meters, for the reason explained in Lee and Ha (2012). For desired rotational potential, we set $k_p^\phi = k_p^\theta = k_p^\psi = 5.5$. The desired damping along the thrust direction is set to $\bar{k}_T = 10$ and for the rotational dynamics it is $\bar{k}_R^\phi = \bar{k}_R^\theta = \bar{k}_R^\psi = 50$. The proxy position is set to $p_z = -0.2$ [m]. Fig.4.5 shows the results for different desired mass and inertia values. By judging the change of $t_z$, and orientation $\theta$, bigger inertia quickly adapts to the disturbances, while smaller inertia is oscillating, which causes disconnection with the contact surface (blue plot). It is noticed that a smaller mass (red plot) establishes the contact with the surface faster than the bigger mass. This shows how the quadrotor can benefit from the proposed controller, where we shape and dissipate both kinetic and potential energies. Changing the desired mass creates a difference in orientation at steady state, since the total force (with surface reaction) along $\mathbf{z}_W$ creates bigger torque for bigger mass, which is directly related to the length of the tool. This is an important motivation of choosing a reasonable

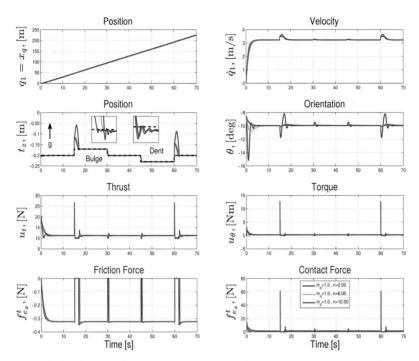

Figure 4.6: Stable contact and sliding along a rough surface, where there are dents and bulges. Zoomed windows in the tooltip position plot highlight the behavior of the system when a bulge and a dent are encountered, respectively. Bigger inertia (red plot) reestablishes the contact with the surface faster, while smaller inertia (blue and green plots) are more vulnerable to the external forces.

$\mathbf{d}_t$ value. Note that the maximum penetration of the red plot to the surface is calculated as 4 [mm], and the final penetration is 0.3 [mm].

Notice the differences in $q_1$ and $\dot{q}_1$ in Fig. 4.5. It is not due to the difference in the desired inertia, but the desired mass assigned to the quadrotors. The one with lower desired mass is controlled with smaller $\mathbf{u}_o$ to scale the gravity effect on it, which causing it to accelerate less than the others along the $\mathbf{x}_W$ when the platform is tilted.

As explained before in (4.54), the external forces are modeled discontinuously. It is seen in the blue plot of Fig. 4.5, the tooltip loses the contact with the surface, yet the controller stabilizes the system anyway. In fact, an advantage of the passivity based controllers is that they stabilize (hybrid) systems, where discontinuities may exist.

In the second case, the quadrotor slides on a rough surface, where there are dents and bulges. For this, we simply change the position of the proxy, $p_z$, and let the quadrotor slide on this new surface. Different from the previous simulation, we set $k_p^\phi = k_p^\theta = k_p^\psi = 10$ and $\bar{k}_T = 15$. The results are shown in Fig. 4.6, where this time the comparison is done for different inertias only, and the desired mass kept at $m_d = 1$ [kg]. The position of proxy is

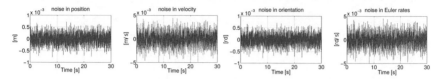

Figure 4.7: Gaussian noise added to the state measurements.

presented as black dashed plot, where it is first shifted 3 [cm] outwards, i.e., representing a bulge, and later 3 [cm] inwards, i.e., representing a dent. Notice that the direction of the gravity is also given in the figure. As it is expected from the outcome of the previous simulations, system with smaller inertia has more compliant reaction to the surface changes. Again, the discontinuity of the external forces does not cause instabilities, thanks to the passivity based controller. One has to notice that by tuning the parameters such as $\mathbf{K}_P$, $\bar{k}_R$ and $\bar{k}_T$, and setting desired mass and inertia values, it is always possible to change the physical behavior of the system depending on the desired objective.

### Using the External Wrench Estimator in the Loop

In this section we present different simulation results to show the capabilities of the nonlinear force observer presented in Sec. 3.3 for different case studies, in which IDA-PBC is used for controlling the physical interaction of the quadrotor system. In order to reproduce realistic scenarios, we added independent Gaussian noises to the measurements of the quadrotor state, based on our laboratory experiences (see Fig. 4.7). The first simulation aims at showing the accuracy of the force observer in an environment, where disturbing forces and torques are acting on quadrotor CoM. In the second simulation we consider the relevant case where a rigid tool is attached to quadrotor CoM and the tooltip is sliding on the surface of a ceiling (same as before and depicted in Fig. 4.4). This case study provides a highly varying external force profile, which is a *non-nominal* situation for using the force observer presented in Section 3.3 (due to the assumption made in (3.8)).

We used the same parameters for the quadrotor as before. In the following results, the legends QC$i$ represent the quadrotor with the $i$-th target dynamics assigned using IDA-PBC (see Sec. 4.2.2 for the controller, and Sec 4.2.5 for its extension). This time, the environment is modeled with no dissipation, which means the aerial drag acting on the body of the quadrotor is not considered.

The aim of the first simulation is to show the accuracy of the force estimation, done by nonlinear force observer proposed in Sec. 3.3. As a case study, we choose an external force/torque profile, where first a 1 [N] of force applied along $+\mathbf{x}_W$; then along $-\mathbf{z}_W$ axis; later 0.5 [Nm] of torque around $+\mathbf{x}_B$; and finally all together at the same time. The external force/torque profile can be seen in the last row of Fig. 4.8. The high level control input $\mathbf{u}_o$ is chosen only for scaling the gravity effect, similar to the previous simulations.

Our first goal is to show the performance of the nonlinear force observer for a quadrotor controlled with IDA-PBC. The first two rows of Fig. 4.8 present the evaluation of quadrotor position, orientation and necessary control inputs when interacting with the external forces. The last row of the same figure shows the estimated forces and torques. It can be seen that force/torque estimation (red plot) is very accurate w.r.t the exact forces (black plot), considering that the measurements are simulated with their noises.

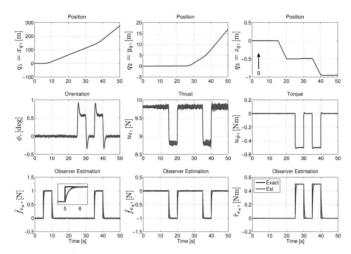

Figure 4.8: IDA-PBC controlled quadrotor for aerial physical interaction and force estimation using nonlinear force observer. The external wrench is directly acting on
the center of gravity of the quadrotor. First 1 [N] of external force along $+\mathbf{x}_W$
is applied between 5 [s] and 10 [s]. Then 1 [N] of external force along $-\mathbf{z}_W$ is
applied between 15 [s] and 20 [s]. Later a constant torque of 1 [Nm] is applied
around $\mathbf{x}_B$ between 25 [s] and 30 [s]. Finally all of these external disturbances
are applied between 35 [s] and 40 [s]. Third column shows the exact external
force and torques (black plots) and their estimations (red plots). The observer
gain for force estimation is set to $c_o = 5$, and for torque estimation to $c_o = 0.5$.

Secondly, we test the observer in an extreme case where there are rapidly varying external
forces. The chosen case study is a challenging one, where a rigid tool is sliding on a rough
surface. This tool is rigidly attached on a quadrotor, placed above the center of gravity
at a position, as before with the distance of $\mathbf{d}_t = [0.2\ 0\ -0.2]^T$ [m], so the surface can be
interpreted as a ceiling. Notice that in Fig. 4.6 we had shown the results of a similar case,
but without the use of a wrench observer. We again apply a constant high-level control
input, $\mathbf{u}_o$, to the system only for scaling the gravity effects, and choose a desired attitude.
Then we let it fly while using IDA-PBC for achieving a stable contact between this uneven
ceiling surface and the tooltip, and slide along the surface for performing such as ceiling
painting, cleaning or surface inspection tasks. The results are shown in Fig. 4.9, where $t_z$
is the tooltip position along $\mathbf{z}_B$ as before. QC1 is assigned with smaller desired inertia;
and damping along the thrust direction, compared to QC2. QC2 shows a stiffer behavior
with respect to the changes on the surface compared to QC1, which has more oscillations
and takes more time to reestablish the contact with the surface. These behaviors are
consistent with the previous results given in Fig. 4.9, where the *exact* knowledge of the
external wrench was used instead of its estimation. The main components of the exact
and estimated wrench are shown in third column of Fig. 4.9. As expected the estimator
cannot precisely track the rapidly varying external wrench. However, the controller shows

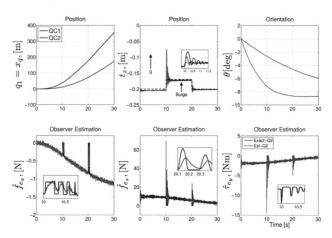

Figure 4.9: Rigid tooltip on a quadrotor interacting with a rough ceiling surface. The blue plot is presenting the behavior of QC1, and the red plot the behavior of QC2. Second row shows the estimation results of the nonlinear wrench observer. The observer gain is set to $c = 20$ for both forces and torques. In case of rapidly varying external force profile, the observer gives a less precise estimation, as expected. However, IDA-PBC controller still stabilizes the interaction, thanks to its ability of preserving the passivity of the controlled system.

the capability to stabilize the physical interaction even when the real and estimated values of the contact force present some discrepancies.

**Quadrotor as 3D Force Effector**

In Sec. 4.2.5 we showed how to use the proposed controller and estimator in a way that the quadrotor can exert an arbitrary constant 3D force on the environment. Here, we present the simulation results of a case where quadrotor is exposed to a constant force (e.g., modeling a constant wind force) which has to be balanced while being subject to other disturbances from the environment at the same time. This is the case in which the quadrotor is applying a desired force in order to balance the external force.

In order to validate the theory in simulation, we consider the case, in which a constant 1 [N] force is applied to quadrotor along the $+\mathbf{x}_W$ axis, continuously. In addition to the constant force, we apply an impulse of a disturbance force of 1 [N] along the $-\mathbf{z}_W$ and an impulse of a disturbance torque of 1 [Nm] around the $\mathbf{x}_B$ axis. Finally all these disturbances are applied in the same time. The controller finds a rotational equilibrium, as explained in Sec. 4.2.5, so that the quadrotor stops accelerating along the 3D axes by balancing both the external force and torques. The results are presented in Fig. 4.10. Two different desired dynamics are shown: QC3 is designed with smaller desired inertia and stiffness in rotational dynamics, compared to QC4, which reacts to external effects faster than QC3. This generates a less travelled distance along the direction of motion for QC4. As it is seen in Fig. 4.10, the quadrotor exerts a counterbalancing force so that the components of the

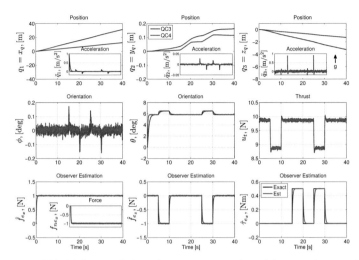

Figure 4.10: IDA-PBC controlled quadrotor is exerting desired forces and counterbalancing external disturbances. The external wrench is directly acting on the center of gravity of the quadrotor. A constant 1 [N] of external force along the $\mathbf{x}_W$ axis is applied during the whole simulation for replicating a constant wind effect. Then 1 [N] of external force along the $-\mathbf{z}_W$ axis is applied between 5 [s] and 10 [s]. Later a constant torque of 1 [Nm] is applied around the $\mathbf{x}_B$ axis between 15 [s] and 20 [s]. Finally all of these external disturbances are applied between 25 [s] and 30 [s] of the simulation. Third row shows the exact external force and torques (black plot); and their estimations (red plot). It can be seen from the first row that the accelerations in all three axes converge to zero at the steady state. The sub-figure presented in the last row is showing the desired force applied by the quadrotor to counterbalance the constant wind effect. The observer gain for force estimation is set to $c_o = 5$, and for torque estimation to $c_o = 0.5$.

accelerations along all translational axes are converging to zero in steady state (sub figures in first row of Fig. 4.10). Another sub-figure in the third row shows that $f_{ne,x}$ rapidly converges to $f_x^* = -f_{e_x}$ in order to counterbalance it. Since no dissipation is present in the environment (no aerial drag is acting on the body of the quadrotor), the system floats with a constant velocity in each direction.

## 4.3.2 Experimental Results

In this section we present the experimental results of a quadrotor VTOL interacting with its environment. The quadrotor setup for the experiments was presented before in Fig. 3.2, where a Force/Torque (F/T) sensor is placed on board providing the external wrench information. For the reasons explained in Sec. 3.4, we performed our experiments using an F/T sensor, instead of a wrench estimator. In the experiments, the flying robot is interacting with its environment through a rigid link, which is attached on top of this F/T sensor, which is rigidly connected to the quadrotor body frame. For more details on how to use this setup for computing the external wrench $\mathbf{w}_{ext} \in \mathbb{R}^6$, see Sec. 3.2. The overall quadrotor setup together with its sensors, hardware and software is explained in detail in Sec. 2.3.

The physical interaction of the flying robot is controlled using IDA-PBC, which is described in Sec. 4.2. Different from the numerical simulations, for the experiments we also implemented a *high-level* position tracking control. The reason is that while IDA-PBC alone can control the physical interaction of the robot by shaping its physical properties, it does not take the experimental limits into account, e.g. the limits of the room the robot is flying in. Without a controller which can track a desired position/velocity trajectory; the quadrotor controlled only with IDA-PBC would float around after an initial contact with its environment, in a stable but undesirable manner. Nonetheless, this is what IDA-PBC in Sec. 4.2 is actually developed for; to allow controlling the physical interaction of the flying robot by shaping its physical properties, while allowing a *high-level* control input (e.g. $\mathbf{u}_o$ of (4.36)) which can be responsible for other purposes, e.g. position tracking (see also Remark 6).

In the following, we briefly explain this position tracker, which is used together with the IDA-PBC controller in our experiments.

### High-level Controller for Position Tracking

A high-level position tracker is used for steering the quadrotor VTOL to a desired trajectory, while letting IDA-PBC shape its physical properties. This tracking controller is developed based on the *near-hovering controller*, presented as in Lee et al. (2013). From the decoupling property of the quadrotor, the rotational dynamics in (2.4) can be computed independently from the translational dynamics given in (2.5). Let us consider a desired position trajectory of the quadrotor as $\mathbf{p}_q^d = [x_q^d \ y_q^d \ z_q^d]^T \in \mathbb{R}^3$, and assume that $\mathbf{f}_{ext} = \mathbf{0}$. Then, following from Lee et al. (2013), and from the third row of (2.5), the thrust input

$$u_{o_t} = -\frac{m_d}{c_\phi c_\theta}(g + \ddot{z}_q^d + k_{d_z}(\dot{z}_q^d - \dot{z}_q) + k_{p_z}(z_q^d - z_q)) \qquad (4.57)$$

ensures the local exponential stability of $(z_q^d - z)$, as long as $c_\phi c_\theta \neq 0$, which is violated only when the quadrotor configuration is in a singularity that we avoid all the time (see Remark 2.) The control gains $k_{d_*} \in \mathbb{R}_{\geq 0}$ and $k_{p_*} \in \mathbb{R}_{\geq 0}$ are used for removing the velocity and position errors along the $*$-axes, respectively, where $* = \{\mathbf{x}, \mathbf{y}, \mathbf{z}\}$. Then from the first two rows of (2.5) we have

$$m_d \begin{bmatrix} \ddot{x}_q \\ \ddot{y}_q \end{bmatrix} = -u_{o_t} \underbrace{\begin{bmatrix} c_\phi c_\psi & s_\psi \\ c_\phi s_\psi & -c_\psi \end{bmatrix}}_{=:\mathbf{Q}(\phi,\psi)\in\mathbb{R}^{2\times 2}} \begin{bmatrix} s_\theta \\ s_\phi \end{bmatrix}, \qquad (4.58)$$

where $\mathbf{Q}$ is always invertible as long as $c_\phi \neq 0$; which means that the system is not in singularity (this is in line with Remark 2). Then choosing the following roll and pitch commands will make $(x_q^d - x_q, \ y_q^d - y_q)$ locally exponentially stable

$$\begin{bmatrix} \bar{\theta}_c = s_{\theta^d} \\ \bar{\phi}_c = s_{\phi^d} \end{bmatrix} = -\frac{m_d \mathbf{Q}^{-1}}{u_{o_t}} \begin{bmatrix} \ddot{x}_q^d + k_{d_x}(\dot{x}_q^d - \dot{x}_q) + k_{p_x}(x_q^d - x_q) \\ \ddot{y}_q^d + k_{d_y}(\dot{y}_q^d - \dot{y}_q) + k_{p_y}(y_q^d - y_q) \end{bmatrix}. \qquad (4.59)$$

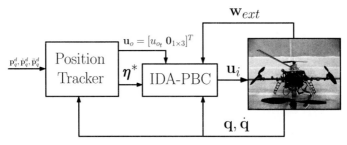

Figure 4.11: Sketch of the control framework used for the experiments. The position tracker is developed based on the near-hovering controller, sending the desired attitude equilibrium $\boldsymbol{\eta}^*$ and the high-level control input $\mathbf{u}_o$ to the IDA-PBC controller.

In this step of the computations, let us define some maximum boundaries to both roll and pitch commands, preventing the system coming close to its singularities. In our experiments, we choose $\phi_c^{max} = \theta_c^{max} = \sin(r_l)$, where $r_l = 0.52326$ [rad]. Then let us implement the following soft saturation for both roll and pitch commands;

$$
\begin{aligned}
\phi_c &= \frac{2\phi^{max}}{\pi} \arctan\Big(\frac{\bar{\phi}_c}{2\phi_c^{max}}\Big), \\
\theta_c &= \frac{2\theta_c^{max}}{\pi} \arctan\Big(\frac{\bar{\theta}_c}{2\theta_c^{max}}\Big).
\end{aligned}
\tag{4.60}
$$

Then, the desired roll and pitch angles to steer the system to the desired $\mathbf{x}$ and $\mathbf{y}$ configurations are

$$
\begin{aligned}
\phi^* &= \arctan(\phi_c), \\
\theta^* &= \arctan(\theta_c).
\end{aligned}
\tag{4.61}
$$

Now, remember that in (4.50), we showed how one can change the desired attitude equilibrium $\boldsymbol{\eta}^* = [\phi^* \ \theta^* \psi^*]^T \in \mathbb{R}^3$, which shapes the desired potential energy of the system as in (4.25). Then using a desired rotational potential energy $\bar{V}_d$ in form of (4.50), where for the desired attitude equilibrium $\phi^*, \theta^*$ are available from (4.61), and $\psi^*$ is chosen any arbitrary number, e.g. $\psi^* = 0$, and placing this desired potential energy $\bar{V}_d$ in (4.25); we make sure that the IDA-PBC controller can steer the system to a desired $\mathbf{x}_W - \mathbf{y}_W$ configuration using the control input in (4.36). Moreover, by choosing the high-level control input as $\mathbf{u}_o = [u_{o_t} \ \mathbf{0}_{1\times3}]^T \in \mathbb{R}^4$, and implementing it in (4.36), we can let the quadrotor setup track a trajectory along the $\mathbf{z}_W$ axis. A sketch of this control scheme is depicted in Fig. 4.11 for fixing the ideas.

**Remark 7.** *Notice, also from Fig. 4.11, that the high-level control input $\mathbf{u}_o$ is providing only the additional thrust input for tracking $z_q^d$ and its derivatives. Other desired trajectories along the underactuated directions, i.e. $\mathbf{x}_W$ and $\mathbf{y}_W$, are tracked using solely the control inputs generated by IDA-PBC, i.e. $\mathbf{u}_i$. However to generate this input, we actively compute a new desired attitude $\boldsymbol{\eta}^*$, which is done using the near-hovering scheme from Lee et al. (2013).*

**Shaping the Inertia**

As explained in Sec. 4.2, IDA-PBC is a powerful control method, controlling the physical interaction of a quadrotor VTOL by shaping its physical properties, through passivation. To test this in real experiments, we first bring the quadrotor to a hovering condition using the controller presented in the previous section and depicted in Fig. 4.11. Then the flying system is disturbed with an external interaction on top of its tool tip, as shown in the top of Fig. 4.12. We repeat this twice; first the IDA-PBC controller is tuned for a smaller desired inertia $\mathbf{N} = \text{diag}([0.008, \ 0.008, \ 0.0274]) \in \mathbb{R}^{3 \times 3}$, and then when it is tuned for a bigger desired inertia $\mathbf{N} = \text{diag}([0.03, \ 0.03, \ 0.0274]) \in \mathbb{R}^{3 \times 3}$, only around the $\mathbf{x}_B$ and $\mathbf{y}_B$ axes. Remember from Fig. 3.2 that the mass of the real system is $m_q = 1.49$ [kg], and its rotational inertia is $\mathbf{M}_{qr} = \text{diag}([0.01708, \ 0.0172, \ 0.0274]) \in \mathbb{R}^{3 \times 3}$.

The results are given on the bottom of Fig. 4.12. For brevity, we only show the response of the second order rotational dynamics to the external torque around the $\mathbf{y}_B$ axis. In the figure, superscript $*^s$ stands for the measurements of the *small desired inertia* case, while $*^b$ for the *bigger desired inertia* one. Notice that the external torques for both cases (i.e. $\tau_{e_y}^s, \tau_{e_y}^b$, depicted with black solid and dashed magenta lines, respectively) are the same. However the pitch orientations (i.e. $\theta^s, \theta^b$, depicted with blue solid and red dashed lines, respectively) are different from each other. Due to the position tracker implemented together with IDA-PBC (see Fig. 4.11), in both cases quadrotor comes back to its equilibrium after the disturbances. This creates a virtual rotational spring effect, making the system oscillate around its equilibrium until it reaches to a region of attraction. Notice the difference between the settling times of the two different cases; when the desired inertia is bigger, it takes longer for the system to reach its steady state than when the desired inertia is smaller. This is in line with the fact that for a rotational mass-spring-damper system with constant spring[2], it would take longer for it to come back its equilibrium when the rotational mass is greater.

**Sliding on an Uneven Ceiling Surface**

Here we present some preliminary experimental results of the quadrotor+rigid tool setup, sliding on an uneven ceiling surface. The purpose of the experiments is to show that by changing the desired inertia of the system using IDA-PBC, we can change the performance of the aerial physical interaction task, e.g. letting the quadrotor slide on the ceiling surface with a better contact profile.

The quadrotor+rigid tool setup is controlled using the method depicted in Fig. 4.11, where the system is steered via joystick commands, which are provided by a human observer. Although the physical interaction is controlled autonomously, by bringing the human in the loop we aim to bring some level of security to the system for avoiding an unexpected crash, and also pave the way for future human in the loop experiments for APhI. The latter one can be studied more intensively by replacing the joystick controller with a haptic device, which allows bilateral control of the robot (Franchi et al. (2012b)).

The results of the experiments are given in Fig. 4.13. There, we compare two cases: quadrotor controlled with a small desired inertia, i.e. $\mathbf{N} = \text{diag}([0.004, \ 0.004, \ 0.0274]) \in \mathbb{R}^{3 \times 3}$, and with a big desired inertia, i.e. $\mathbf{N} = \text{diag}([0.014, \ 0.014, \ 0.0274]) \in \mathbb{R}^{3 \times 3}$. Notice

---

[2]Note that the spring effect is due to the choice of the rotational desired inertia $\bar{V}_d$, damping is due to the damping injection implemented inside of the IDA-PBC, and the angular mass is the desired inertia $\mathbf{N}$.

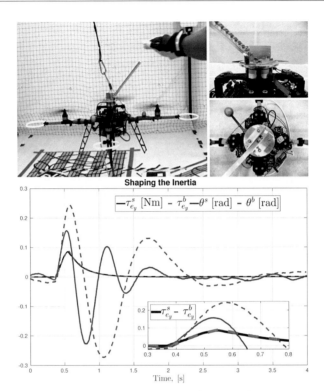

Figure 4.12: – *Top:* Quadrotor equipped with an F/T sensor (see details of the setup in Sec. 3.2) is about to be disturbed by an external interaction from the tip point of the rigid tool, during the hovering condition.

– *Bottom:* System response (second order rotational dynamics) to the external disturbances around the $\mathbf{y}_B$ axis. Two cases are compared: system with bigger desired inertia (denoted with superscript $*^b$) and the one with smaller desired inertia (denoted with $*s$). IDA-PBC is used to assign the desired inertial properties, together with a high-level position controller described in Sec. 4.3.2.

that the desired inertias are assigned only around $\mathbf{x}_B$ and $\mathbf{y}_B$ axes, while for the rotations around $\mathbf{z}_B$ it is same as the original system. On the top of Fig. 4.13 several snapshots from the experiments are given, where; (a) the quadrotor+rigid tool is first time in contact with the ceiling surface, (b) it is sliding on the even part of the ceiling, (c) just before the dent, (d) right after the dent, (e) just before a bulge which is built smoothly, (f) right after the bulge[3]. On the bottom of the figure the results are given, where blue solid curves

---

[3]Notice that the experiments are performed in a limited environment, since the artificial ceiling we have built has a limited size (1.73 [m] in longitudinal). On the other hand, this was not the case for the simulations (e.g. Fig 4.9), where the quadrotor was able to slide on a surface for hundreds of meters.

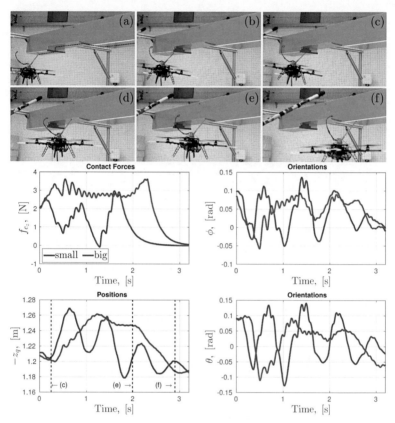

Figure 4.13: – *Top:* A series of snapshots (from (a) to (f)) from the experiments. The quadrotor setup as shown in Fig. 3.2 is sliding on a blue-colored uneven ceiling surface. The tip of the rigid tool is in contact with the ceiling, and its bottom is rigidly attached to the F/T sensor and the quadrotor body frame. The overall system is secured with a *slack* cable connected to a stick, for avoiding any dangerous crashes.
– *Bottom:* Experimental results for a quadrotor+rigid tool sliding on an uneven surface. Results for the system with the smaller desired inertia are depicted with blue curves, and the one with bigger desired inertia with red curves. Three important time instants for $z_q$ are highlighted with black dashed vertical lines; the moment before the dent (c), at the end of the dent and before the bulge (e), and the moment at the end of the bulge (f). Clearly, the system with bigger desired inertia follows the profile of the ceiling better than the one with the smaller desired inertia.

stand for the response of the system with smaller desired inertia, and the red solid curves for the one with bigger desired inertia. The contact forces acquired from the F/T sensor along the **z**-axis are given as the first plot, and below it the $z_q$ position of the quadrotor. Two plots in the second column show the roll ($\phi$) and pitch ($\theta$) values. Notice that the system with a bigger desired inertia (red) preserves its contact with the ceiling much better than the one with a smaller desired inertia (blue), despite the uneven profile of the surface. A smaller desired inertia, in this case $\mathbf{N} = \mathrm{diag}([0.004,\ 0.004,\ 0.0274]) \in \mathbb{R}^{3\times3}$, causes more oscillations for the system along the $\mathbf{z}_W$ axis ($z_q$, up and down) and also around its rotational axes (see $\phi$, $\theta$). When we implement IDA-PBC with a bigger desired inertia, i.e. $\mathbf{N} = \mathrm{diag}([0.014,\ 0.014,\ 0.0274]) \in \mathbb{R}^{3\times3}$, these oscillations are reduced and the contact with the surface during sliding is much better (see the plot of the contact forces in Fig. 4.13). This result is in line with the numerical simulations.

Note that even when the system is controlled for a bigger desired inertia, some small oscillations appear during the contact. This can be further improved by shaping the dissipation of the system, changing e.g. $k_T$, as explained in Sec. 4.2.4.

## 4.4  Discussions

Controlling the aerial robots when they are physically interacting with their environment is not a trivial task. In this chapter we approached this problem from energetic point of view, and adopted the IDA-PBC method for quadrotor VTOLs. This powerful method allowed us *reshaping* the physical properties of a quadrotor, e.g. assigning to it a desired inertia, potential energy or dissipation; while it ensured the passivity of the controlled system.

We have to note that the IDA-PBC method presented here is a *low-level* framework, helpful for achieving a stable APhI of the quadrotors (see Sec. 4.2 and also Remark 6). It can be used to turn the quadrotors into 3D force effectors, as explained in 4.2.5; however it is not developed for, e.g. position or force/torque tracking purposes. It elegantly renders the nonlinear quadrotor dynamics into a stable, desired behavior, especially when the system is interacting with its environment. How it assigns desired dynamics to a quadrotor is shown overall in Sec. 4.3, both numerically and experimentally.

Nonetheless, IDA-PBC can be used together with a *high-level* controller, which enters to the new controlled system via a new input signal $\mathbf{u}_o$, as depicted in Fig. 4.1 and shown in (4.36). As given in Proposition 4, the controlled system with new desired dynamics will be also cyclo-passive (or passive with a condition on lower bound of the system energy) w.r.t. this new input. This high-level new input can be used for position tracking (see Fig. 4.11), or even for force/torque tracking. We consider the latter one in the scope of our future studies.

One note is that during the preliminary experiments, our goal was to show that using IDA-PBC, one can reshape the physical properties of a quadrotor VTOL when it is interacting with its environment. The results given in both Fig. 4.12 and Fig. 4.13 are clearly showing that, by shaping the desired inertia of the system we can achieve different interactive behaviors. This can lead to a better APhI depending on the task, as shown in Fig. 4.13 when the quadrotor sliding on an uneven surface, which is in line with the results of Fig. 4.9. Sliding task, which can also be interpreted as surface inspection, cleaning or painting, can further be improved by, e.g. tuning the dissipative parameters of the controller ($\mathbf{K}_v$ of (4.36)), decreasing the desired mass $m_d$, or implementing a force/torque tracking controller providing a high-level control input to the IDA-PBC controller. This as well is in the scope of our future studies.

# Chapter 5

# Design of a Compliant Actuator for Aerial Physical Interaction and Manipulation

The control of the flying robots during physical interaction or manipulation is a challenging problem, not only in terms of estimation and control, but also in terms of designing new tools or actuators. For this reason, this chapter is dedicated to the design, identification and control of a light-weigh elastic-joint arm, to be used on board of a small-size aerial robot, e.g. the quadrotor in Fig. 2.2. The elastic-joint arms are known with their intrinsic stability when the output shaft of the arm is physically interacting with the environment. Here we show how to use this ability for Aerial Physical Interaction (APhI), by controlling an in-home built arm on the flying robot (quadrotor). Moreover elastic arms are favorable for explosive movement tasks, e.g. throwing, due to the potential energy stored in the elastic components. Such manipulation task has never been considered for aerial robots before and in this chapter we investigate its possibility with experimental results.

This chapter can be considered as a bridge from aerial physical interaction to aerial manipulation, where we propose a new light-weight elastic joint-arm design, that in theory is capable of achieving both tasks. It can be seen, that different from Chapter 4, here we approach to the problem of APhI from the design point of view. We note that the content of this chapter is published in Yüksel et al. (2015).

## 5.1 Introduction

The control of flying robots during physical interaction is a challenge in terms of designing new tools or actuators, as well as developing powerful algorithms to allow the exertion of forces and torques on the environment, while stabilizing the overall system and protect the expensive hardware. In Chapter 4 we have presented a control method addressing this problem.

In parallel, other studies presented useful designs for improving the performance of physical interaction tasks, by introducing novel tools. In Kondak et al. (2013), a light-weight industrial arm is attached to a small-size helicopter, under the consortium of ARCAS (2011-2015). Smaller scale designs are also presented in the literature. In Kim et al. (2013) a 2D rigid arm for aerial manipulation is introduced. A tool for surface inspection using a flying robot is developed and presented in Fumagalli et al. (2012b). Besides many different joints and actuated tools, passive ones are also used for physical interaction, as in Gioioso et al. (2014). See also Sections 1.2 and 1.3 for a broad literature review.

In APhI, it is necessary to always guarantee a safe (i.e., non-destabilizing) behavior of the system during any contact with the environment, that can be either desired or unforeseen (e.g., the robot moving in an unknown and/or hostile area). Furthermore, it should be possible to implement explosive motions for useful aerial interaction tasks, like aerial repairing or fixing. The use of flexible joints has proven in the literature to be successful for the implementation of interactive tasks, so far for the grounded robotic arms or humanoid robots. The elasticity of the joints can in fact be exploited for achieving an intrinsic safe behavior of the system and for amplifying the mechanical performance of a rigid arm by exploiting the energy stored into the elastic element, as also described in Braun et al. (2012) and in Braun et al. (2013). In this chapter (and later in the following in Chapter 6 as well), we explore the possibility of exploiting the benefits of elastic joints in aerial physical interaction, which was not done before to the best of our knowledge. The goal of this chapter is to conduct a preliminary work to start filling this gap. For this, we present a novel design of a light weight flexible-joint arm that can be mounted on a small size aerial vehicle for achieving an intrinsically safe APhI and that allows to exploiting the joint elasticity for implementing aerial explosive tasks (e.g., aerial hammering and aerial throwing). The proposed arm is then mounted on a quadrotor and its benefits for APhI are experimentally validated.

## 5.2  Design of an Elastic-Joint Arm for Aerial Robots

Let us describe the light-weight elastic-joint arm, designed in-home, to be used for APhI together on board of a small size flying robots, e.g. the quadrotor platform shown in Fig. 2.2. The flexible-joint arm consists of several parts. The rigid parts, with the exception of the actuators, are CAD modeled and 3D printed. As can be seen in Fig. 5.1, a rigid pulley is attached to the shaft of the servo motor, which is connected to a second rigid pulley via two elastic elements (see also Fig. 5.2 for its realization). The first pulley is referred to as the *active pulley* and the second one as the *passive pulley*. The passive pulley is attached to a rigid link, whose objective is to interact with the environment. The position of the active pulley is measured by the encoder of the servo motor. The measurements for the arm motion are collected with a magnetic encoder attached to the passive pulley. The magnet is placed on a cylindric part attached to the center of the passive pulley, and the encoder is placed on a fixed surface.

A second linear servo motor can let the active pulley slide along the bottom surface in order to regulate the distance between the two pulleys. At present however this feature is not used and the distance has been fixed with a rigid connection. The reason for introducing a second linear motor in the design is to have room for future improvements, such as changing the design from an elastic actuator to a variable stiffness actuator by using, e.g., nonlinear elastic components Ham et al. (2009). Finally, a rigid box covering the mechanism except the rigid arm is adopted for protection of the hardware components.

The elastic components between two rigid pulleys are chosen as linear springs (see in Figs. 5.2 and 5.3). The two springs are used as antagonistic pairs, i.e., in case one of the springs contracts, the other one relaxes, and viceversa. This is a natural way of designing a flexible arm, similar to biceps and triceps in the human arm; where the muscels cannot push, but only pull (see also Ham et al. (2009)). The advantage of using antagonistic pairs is that, they allow the full elastic behavior in both directions of the rotation.

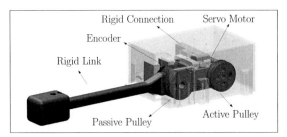

Figure 5.1: CAD Model of the flexible-joint arm.

| Link Inertia | $J_1$ | 0.0019 | [kgm²] |
|---|---|---|---|
| Spring Constant | $k_e$ | 0.3374 | [Nm/rad] |
| Natural Frequency | $\omega_n$ | 13.3 | [rad/s] |
| Motor-side Dissipation | $\mu_m$ | 0.0364 | [Ns/rad] |
| Link-side Dissipation | $\mu_l$ | 0.0048 | [Ns/rad] |

Table 5.1: Estimated physical parameters of the flexible-joint arm

Besides the mechanics of the system, the electronics and communication with the arm is another part of the design. This is explained in Section 5.4 in detail.

## 5.3 Identification and Control

We have designed the elastic-joint arm in Fig. 5.1 from stretch. This means that the physical parameters of the flexible arm are mostly unknown. Here we show how to acquire these parameters that will be later used for controlling the flexible-joint arm motion. Additionally, we derive the models of the link and of the motor that will be exploited for evaluating the dynamic behavior of the system during fast motions.

### 5.3.1 Parametric Identification

The flexible-joint arm consists of the motor-side and the link-side dynamics. Both are of second order as shown in Ozparpucu and Haddadin (2013). However, if the motor dynamics is considerably faster than the link dynamics, it is possible to assume that the motor velocity can be controlled instantaneously, as also shown in Ozparpucu and Haddadin (2013) and in Haddadin and Krieger (2012). This means that the servo motor behaves as a perfect velocity source. Under this assumption, the linear dynamics of the flexible-joint arm can be described as:

$$J_1\ddot{\theta}_1 + \mu_l\dot{\theta}_1 + k_e\theta_1 = \mu_m\dot{\theta}_m + k_e\theta_m \tag{5.1}$$

$$\theta_m = \int_0^t \dot{\theta}_m^d dt + \theta_m^0, \tag{5.2}$$

where $\theta_m \in \mathbb{R}$ and $\theta_1 \in \mathbb{R}$ are motor side and link side orientations, respectively, $J_1 > 0$ is the inertia of the rigid link, $k_e > 0$ is the linear spring constant, and $\mu_m > 0$ and $\mu_l > 0$

Figure 5.2: The light-weight flexible-joint arm.

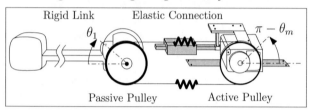

Figure 5.3: Schematic figure of the flexible-joint arm with notations used Chapter 5. The linear motor is grayed out and inactive for the current design. A future development includes activating this part too, for upgrading the current elastic actuator design to a variable stiffness actuator one.

represent the dissipations (viscous elements) of the motor side and link side, respectively. The desired motor velocity is shown with $\dot{\theta}_m^d$, and $\theta_m^0$ is the initial motor position.

The validity of the assumption of considering motor as a velocity source depends on the load attached to the motor. Since we are introducing a novel design, it is in our interest to see whether the servo motor can be used as velocity source when the elastic and rigid parts are also attached.

Considering as input the desired motor velocity $\dot{\theta}_m^d$, and as output the measured motor velocity $\dot{\theta}_m$, the transfer function, in the Laplace domain, is expected to be of the first order (a servo motor dynamics), namely

$$\frac{\dot{\Theta}_m(s)}{\dot{\Theta}_m^d(s)} = e^{-s\tau_d} \frac{a}{s+b}, \tag{5.3}$$

where $\tau_d$ represents the system delay.

In order to validate the model in (5.3) and to estimate the parameters $a$, $b$, and $\tau_d$ we performed some experiments and used the nonlinear least squares method (more details can be found in Ljung (1986)). In Fig. 5.4 the frequency response of the motor is shown for different step inputs as desired velocity profile and with different conditions: loaded (springs and the arm is attached to the motor) and unloaded cases (motor output shaft is free of any load). The output is the measured motor velocity that is retrieved using the motor encoder. The different plot colors correspond to the pulse trains of different frequencies. Solid lines are corresponding to conditions where no load is attached to the motor. Dashed lines represent the cases when the motor is connected to the arm though the elastic interconnection. As it can be seen, the frequency response of the system stays

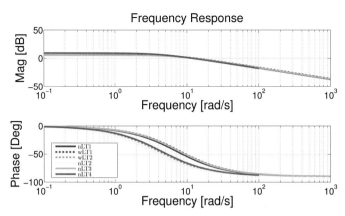

Figure 5.4: Frequency response of the motor for different step inputs. The input is the desired motor velocity, and the output is the measured motor velocity. The dashed lines depicts the conditions 'with load' (wL: springs, passive pulley and the rigid arm are connected to the motor), and the solid lines show the ones with 'no load' (nL: only the active pulley is connected to the motor). Different colors correspond to pulse trains of different frequencies (T1, T2, T3, and T4).

almost the same in the frequency range of interest, which is around the natural frequency of the total system (this will be identified in the next step). Hence we can consider the motor as a velocity source even when it is connected to its load. A good fit for (5.3) considering different conditions is found for the parameters $a = 13.79$ and $b = 7.175$, and $\tau_d = 0.2$s.

The parameters of the flexible-joint arm dynamics, given in (5.1), have to be estimated as well. Let us first write the transfer function of the system dynamics for motor position $\theta_m$ as input and link position $\theta_1$ as output. Denoting with $\omega_n$ and $G$ the natural frequency and the low-frequency gain of the system, respectively, we obtain:

$$\frac{\Theta_1(s)}{\Theta_m(s)} = e^{-s\tau_d} K \frac{s + \mu}{s^2 + 2\xi\omega_n + \omega_n^2}, \tag{5.4}$$

where $K = GD_\theta/J_1$, $\mu = k_e/\mu_m$, $\omega_n = \sqrt{k_e/J_1}$, and $\xi = \mu_l/2\sqrt{k_e J_1}$.

The moment of inertia for the rigid link is computed as $J_1 = 0.0019$ [kgm$^2$] from the CAD model shown in Fig. 5.1. Although the inertia is easy to compute from the geometry of the system, the dissipative parameters such as the damping and frictions are hard to retrieve from a simulation. For this reason, we again used nonlinear least squares Ljung (1986) to identify the system parameters. By choosing motor position $\theta_m$ as input and link position $\theta_1$ as output we computed the frequency response of the system for different step input profiles. The results are presented in Fig. 5.5. The best fit for (5.4) is

$$\frac{\Theta_1(s)}{\Theta_m(s)} = e^{-0.2s} 8 \frac{s + 9.2634}{s^2 + 2.573s + 197.3}. \tag{5.5}$$

The physical parameters corresponding to this transfer function are shown in Table 5.1.

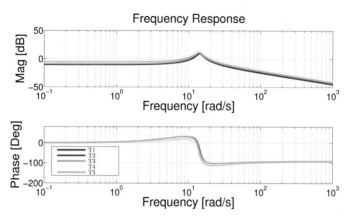

Figure 5.5: Frequency response of the flexible-joint arm system for different step inputs.
The input is the motor position and the output is the measured link position.
Different colors present the trials with different step input profiles in motor
velocities.

In Figs. 5.4 and 5.5 it is clear that the both motor and link can be operated around the
natural frequency of the flexible-joint arm, $\omega_n$, when the servo motor is considered as a
perfect velocity source. The natural frequency of the system is especially interesting for
us, since we would like to test the new design during fast movement tasks, as explained in
Section 5.3.2 in detail.

Now, by considering the flexible-joint arm as a second-order LTI system (which is actually
the case considering (5.1) and (5.2)), where the desired velocity is the input, and the motor
position and the link position are the outputs, we have

$$\frac{\Theta_m(s)}{\dot{\Theta}_m^d(s)} = e^{-s\tau_d}\frac{a}{s(s+b)} \tag{5.6}$$

and

$$\frac{\Theta_1(s)}{\dot{\Theta}_m^d(s)} = \frac{e^{-s\tau_d}(Kas + K\mu a)}{s^4 + (2\xi\omega_n + b)s^3 + (\omega_n^2 + 2\xi\omega_n b)s^2 + \omega_n^2 bs}, \tag{5.7}$$

where both (5.6) and (5.7) can be used to evaluate motor and link motions during explosive
movements of the system.

## 5.3.2 Control of the Flexible-joint Arm

Due to its compliance, a flexible-joint arm is intrinsically safe and an unforeseen interaction
with the environment is absorbed by the arm and it perturbs only slightly the motion of
the aerial vehicle it is mounted on. Another interesting interactive feature of flexible-joint
arms is the possibility of achieving fast, or even explosive motions. Such movements can
be transformed to hammering or throwing tasks using flying robot, which can be useful
especially for construction works in relatively high buildings where a ground robot cannot
reach.

By explosive movement, we mean the amplification of the link velocity using elastic components of the flexible-joint arm. A rigid arm, directly connected to the actuator without any elastic element, will always have the velocity of the motor. However, a flexible-joint arm can reach higher velocities than what actuator can provide, thanks to its elastic components.

It is shown in the literature (e.g. in Haddadin and Krieger (2012)) that the elastic components, e.g., springs, can be used to amplify the velocity of the actuation source. Let us then use the unconstrained optimal control strategy presented there to maximize the flexible-joint arm velocity in a specified final time $t_f$. Hence the cost function to be maximized is

$$J = \dot{\theta}_1(t_f). \tag{5.8}$$

Since we identified the damping factors relatively low (see Table. 5.1), we can neglect them. Hence, the optimal control policy presented in Ozparpucu and Haddadin (2013) and in Haddadin and Krieger (2012) for unconstrained undamped mass-spring system is suitable for our purpose. The optimal controller is then

$$u^* = \begin{cases} \dot{\theta}_m^{max}, & \sin(\omega_n(t_f - t)) > 0 \\ \dot{\theta}_m^{min}, & \sin(\omega_n(t_f - t)) < 0, \end{cases} \tag{5.9}$$

where the optimal control input is $u^* = \dot{\theta}_d^*$. Note that we consider the servo motor as a perfect velocity source (see (5.2)) which is however constrained, i.e., $u^* \in [\dot{\theta}_m^{min}, \dot{\theta}_m^{max}]$. This type of controller is called *bang-bang* controller. In our case no state constraints are considered.

## 5.4 Experimental Results

The flexible-joint arm is a combination of 3D printed rigid parts, elastic components, actuator, and measurement units (sensors) with their electronics. The 3D printed parts and elastic elements are introduced previously in detail in Sec. 5.2. In this section we present the actuator, i.e., the servo motor, the sensors, and its electronics. Moreover, we briefly recall the flying platform and present the setup for the experiments. Finally we present the experimental results.

### Control of the Aerial Robot

In this part we give an overview of the controller used for quadrotor, to achieve a stable flight while the flexible-joint arm is operating. In addition, we expect from the controller to give a good tracking results when a human operator is controlling its trajectory. In real-life cases, where flying robots are required to be interacting with their environment (such as in outdoor scenarios) we believe that controlling the flying system with human input is as important as fully autonomous trajectory control. For this reason, we choose to control the quadrotor using so called near hovering control, as deeply explained in Lee et al. (2013).

Briefly, the controller is developed for teleoperation tasks using quadrotors, where it allows operator more focusing on high-level tasks, while low-level ones are hidden from the operator such as underactuation of the translational dynamics. The goal of the controller is to make quadrotor follow a smooth trajectory in translational motion, using the inputs given

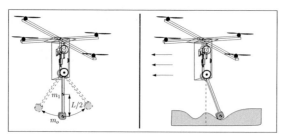

Figure 5.6: Flexible-joint arm is attached to the quadrotor system. Two experimets are designed; explosive arm movement (on the left) and aerial physical interaction (on the right). The second one is a task, where flexible-joint arm is sliding on a sloped surface, while quadrotor is performing a stable flight. The rigid link mechanism consists of a mass $m_o = 0.044$ [kg] at the tip of the link, in a distance of $L = 0.18$ [m] to the link center of mass, where $m_1 = 0.044$ [kg].

by the human operator, while trying to keep hovering configuration, i.e., small pitch and roll angles, as much as possible. In the experiments, we used this controller for controlling the translational trajectory of the quadrotor, where the novel flexible-joint arm is placed on it, as depicted in Fig. 5.6. The main limitation of the design is the weight and size of the flexible-joint arm, since it is planned to be mounted on a small-scale quadrotor. For this reason we chose Dynamixel AX-12A servo motor[1], which is both velocity and position controllable. The motor provides velocity, position, load, and temperature measurements. The communication with servo motor is done using serial communication block of Simulink-Matlab. The position and velocity measurements of the passive pulley (so the rigid link) are acquired using 10-bit AEAT-6010 magnetic encoder[2]. The encoder is fixed to the body of the actuator, whereas the rotating part (magnet) is directly connected to the passive pulley. The encoder readings are acquired using serial channel of Arduino-Uno[3], and transferred to another serial communication block in the same simulink file where the servo motor is controlled.

**Experimental Setup**

The experiments are conducted on the flying platform, which is a quadrotor UAV (see Sec. 2.3). The maximum payload of the quadrotor is about 2 [kg], hence we designed the flexible-joint arm as light as possible (total mass of the flexible-joint arm is 0.36 [kg]). The maximum reachable rotation angle of the flexible-joint arm is limited to $\theta_1 \in [-1, 0.55]$ [rad]. The experiments are done indoor, and the state estimation of the quadrotor is done as explained in Sec. 2.3. The total system weight is 1.36 [kg]. The communication with the on board electronics of the quadrotor and the MoCap is done on an Ubuntu 12.04 machine, using ROS-fuerte (Quigley et al. (2009)) and Telekyb softwares, which is developed for controlling the aerial robots (see Grabe et al. (2013)).

The quadrotor is controlled by a human operator, using a standard joypad connected

---

[1]http://www.robotis.com/xe/dynamixel_en

[2]http://www.avagotech.com/docs/AV02-0188EN

[3]http://arduino.cc/en/Main/arduinoBoardUno

Figure 5.7: Explosive movement of the flexible arm during flight on the quadrotor. On the left picture arm swings right, and on the right picture it swings left. Results of this experiment is given in Fig. 5.8.

to the Ubuntu machine. Besides the operator, another person was holding the security stick, which is connected to the quadrotor from top. The connection is done with ropes for security reasons, which have no tension the during flight.

**Experiments and Results**

We have designed two experiments for showing the effectiveness of the elastic joint-arm on a quadrotor; fast movement of the arm, and stable aerial physical interaction task (see Fig. 5.6 for a sketch of these two tasks).

For the first experiment, the quadrotor is actively keeping a hovering condition, and the optimal controller presented in (5.9) is applied to the flexible arm, by choosing $t_f = 3$ [s]. The motion of the arm is shown in Fig. 5.7, with two snapshots from the real experiment.

The results are shown in Fig. 5.8. The control input as desired motor velocity is shown in the figure with blue dashed line. The measurements are depicted with black dots, in 10 [Hz] from both motor and link encoders. Here we also plotted the evolution of the identified transfer functions in 1 [kHz], which is presented with solid red plots. As seen in the lower right plot of Fig. 5.8 the maximum link velocity is reached in around $t_f = 3$ [s], which is almost five times more than the servo motor velocity. Thanks to the storage of elastic energy due to the flexible joint, the motion of the link is amplified, which provides some preliminary results on explosive movement tasks using aerial robots. Such tasks can be imagined as aerial hammering or throwing.

The second experiment is for the aerial physical interaction task using flexible-joint arm on the quadrotor. A platform with a slopped/uneven surface (blue colored surface in Fig. 5.10) is designed for this experiment. The goal is to have a stable flight, while quadrotor equipped with the arm is sliding on the blue surface in a stable manner. For this experiment, we have set the proportional gain of the quadrotor position controller (from Lee et al. (2013)) to zero along the $\mathbf{x}_W$-axis. The reason is to let the quadrotor drift along the direction of the motion, while the flexible-joint arm is interacting with the environment (see Fig. 5.10). The flexible-joint arm is position-controlled in this case, meaning that we send the desired motor position and velocity values at will. The results are given in Fig. 5.9. First column of the figure shows the positions of the quadrotor in three Cartesian axes, with blue solid lines. The gray areas are representing the hovering condition of the quadrotor, meaning that its desired position is fixed. After the first grayed area, quadrotor starts to descend and approaches to the blue platform so that the flexible arm touches the surface.

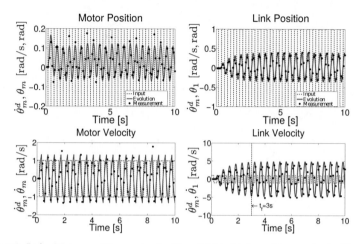

Figure 5.8: Left: Motor position and velocities. Right: Link position and velocities. Blue dashed line represents the optimal control input computed using (5.9) for $t_f = 3$ [m] and $\omega_n = 13.3$ [rad/s]. Black dots are the measured motor position and velocities using embedded servo motor encoder in 10 [Hz]. Red plot are for the corresponding evolutions of the identified transfer functions.

Later the quadrotor flies along the $+\mathbf{x}_W$ direction while trying to keep contact with the uneven surface (see Fig. 5.10). After holding in hovering position in the second grayed area, it flies back along the $-\mathbf{x}_W$ direction by keeping the contact. Finally it arrives to initial position, ascends, and stays in hovering condition as in the beginning. The pitch angles (rotation around the $\mathbf{y}_B$ axis) are given as black solid lines in Fig. 5.9. The near-hovering controller is keeping the system and the interaction stable during the whole flight. The red plots in the same figure are showing the motor positions ($\theta_m$), and the link positions ($\theta_1$) of the joint arm. Depending on the direction of the flight, different motor positions are set to the servo motor. The changes in the link position is clearly seen in Fig. 5.9 (bottom-right), which is following the profile of the interaction surface. The motor positions, on the other hand, are fixed unless it is not commanded. Thanks to the flexibility of the joint the arm, the overall system can safely interact with an unknown environment. Furthermore, since the elasticity of the joint absorbs the impact with the environment, the motion of the quadrotor is left almost unperturbed by the unforeseen impact with the environment.

Please note that this setup is further developed by; *i)* implementing the elastic-arm control in ROS environment *ii)* bringing the Leap-Motion hand tracking sensor[4] as a new user interface in the loop. The details of this practical developments are avoided due the brevity reasons. Nonetheless, in Fig. 5.11 we show a snapshot from an experiment, in which a leap-motion sensor is used to operate the arm motion for a hammering task, based on human hand tracking, using ROS-groovy interface.

---

[4]https://www.leapmotion.com/

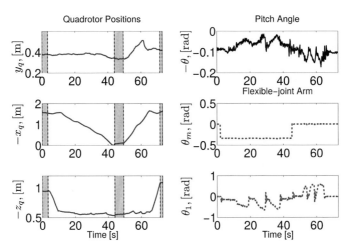

Figure 5.9: Results of aerial physical interaction with flexible-joint arm. On the left side three Cartesian coordinates are figured (blue solid lines) where the grayed areas are representing the hovering case of the quadrotor. Black solid figure on the right shows the pitch angles, and red dashed plots are measurements of motor angles ($\theta_m$) and link angles ($\theta_1$) for the flexible arm during the flight. A snapshot from this experiment is given in Fig. 5.10.

## 5.5 Discussions

In this chapter we have studied the design of a light-weight elastic joint-arm, for using in the APhI tasks of the flying robots. By developing such a tool, we approached to APhI problem from modeling and design point of view. After realizing the arm and identifying its physical parameters, we showed how it can be controlled, and used on board of an available quadrotor setup.

The work explained in this chapter is paving the way of using compliant actuators for APhI and maybe for aerial manipulation as well. Here it is shown that due to their nature and design, actuators with elastic elements have stabilizing effect during APhI. Morover they can even be further benefited for performing explosive movement tasks, and such movements can be turned into to hammering or throwing tasks. Enabling the flying robots for such tasks can be extremely effective especially for construction works in relatively high buildings, where a ground robot cannot reach.

Since the overall setup (shown during flight in Fig. 5.10) is a binomial of a flying base and a manipulating arm, it is actually an *aerial manipulator*. However in this chapter we did not consider this aerial manipulator as a full system, namely the control employed for the experiments is unaware of the dynamical effects of the arm on the quadrotor, and of course vice versa. Both systems consider these effects as disturbances, and try to cancel each other.

A future development of the design presented here is activating the linear motor (see the grayed out part of Fig. 5.3), which changes the distance between the passive and the active

Figure 5.10: Snapshots from the experiments for aerial physical interaction using elastic-joint arm. Three figures represent the evolution of the motion along the $\mathbf{x}_B$-axis, as mentioned in Fig. 5.9, where the results are shown. The three axes of the world frame are depicted with three different colors in the last figure.

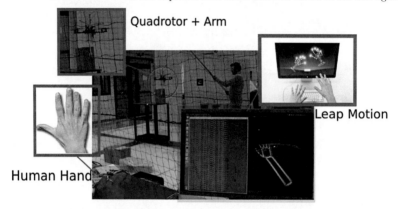

Figure 5.11: During the experimental tests of the elastic-joint arm, we have used it with different user interfaces. In this figure, the communication with the arm using a hand tracking sensor, a.k.a *Leap Motion*, is shown. The desired positions and the velocities of the hand palm is acquired using the Leap Motion sensor, and sent to the actuator of the arm via ROS interface.

pulleys where the springs are located. In case of the nonlinear elastic components, the change in the position of this linear motor would vary the stiffness of the elastic actuator design, turning it a variable stiffness one.

Another improvement is considering the dynamic model of the aerial manipulator, taking the couplings between the flying base and the manipulating arm into account. In fact, this is exactly what we do in the following chapter; we study the aerial manipulators as a full system, consisting of a flying platform and manipulating arm(s), and propose controllers that are aware of the dynamical coupling between those systems.

# Chapter 6

# Control of Aerial Manipulation using Differential Flatness and Exact Linearizability

In Chapter 5 we presented a novel flexible-joint arm as a powerful tool for aerial physical interaction and manipulation. However, the control of the aerial robot equipped with this arm was independently done, i.e. neither the arm nor the quadrotor was aware of each others dynamics.

In this chapter we intensively study the system dynamics of different aerial manipulators, when they consist of a flying platform and one or multiple number of manipulating arms. In fact, here we show that such systems are *differentially flat*, and for them there exist an *exact linearizing controller* (or Dynamic Feedback Linearizing - DFL- controller). Hence, we first briefly recall the concept of differential flatness in Sec. 6.1.1. Our first set of objectives, as defined in Sec. 6.2.1, is to exploit the differential flatness property of some types of aerial manipulators, which are binomial of a flying platform and a single joint-arm, whose dynamics are constrained in a plane. To achieve this, we formalized our methodology in Sections 6.2.2 and 6.2.3. Note that Section 6.2 is entirely dedicated to the aerial manipulators with a single-joint arm, which can be rigidly or elastically actuated. There we show that the attachment point of the manipulating arm has a tremendous effect on the differential flatness property of the system, as well as the design of the tracking control problem. Briefly, when the arm is attached to the CoM of the flying robot, then the end-effector positions become part of the flat outputs, and this has a practical importance for us. We validate our theoretical results first numerically in Sec. 6.2.9, where we present advantages of using rigid or elastic actuation over each other for different aerial manipulation tasks. Then in Sec. 6.2.10 we present the preliminary experimental results for tracking control of a quadrotor equipped with an arm that is actuated via a Variable Stiffness Actuator (VSA).

Let us stress again, that the end-effector position of the aerial manipulator is part of the flat outputs of the overall system, when the arm is attached to the CoM of the flying platform. This is important because most of the aerial manipulation tasks require the control of the end-effector positions. In light of this result, in Sec. 6.3 we consider an aerial manipulator design, where the flying robot is equipped with a generic number of manipulating arms, each may have any numbers of DoF, with either rigid or compliant actuators. The system dynamics, their differential flatness property and a proper exact linearizing controller is explained in Sec. 6.3. Since so far the system motion was constrained to a plane, its extension to 3D is required for performing practical applications. For this reason in Sec. 6.4 we investigate control methods developed in 3D, which can benefit from

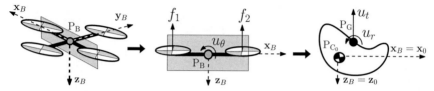

Figure 6.1: Conceptual sketch of a PVTOL as a quadrotor on a plane. First figure on the left is the sketch of a quadrotor, which was depicted before in Fig. 2.1. Second figure is the quadrotor in $\mathbf{x}_B - \mathbf{z}_B$ plane, where the Center of Actuation (CoA) and the Center of Mass (CoM) of the PVTOL are at the same point, $P_B$. Third figure is a generic PVTOL we consider in Chapter 6, where the CoA ($P_G$) is different from the CoM ($P_{C_0}$) of the PVTOL. Note that the pitch torque of the quadrotor $u_\theta$ given in (2.3) is equivalent to the PVTOL torque $u_r$.

the differential flatness property of the aerial manipulators discovered in 2D.

We note that the content of this chapter has been partially published in different papers. Some parts of Sec. 6.1, and Sec. 6.2 are available online in Yüksel and Franchi (2016) and in Yüksel et al. (2016b); Sec. 6.3 in Yüksel et al. (2016a); and Sec. 6.4 in Tognon et al. (2017).

# 6.1 Introduction

In this chapter we study the system properties and control of an *aerial manipulator*, a system consisting of a flying robot and one or multiple manipulating arms. A broad literature review for the existing aerial manipulators is given in Section 1.3. In particular, here we consider the motion of a quadrotor VTOL in a vertical plane, e.g. $\mathbf{x}_B - \mathbf{z}_B$ plane as depicted in Fig. 6.1. The system in this plane can be called as a *Planar Vertical Take-off and Landing* (PVTOL) aerial robot, which is actually a quadrotor VTOL, but its motion is constrained in a plane.

A PVTOL aerial platform, similar to previous studies (see, e.g., Lupashin and D'Andrea (2013); Lupashin et al. (2010)), does not only capture the nonlinear features and the underactuation of a 3D system, but also allows generalizing the obtained results in a later stage (e.g. see Sec. 6.2.10 and Sec. 6.4). Furthermore, many practical aerial problems are, fundamentally, 2D problems immersed in a 3D world (as, e.g., the aerial grasping problem addressed in Thomas et al. (2013)).

We note that this is a long chapter, including various aerial manipulator designs, and their striking nonlinear system properties. By aerial manipulator, we now mean a PVTOL, which is equipped with one or multiple manipulating arms. In particular, here we are going to show how the differential flatness property of such systems changes with; *i)* the attachment point of the arm, *ii)* the actuation type of the arm, *iii)* and the DoF as well as the number of the manipulating arms.

Although the control of aerial robots (Mistler et al. (2001), Koo and Sastry (1999)) and the control of the fixed-base manipulators with: rigid (Siciliano et al. (2009)), elastic (De Luca and Book (2008)), or mixed (De Luca (1996)) joint types have been studied in the literature separately, the analysis and control of systems consisting of these two is still an open (and rapidly growing) topic. Especially, the control of aerial manipulators with elastic arms has not been addressed so far and is not yet throughly understood, although benefiting from elastic components of a ground manipulator has been studied before for explosive movement tasks, e.g. in Braun et al. (2012). Here, we will try to set some light to their system analysis by studying their differential flatness property, which can be very effective in the trajectory planning phase, or their control.

Let us first recall some known facts on the differential flatness and the Dynamic Feedback Linearization (DFL).

## 6.1.1 Relative Degree, Exact Linearization and Differential Flatness

We start with some preliminaries on the relative degree of a system, and its connection to the differential flatness. To fix the ideas better, first consider the following *Single-Input-Single Output* (SISO) nonlinear system;

$$\begin{aligned} \dot{\mathbf{x}} &= f(\mathbf{x}) + g(\mathbf{x})u \\ y &= h(\mathbf{x}) + z(\mathbf{x})u, \quad \mathbf{x} \in \mathbb{R}^n, \quad u, y \in \mathbb{R}, \end{aligned} \tag{6.1}$$

where $\mathbf{x} \in \mathbb{R}^n$ is the state and $u, y \in \mathbb{R}$ are the input and the output of the system, respectively. If $z(\mathbf{x} \equiv \mathbf{0})$ is in the neighborhood of $\mathbf{x}_0$, then by differentiating the output we can write

$$\dot{y} = \frac{\partial h}{\partial \mathbf{x}} \dot{\mathbf{x}} = L_f h(\mathbf{x}) + L_g h(\mathbf{x}). \tag{6.2}$$

Then let us give the following definition,

---

**Definition 4** (Relative degree of SISO Systems-Sepulchre et al. (1997)). *The relative degree of the nonlinear system* (6.1) *at the origin* $\mathbf{x} = \mathbf{x}_0$ *is the integer* $r$, *s.t.*

*(i)* $L_g L_f^k h(\mathbf{x}) \equiv 0$, *for* $k = \{0, \cdots, r - 2\}$ *and* $\mathbf{x}$ *in a neighborhood of* $\mathbf{x} = \mathbf{x}_0$;

*(ii)* $L_g L_f^{(r-1)} h(\mathbf{x}_0) \neq 0$.

---

In words, for a nonlinear system, the *relative degree* of the *output* w.r.t the *input* is the differential order of the output at which the input explicitly appears for the first time.

Now for a *Multi-Input-Multi-Output* (MIMO) nonlinear system;

$$\begin{aligned} \dot{\mathbf{x}} &= f(\mathbf{x}) + g(\mathbf{x})\mathbf{u} \\ \mathbf{y} &= h(\mathbf{x}) + z(\mathbf{x})\mathbf{u}, \quad \mathbf{x} \in \mathbb{R}^n, \quad \mathbf{u}, \mathbf{y} \in \mathbb{R}^m \end{aligned} \tag{6.3}$$

assign to each output $y_i$ an integer $r_i$ which is the number of differentiations of $y_i$ needed for at least one of the inputs to appear first time explicitly, and $i = \{1, \cdots, m\}$. Then let us give the following definition,

---

**Definition 5** (Relative degree of MIMO Systems-Sepulchre et al. (1997)). *The nonlinear system* (6.3) *has a relative degree* $\{r_1, \cdots, r_m\}$ *at the origin* $\mathbf{x} = \mathbf{x}_0$, *if*

*(i)* $L_{g_j} L_f^k h_i(\mathbf{x}) \equiv 0, \, \forall \, 1 \leq i, j \leq m, \, \forall \, k < r_i - 1$ *and* $\forall \mathbf{x}$ *in a neighborhood of* $\mathbf{x} = \mathbf{x}_0$,

*(ii)*

$$\left[ \frac{\partial y_i^{(r_i)}}{\partial u_j} \right]_{1 \leq i,j \leq m} = \begin{bmatrix} L_{g_1} L_f^{(r_1-1)} h_1(\mathbf{x}) & \cdots & L_{g_m} L_f^{(r_1-1)} h_1(\mathbf{x}) \\ \vdots & \ddots & \vdots \\ L_{g_1} L_f^{(r_m-1)} h_1(\mathbf{x}) & \cdots & L_{g_m} L_f^{(r_m-1)} h_m(\mathbf{x}) \end{bmatrix} \quad (6.4)$$

*is nonsingular at* $\mathbf{x} = \mathbf{x}_0$.

---

Notice that the second condition of Definition 5 is a generalization of the second condition of Definition 4. In these both definitions, the relative degree of the *square* systems is considered, i.e. the number of input is equal to the number of outputs. For underactuated systems, as will be discussed in Section 6.2.2, for computing the relative degree, we will make sure that the number of the chosen outputs are equal to the number of inputs. Note that if $\{r_1 = \cdots = r_m\}$, then the outputs of the system (6.3) has a *uniform* relative degree $r_1$ w.r.t. its outputs. Moreover we call $r = \sum_{i=1}^{m} r_i$ as the *total relative degree* of the system in (6.3), which (later to be shown) has a significant impact for both the *differential flatness* and the *exact linearization* of a nonlinear system. Let us shortly define the differential flatness;

---

**Definition 6** (Differential Flatness-Fliess et al. (1995); Murray et al. (1995)). *The system in* (6.3) *is said to be* differentially flat, *if there exists an output* $\mathbf{y}$ *(called flat output), such that the states* $\mathbf{x}, \dot{\mathbf{x}}$ *and the control inputs* $\mathbf{u}$ *can be expressed as an algebraic function of* $\mathbf{y}$ *and a finite number of their derivatives.*

---

We note that this definition is repeated later in Definition 8 for the aerial manipulator system studied in Sec. 6.2. Furthermore, for the same aerial manipulator we describe the *exact linearization* in Definition 7. Although differential flatness is a geometric property of a nonlinear system and exact linearization is a control approach[1], they both imply each other because of the Fact 1. This means, that the differentially flat outputs of a nonlinear system is also its exactly linearizing outputs. As elegantly stated in Isidori (2013), when the exactly linearizing (or flat) outputs are used for constructing a dynamic feedback linearization (DFL) controller, then the zero dynamics of the system becomes trivial.

Now this in mind, in Definition 7 it is given that for the underactuated systems, two conditions are required for an output to be a differentially flat output; *i)* the total relative degree of the system matches with the the total number of states, *ii)* there is an invertible decoupling matrix in form of (6.4). The second condition allows us finding a control in form of (6.9), which brings the system to a linear controllable form[2]. The first condition

---

[1]See overall Sec. 6.2.2 for an informal description of *exact linearizability via dynamic feedback* for the aerial manipulators considered in Section 6.2.

[2]Then, a linear controller satisfying *Hurwitz* criteria using pole-zero cancellation can steer the output dynamics in a stable manner, if the zero dynamics of the system is stable.

makes sure that there are no unstable zero dynamics (or they are trivial). It is important to realize that the *zero dynamics* of the nonlinear system has a fundamental role for the feasibility of the exact linearization and output regulation (see Sepulchre et al. (1997)). If it is unstable, i.e. there are states that are not observable from the output of the system, then this pole-zero cancellation can potentially destabilize the whole system. However, thanks to the matching condition of the total relative degree (fist condition of Definition 8), after an exactly linearizing DFL control[3], there will be no unobservable states, hence no unstable zero dynamics (or they become trivial as stated in Isidori (2013)).

In summary, when these two important conditions are met, we can say that the output $\mathbf{y}$ of a nonlinear system is an exactly linearizing output via dynamic feedback, and the linearized dynamics is exact of the nonlinear one, which can be stabilized using a pole-zero cancellation method (under the Hurwitz condition). Furthermore, from the Definition 8 and the Fact 1, this means that the system is differentially flat for the output $\mathbf{y}$.

Clearly, the choice of using DFL controller has its own advantages, or disadvantages as discussed in Sec. 1.4.2. Hence, using it or another controller enjoying the flatness property of the system (e.g. the one presented in Sec. 6.4.2) depends on the convenience.

In the following, we will use these concepts for addressing the output tracking control of the aerial manipulators. The considered system is a PVTOL equipped with a robotic arm (see Fig. 6.2). Note that different from the previous chapters, here we do the following change of coordinates; *i)* For the convenience of the further computations, the frame fixed at the CoM of the quadrotor, $\mathcal{F}_B : \{P_B, \mathbf{x}_B, \mathbf{y}_B, \mathbf{z}_B\}$ in 3D, will be called as $\mathcal{F}_0 : \{P_{C_0}, \mathbf{x}_0, \mathbf{z}_0\}$ on the $\mathbf{x}_B - \mathbf{z}_B$ plane, *ii)* similarly the world frame on the plane is depicted with $\mathcal{F}_W : \{P_W, \mathbf{x}_W, \mathbf{z}_W\}$, *iii)* and the torque around $\mathbf{y}_B$, called $u_\theta$ before, is denoted now by $u_r$ (see Fig. 6.1).

# 6.2 Aerial Manipulators with Single Joint-Arm

In this section we consider a Planar-VTOL (PVTOL) aerial robot equipped with a joint-arm. This joint arm can be connected to its actuator either rigidly or via some elastic elements (see Chapter 5 for an example). Here, we address the design and the control problems of the aerial manipulators (binomial of a PVTOL and a joint arm), and propose four new nonlinear controllers for their four most important design cases given by the possible combinations of the joint nature (rigid or elastic) with the kinematics of the platform-arm combination (generic attachment or CoM attachment), as summarized in Table 6.1.

Before going into the details, let us motivate the work done for this section. We aim at *i)* studying both the aerial vehicle and the manipulator together as one system, the *aerial manipulator, ii)* extending the preliminary insights shown in Chapter 5 and rigorously laying the foundations of the topic addressed there, and *iii)* comparing different designs of aerial manipulators for different tasks. We reach our goals by achieving the Objectives 1, 2 and 3 in Section 6.2.1. To do so, we rigorously analyze the exact output tracking and differential flatness properties of different aerial manipulator designs, and propose dynamic feedback linearization (DFL) control for each of them. A summary of the theoretical results can be found in Table 6.2.

To the best of our knowledge, such an extensive study for aerial manipulators using differential flatness and exact linearizability has not been presented before. Furthermore,

---

[3]i.e, that the flat (or exactly linearizing) outputs are used in DFL control.

controlling the motion of a robot composed by an elastic-joint arm attached to a flying vehicle has not been studied yet, except the work done in Chapter 5. Here, we fill this gap, by showing how we can independently and dynamically control the orientation of the elastic-joint arm together with the position of the PVTOL. By analyzing and proposing several controllers for robots of this novel type we aim at paving the way for the use of flexible-joint manipulators, which are able to benefit of the compliance advantages, also in the aerial physical interaction and manipulation field.

We exhaustively study the four different cases of Table 6.1 because each one is interesting for a different reason. First of all, the cases in which the joint is attached to any point of the PVTOL (shortly named 'Case RG' and 'Case EG' later) are interesting for their generality because they cover any possible real case. We prove that, in this case, the center of mass of the whole system (VTOL + arm) is a part the flat outputs. On the other hand, we prove that the end-effector position is not in general a part of the flat outputs, except for the cases in which the joint is attached to the CoM of the PVTOL, (shortly named 'Case RC' and 'Case EC' later). This fact brings strong advantages for the motion planning and control of the end-effector position because the whole state and the input can be computed analytically from a sufficiently smooth trajectory of the end-effector and the corresponding controllers are computationally simpler and do not generate a zero dynamics (see also Sec. 6.1.1).

The rigid-joint cases (Case RG and Case RC) are found to be more suitable for tasks such as aerial grasping (see Sec. 6.2.9) or trajectory tracking. The reason is that elastic-joint need more effort than rigid ones for these tasks, since the motor has to fight against the tendency of the spring to oscillate at its natural frequency. On the other hand the elastic-joint cases (Case EG and Case EC) are favorable for tasks in which one has to achieve high-speed link velocities such as aerial throwing (see Sec. 6.2.9). This is because of the ability of the elastic components to store potential energy and release it in the form of kinetic energy.

Summarizing, the main contribution of this section is that we systematically provide *i)* a set of exact linearizing (i.e., flat) outputs for the all the four cases, *ii)* the explicit algebraic map from the flat outputs to the states and the control inputs, *iii)* a nonlinear controller for each case with formal proofs, *iv)* the formalization of an optimal control problem for aerial manipulators using differential flatness property, *v)* an extensive set of realistic numerical tests that shows its practicability with real robots, *vi)* a comparison between the rigid-joint and the elastic-joint cases that shows the benefits and drawbacks of each choice, *vii)* a numerical study on the robustness of the controller for the coinciding cases (Case RC and case EC) when the coinciding assumption is not exactly verified[4], *viii)* preliminary experiments validating our theory for a quadrotor setup equipped with an arm actuated via a VSA.

In Section 6.2.1 we describe the kinematics of such system, which is sketched in Fig. 6.2. We first start with the case, in which the arm is rigidly actuated, and can be placed at any point on the PVTOL body (Sec. 6.2.4). Then in Sec. 6.2.5 we study the same case when the arm is attached to the CoM of the PVTOL, which enormously increases the capabilities of the controller. Later we do the same for the case when the arm is compliant, in Sec. 6.2.6 and Sec. 6.2.7, respectively. See Table 6.1 for the summary of these cases. We note that the results of these cases are summarized later in Table 6.2. Later in Sec. 6.2.8

---

[4]Meaning that the attachment point of the arm and the PVTOL CoM are not coinciding, but the controllers are not aware of that.

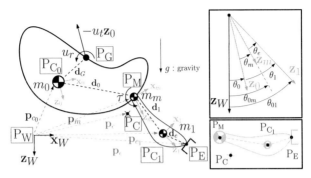

Figure 6.2: – *Left:* a sketch of the mobile aerial manipulator, composed by a PVTOL equipped with a 1-link arm. Notice the offsets between:

*i)* the center of mass (CoM) of the PVTOL ($P_{C_0}$),

*ii)* the center of actuation of the PVTOL ($P_G$), and

*iii)* the attachment point of the link ($P_M$), around which the motor rotates and an either rigid or elastic joint is placed.

– *Right:* above, the relative and absolute angles of the rigid bodies. The lengths of the **z** axes are made different just for illustration purposes. Below, the location of the important points around the arm.

we show how to use the powerful flatness property for optimal trajectory planning of the aerial manipulators. Finally the extensive numerical results are presented in Sec. 6.2.9; and preliminary experiments in Sec. 6.2.10.

## 6.2.1 Nomenclature and Objectives

The considered mobile aerial manipulator is composed by a generic model of a PVTOL with an attached 1-link arm, as depicted in Fig. 6.2 (left). We denote with $\mathcal{F}_W : \{P_W, \mathbf{x}_W, \mathbf{z}_W\}$ and $\mathcal{F}_0 : \{P_{C_0}, \mathbf{x}_0, \mathbf{z}_0\}$, the world (inertial) frame and the frame attached to the PVTOL, respectively, where $P_{C_0}$ is the Center of Mass (CoM) of the PVTOL (without the arm). Since all the motions are in a plane, both the motor and the joint of the arm rotate about an axis parallel to $\mathbf{z}_W \times \mathbf{x}_W$ and passing through a point $P_M$. We then define the *motor frame* as $\mathcal{F}_M : \{P_M, \mathbf{x}_m, \mathbf{z}_m\}$ that is rigidly attached to the motor output shaft. The joint can be either rigid (cases considered in Sections 6.2.4 and 6.2.5) or elastic (cases considered in Sections 6.2.6 and 6.2.7). We define also a *link frame* $\mathcal{F}_1 : \{P_{C_1}, \mathbf{x}_1, \mathbf{z}_1\}$, where $P_{C_1}$ is the CoM of the link. Finally we denote with the points $P_E$ and $P_C$ the end-effector of the arm and the CoM of the whole system (PVTOL+motor+link), respectively.

Given an angle $\theta_* \in \mathbb{R}$ between the **z**-axes of two frames, defined in Fig. 6.2 (right) we define:

$$\mathbf{R}_* = \begin{bmatrix} \cos\theta_* & \sin\theta_* \\ -\sin\theta_* & \cos\theta_* \end{bmatrix} \in \mathrm{SO}(2). \tag{6.5}$$

Therefore, the orientations of $\mathcal{F}_0$ in $\mathcal{F}_W$, $\mathcal{F}_M$ in $\mathcal{F}_0$, $\mathcal{F}_1$ in $\mathcal{F}_0$, and $\mathcal{F}_1$ in $\mathcal{F}_M$ are expressed by $\mathbf{R}_0(\theta_0)$, $\mathbf{R}_m(\theta_m)$, $\mathbf{R}_1(\theta_1)$, and $\mathbf{R}_e(\theta_e)$, respectively. The absolute motor angle is $\theta_{0m} = \theta_0 + \theta_m$

83

|  | Rigid Joint | Elastic Joint |
|---|---|---|
| $P_{C_0}$ and $P_M$ are *generic* | Case **RG**, Sec.6.2.4 | Case **EG**, Sec.6.2.6 |
| $P_{C_0}$ and $P_M$ are *coinciding* | Case **RC**, Sec.6.2.5 | Case **EC**, Sec.6.2.7 |

Table 6.1: Summary of the aerial manipulator categories considered in Section 6.2.

and absolute link angle is $\theta_{01} = \theta_0 + \theta_1$, as depicted in Fig. 6.2 (right). The angle $\theta_e = \theta_1 - \theta_m = \theta_{01} - \theta_{0m}$ is constantly zero if the joint is rigid and can be nonzero if the joint is elastic.

The constant positions of $P_M$ and $P_G$ in $\mathcal{F}_0$ are denoted with $\mathbf{d}_0 = [d_{0_x} \ d_{0_z}]^T \in \mathbb{R}^2$ and $\mathbf{d}_G = [d_{G_x} \ d_{G_z}]^T \in \mathbb{R}^2$ respectively. The constant position of $P_M$ in $\mathcal{F}_1$ is denoted with $-\mathbf{d}_1 = [-d_{1_x} \ -d_{1_z}]^T \in \mathbb{R}^2$. Finally, the vector $\mathbf{d}_e = [d_{e_x} \ d_{e_z}]^T \in \mathbb{R}^2$ denotes the constant position of the end-effector $P_E$ in $\mathcal{F}_1$.

The (time-varying) positions of $P_C$, $P_{C_0}$, $P_M$, $P_{C_1}$ and $P_E$ in $\mathcal{F}_W$ are denoted with $\mathbf{p}_c = [x_c \ x_c]^T \in \mathbb{R}^2$, $\mathbf{p}_0 = [x_0 \ z_0]^T \in \mathbb{R}^2$, $\mathbf{p}_m = [x_m \ z_m]^T \in \mathbb{R}^2$, $\mathbf{p}_1 = [x_1 \ z_1]^T \in \mathbb{R}^2$, and $\mathbf{p}_e = [x_e \ z_e]^T \in \mathbb{R}^2$, respectively.

The mass and moment of inertia of the PVTOL, motor, and link are denoted with $m_0 \in \mathbb{R}_{>0}$, $J_0 \in \mathbb{R}_{>0}$; $m_m \in \mathbb{R}_{>0}$, $J_m \in \mathbb{R}_{>0}$; $m_1 \in \mathbb{R}_{>0}$, $J_1 \in \mathbb{R}_{>0}$, respectively. As before, the symbol $g \in \mathbb{R}^+$ denotes the gravitational constant.

The PVTOL is actuated by means of: *i)* a total *thrust force* $-u_t \mathbf{z}_0 \in \mathbb{R}^2$ applied at $P_G$, where $u_t$ is its (signed) intensity and its direction $\mathbf{z}_0$ is constant in $\mathcal{F}_0$, and *ii)* a *total torque* (moment) $u_r(\mathbf{z}_0 \times \mathbf{x}_0) \in \mathbb{R}^2$ applied at $P_G$, where $u_r \in \mathbb{R}$ is its (signed) intensity.[5] Furthermore, a motor is attached to the PVTOL and applies a torque $\tau(\mathbf{z}_m \times \mathbf{x}_m) \in \mathbb{R}^2$ at $P_M$ to the joint, where $\tau \in \mathbb{R}$ is its (signed) intensity. The three inputs of the system are gathered in the vector $\mathbf{u} = [u_t \ u_r \ \tau]^T \in \mathbb{R}^3$ and shortly denoted in the following as *thrust, PVTOL torque* and *motor torque*, respectively.

We note that in this chapter, the pitch torque of the quadrotor $u_\theta$ given in (2.3) is equivalent to the PVTOL torque $u_r$, due to the choice of the plane as shown in Fig. (6.1).

Let us then consider the following cases: *i)* $P_{C_0}$ and $P_M$ are *generic*, i.e., there exist an arbitrary offset $\mathbf{d}_0 \neq \mathbf{0} \in \mathbb{R}^2$ between each other; or *ii)* *coinciding*, i.e., $\mathbf{d}_0 = \mathbf{0}$. Moreover, for each of the previous cases we consider the case in which the connection between the PVTOL and 1-link arm is either *rigid* or *elastic*. Hence, four Cases are investigated in total, summarized in Table 6.1.

Clearly, Case RC and Case EC are sub-cases of Case RG and Case EG, respectively. Nevertheless we shall show that they deserve a special treatment because new properties appear in those cases that significantly increase the capabilities of the platform. Notice that in all cases the position of $P_G$ can be any, i.e., $\mathbf{d}_G \in \mathbb{R}^2$ (while in the literature is typically assumed $P_G \equiv P_{C_1}$, i.e., $\mathbf{d}_G = \mathbf{0}$).

Like for similar mechanical systems, the robot dynamics can be expressed, using the Lagrange's equation, as

$$\ddot{\mathbf{q}} = \mathbf{M}^{-1}(\mathbf{q})\big(\mathbf{G}(\mathbf{q})\mathbf{u} - \mathbf{c}(\mathbf{q}, \dot{\mathbf{q}}) - \mathbf{g}(\mathbf{q}) + \mathbf{f}_E(\mathbf{q}) + \mathbf{f}_{ext}\big), \qquad (6.6)$$

---

[5]For example, in the case of a planar birotor, $P_G$ would be the center of two coplanar propellers, $u_t$ the sum of the forces provided by each propellers and $u_r$ their difference times the distance from $P_G$, see, e.g., Mahony et al. (2012).

where $\mathbf{q} \in \mathbb{R}^n$ are the considered generalized coordinates ($n = 4$ for the rigid-joint cases whereas $n = 5$ for the elastic-joint cases), $\mathbf{M} \in \mathbb{R}^{n \times n}$ is the generalized mass and inertia matrix, $\mathbf{G} \in \mathbb{R}^{n \times 3}$ is the control input matrix, $\mathbf{c} \in \mathbb{R}^n$ stands for the centrifugal/Coriolis forces, $\mathbf{g} \in \mathbb{R}^n$ represents the gravitational forces, and $\mathbf{f}_E \in \mathbb{R}^n$ represents the forces due to the potential energy stored in the elastic joints (in the rigid-joint cases $\mathbf{f}_E = \mathbf{0}$). Finally, the external forces are denoted with $\mathbf{f}_{ext} \in \mathbf{R}^n$, and represent the force and torques applied to the system from the external environment. We shall specify the elements of (6.6) for each Case of Table 6.1 in the following sections.

Notice that in all the four cases the system is underactuated, because only three inputs are available for a system whose configuration space is either 4- or 5-dimensional.

For the PVTOL aerial manipulator considered in this section, we want to achieve the following objectives:

**Objective 1.** *(Trajectory generation) formally discover, if it exists, an exact linearizing (i.e., differentially flat) output and explicit the algebraic map from the flat output to the state* $\mathbf{q}, \dot{\mathbf{q}}$ *and the input* $\mathbf{u}$.

**Objective 2.** *(Control) Find the domain in which the decoupling matrix* $\bar{\mathbf{G}}$ *is invertible (see Sec. 6.2.2) and therefore a dynamic exact feedback linearization control is applicable.*

**Objective 3.** *(Rigid–Elastic Comparison) Compare the different cases, especially the elastic versus the rigid case in order to discover pros and cons of the two architectures.*

Note that it is a particularly challenging task to achieve the aforementioned Objectives for a PVTOL aerial manipulator due to the nonlinearity and underactuation of the system, and the presence of dynamical couplings between the floating base and the rigid- or elastic-joint arm.

## 6.2.2 Review of Exact Output Tracking

Let us briefly and rather informally recap some well known concepts in nonlinear control, see, e.g., Isidori (1995) for a rigorous introduction to the topic. In the following we refer to a system in the form of (6.6).

Let be given an output $\mathbf{y}(\mathbf{q}) = [y_1 \; y_2 \; y_3]^T \in \mathbb{R}^3$ that is function of $\mathbf{q}$ and has the same size of $\mathbf{u}$ (i.e., three), and let us ask whether it is possible to make $\mathbf{y}$ track a desired trajectory $\mathbf{y}^d(t)$ whose derivatives are known and bounded, while maintaining the state $(\mathbf{q}, \dot{\mathbf{q}})$ and the input $\mathbf{u}$ bounded and with a known evolution that depends only on $\mathbf{y}^d(t)$ and its derivatives.

It is evident that finding an output possessing this strong property is very useful in practice. In fact, this is known as the *exact tracking control problem* and it is solvable if and only if $\mathbf{y}$ is an *exact linearizing output via dynamic feedback* for the system (6.6), which is defined in the following:

**Definition 7.** *An output* $\mathbf{y}$ *is an exactly linearizing output via dynamic feedback for* (6.6) *if it is possible to find* $s_1, s_2, s_3 \in \mathbb{N}_{\geq 0}$ *such that if one considers* $\bar{\mathbf{u}} = [u_1^{(s_1)} \ u_2^{(s_2)} \ u_3^{(s_3)}]^T \in \mathbb{R}^3$ *and* $\bar{\mathbf{x}} = [\mathbf{q}^T \ \dot{\mathbf{q}}^T \ u_1 \cdots u_1^{(s_1-1)} \ u_2 \cdots u_2^{(s_2-1)} \ u_3 \cdots u_3^{(s_3-1)}]^T \in \mathbb{R}^{\bar{n}}$ *as the new input and the new state of the system, respectively[a], then the the dynamics of* $y_1$, $y_2$, *and* $y_3$ *can be written as*

$$\bar{\mathbf{y}} = \begin{bmatrix} y_1^{(r_1)} & y_2^{(r_2)} & y_3^{(r_3)} \end{bmatrix}^T = \underbrace{\bar{\mathbf{f}}(\bar{\mathbf{x}})}_{\in \mathbb{R}^3} + \underbrace{\bar{\mathbf{G}}(\bar{\mathbf{x}})}_{\in \mathbb{R}^{3 \times 3}} \bar{\mathbf{u}}, \tag{6.7}$$

*where the following conditions are verified*

1. *the total relative degree* $r = r_1 + r_2 + r_3$ *matches with the dimension* $\bar{n}$ *of the augmented state, i.e.,*

$$r = r_1 + r_2 + r_3 = 2n + s_1 + s_2 + s_3 = \bar{n} \tag{6.8}$$

2. *the decoupling matrix* $\mathbf{G}(\bar{\mathbf{x}})$ *is invertible for some* $\bar{\mathbf{x}}$.

---

[a]Notice that $u_i^{(0)} = u_i$ and that $u_i^{(-1)}$ means that $u_i$ does not belong to $\bar{\mathbf{x}}$.

The name 'exactly linearizing' comes from the fact that if $\mathbf{y}$ is an exactly linearizing output, then input transformation

$$\bar{\mathbf{u}} = \bar{\mathbf{G}}^{-1}(\boldsymbol{u}_v - \bar{\mathbf{f}}) \tag{6.9}$$

brings the system in the fully linear controllable form via dynamic feedback (see Isidori (1995))

$$\bar{\mathbf{y}} = \boldsymbol{u}_v, \tag{6.10}$$

which is equivalent to system (6.6) thanks to the matching condition on the relative degree (i.e., thanks to the absence of an internal dynamics).

Furthermore, the transformation (6.9) can also be used in a control scheme as inner linearizing control loop on top of which any linear pole placement or LQR control strategy can be employed (see e.g. Ogata (2010)) for the transformed system (6.10). However, the existence of an exact linearizing output is a general property of the system that is not necessarily related to the need of controlling it. In fact a system is said to be *exactly input-output linearizable via dynamic feedback,* if it admits (at least) one exactly linearizing output, i.e., it exists a state and input change of coordinates (possibly including a state extension) which brings it to the simpler equivalent form (6.10).[6]

A similar concept introduced later in the literature (see, e.g., Fliess et al. (1995); Murray et al. (1995)) is the one of *differentially flat system.*

---

[6]Note that here the obtained linear system is the same as the original one, i.e., the linearization is 'exact', and must not be confused with the linear approximation of a nonlinear system based on Taylor expansion and truncation at the first order.

---

**Definition 8.** *The system* (6.6) *is differentially flat if it exists an output* $\mathbf{y} = \begin{bmatrix} y_1 & y_2 & y_3 \end{bmatrix}^T$ *(called* flat output*) such that* $\mathbf{q}$, $\dot{\mathbf{q}}$, *and* $\mathbf{u}$ *can be expressed as an algebraic function of* $y_1$, $y_2$, $y_3$ *and a finite number of their derivatives.*

---

The presence of a flat output allows knowing in advance (algebraically) the nominal state and input trajectories along which the system will evolve while tracking a desired output trajectory. Therefore it turns to be very useful in the planning and trajectory generation phase. Knowing that an output is flat allow also using some flatness-based tracking control techniques, see, e.g., Martin et al. (2003) and Tognon et al. (2017).

The following important equivalence fact holds (see, e.g., De Luca and Oriolo (2002); Martin et al. (2003)):

---

**Fact 1.** *Differential flatness is equivalent to exact input-output linearizability via dynamic feedback in an open and dense set of the state space and an output is flat if and only if is exactly linearizing.*

---

**Remark 8.** *Thanks to the Fact 1, and results of e.g. De Luca and Oriolo (2002), the decoupling matrix* $\bar{\mathbf{G}}(\bar{\mathbf{x}})$ *of Definition 7 is invertible in an open and dense set of the state space, if the chosen outputs of the system are exact linearizing (or differentially flat).*

---

### 6.2.3 Methodology

In this section, we describe the main steps of the method that we will shall employ in order to achieve the aforementioned Objectives 1 and 2.

Regarding Objective 1, in order to understand if an output is flat, one has to find an appropriate algebraic transformation. However, this is clearly not practical criteria, because it requires a priori knowledge if the output is flat or not, for a successful trial. On the other hand, Definition 1 provides a systematic way to assess whether an output is exactly linearizing or not. Moreover if it is, then one also finds a linearizing controller together with the differentially flat outputs, using Fact 1. Therefore, we will achieve Objectives 1 and 2 using the following method:

Given an output $\mathbf{y}$:

1. we define the generalized coordinates $\mathbf{q}$ starting from $\mathbf{y}$ and adding one coordinate more in the rigid case and two more in the elastic case;

2. we compute $\mathbf{M}$, $\mathbf{c}$, $\mathbf{g}$, and $\mathbf{f}_E$ in (6.6) which makes possible to write down the exact dependency of each entry of $\ddot{\mathbf{y}}$ from each entry of $\mathbf{q}$, $\dot{\mathbf{q}}$, and $\mathbf{u}$, i.e.,

$$\ddot{y}_1 = f_1(\mathbf{q}, \dot{\mathbf{q}}, \mathbf{u}), \; \ddot{y}_2 = f_2(\mathbf{q}, \dot{\mathbf{q}}, \mathbf{u}), \; \ddot{y}_3 = f_3(\mathbf{q}, \dot{\mathbf{q}}, \mathbf{u}) ; \tag{6.11}$$

3. using (6.11) and (6.6), we are able to compute the expected dependency of $y_i^{(j)}$ from each entry of $\mathbf{q}$, $\dot{\mathbf{q}}$, $\mathbf{u}, \dot{\mathbf{u}}, \ldots, \mathbf{u}^{(j-2)}$ for any $i = 1 \ldots 3, j > 2$, without exactly computing the derivatives,

4. taking advantage of that we can easily compute, for each choice of $s_1$, $s_2$, $s_3$ what are the expected relative degrees $r_1$, $r_2$, and $r_3$ in (6.7), by just stopping as soon as one among $u^{(s_i)}$, $i = 1, \ldots, 3$ appears for some $j$ in $y_i^{(j)}$, thus having $r_i = j$,

5. we can then check whether a choice exists for $s_1, s_2, s_3$ that possibly satisfies (6.8),

6. if this choice exists then we compute explicitly $\bar{\mathbf{G}}(\bar{\mathbf{x}})$ in (6.7) and we check for its invertibility in a certain domain of $\bar{\mathbf{x}}$ which implies that the output is exactly linearizing and, by virtue of Fact 1, also a flat output. by doing so we achieve then Objective 2.

We note that in Table 6.2 a summary of the result of this methodology applied to the four cases in exam is provided. The proofs of the results will be given in the next sections.

Once we know that an output $\mathbf{y}$ is exactly linearizing we try to derive the algebraic relation described in Definition 8 which certainly exists, thus achieving Objective 1.

Objective 3 is achieved through the Section 6.2 and mainly in Section 6.2.9 with realistic numerical tests.

## 6.2.4  Case RG: Rigid-joint Attached to a Generic Point

In this section we consider the 'Case RG' in which $P_{C_0}$ and $P_M$ are generic, i.e., $\mathbf{d}_0 \in \mathbb{R}^2 \neq \mathbf{0}$, as shown in Fig. 6.2, respectively, and that the arm is attached through a rigid joint (top left case in Table 6.1). Hence, the motor and the link orientations are the same, i.e., $\theta_m = \theta_1$. Notice that $P_G$ (in Fig. 6.2) can be anywhere, as in any other case considered in this chapter.

In order to find an exactly linearizing (i.e., flat) output in this case let us choose some generalized coordinates which show no inertial couplings between translational and rotational part, i.e., $\mathbf{q} = [\mathbf{p}_c^T \; \theta_0 \; \theta_{01}]^T \in \mathbb{R}^4$. With respect to these coordinates the generalized inertia matrix is found, after some algebra, as

$$\mathbf{M} = \begin{bmatrix} m_s \mathbf{I}_2 & * \\ \mathbf{0}_{2\times 2} & \mathbf{M}_r \end{bmatrix} = \mathbf{M}^T \in \mathbb{R}^{4\times 4}, \tag{6.12}$$

where $\mathbf{I}_i$ is $i \times i$ identity matrix, $\mathbf{0}_{i\times j}$ is a zero matrix in $\mathbb{R}^{i\times j}$,

$$\mathbf{M}_r = \begin{bmatrix} m_a & * \\ m_{ab}(\theta_0, \theta_{01}) & m_b \end{bmatrix} \in \mathbb{R}^{2\times 2},$$

$$m_a = \frac{m_0(m_1 + m_m)}{m_s} \|\mathbf{d}_0\|_2^2 + J_0,$$

$$m_b = \frac{m_1(m_0 + m_m)}{m_s} \|\mathbf{d}_1\|_2^2 + J_1 + J_m, \tag{6.13}$$

$$m_{ab}(\theta_0, \theta_{01}) = \frac{m_0 m_1}{m_s} \mathbf{d}_0^T \mathbf{R}_1 \mathbf{d}_1,$$

with $\|\mathbf{d}_*\|_2^2 = \mathbf{d}_*^T \mathbf{d}_*$, $* = \{1, 2\}$ and $m_s = m_0 + m_m + m_1$. The centrifugal/Coriolis and gravitational forces are

$$\mathbf{c}(\mathbf{q}, \dot{\mathbf{q}}) = \begin{bmatrix} 0 \\ 0 \\ \frac{m_0 m_1}{m_s} \mathbf{d}_0^T \bar{\mathbf{R}}_1 \mathbf{d}_1 \dot{\theta}_{01}^2 \\ -\frac{m_0 m_1}{m_s} \mathbf{d}_0^T \bar{\mathbf{R}}_1 \mathbf{d}_1 \dot{\theta}_0^2 \end{bmatrix}, \quad \mathbf{g} = \begin{bmatrix} 0 \\ -m_s g \\ 0 \\ 0 \end{bmatrix}, \tag{6.14}$$

where $\bar{\mathbf{R}}_* = \frac{\partial \mathbf{R}_*}{\partial \theta_*}$. The input matrix is

$$
\mathbf{G}(\mathbf{q}) = \begin{bmatrix} -\sin\theta_0 & 0 & 0 \\ -\cos\theta_0 & 0 & 0 \\ -\frac{m_1+m_m}{m_s}d_{0_x} + d_{G_x} & 1 & -1 \\ -\frac{m_1}{m_s}(d_{1_x}\cos\theta_1 + d_{1_z}\sin\theta_1) & 0 & 1 \end{bmatrix}, \tag{6.15}
$$

and, finally, thanks to the rigid connection, $\mathbf{f}_E = \mathbf{0}_{4\times 1}$.

Replacing $\mathbf{M}$, $\mathbf{c}$, $\mathbf{g}$, $\mathbf{G}$ and $\mathbf{f}_E$ in (6.6) we can derive the explicit dependency of each entry of $\ddot{\mathbf{q}}$, here summarized:

$$
\begin{aligned}
\ddot{x}_c &= f_1(\theta_0, u_t), \quad \ddot{z}_c = f_2(\theta_0, u_t) \\
\ddot{\theta}_0 &= f_3(\theta_0, \theta_{01}, \dot{\theta}_0, \dot{\theta}_{01}, u_t, u_r, \tau) \\
\ddot{\theta}_{01} &= f_4(\theta_0, \theta_{01}, \dot{\theta}_0, \dot{\theta}_{01}, u_t, u_r, \tau).
\end{aligned} \tag{6.16}
$$

We can observe from (6.16) that $u_t$ is the only input appearing in $\ddot{x}_c$ and $\ddot{z}_c$. This implies that if we choose $s_1 > 0$ and include both $x_c$ and $z_c$ in the output, it is possible to let $r$ increase twice as rapidly as $\bar{n}$ when we increase $s_1$, until an input other than $u_t$ appears in the higher order derivatives of $x_c$ or $z_c$ (see Definition 7). Following this intuition, let us consider then $s_1 = 2$ and $s_2 = s_3 = 0$. We then obtain the new control input $\bar{\mathbf{u}} = [\ddot{u}_t \; u_r \; \tau]^T \in \mathbb{R}^3$, new state $\bar{\mathbf{x}} = [\mathbf{q}^T \; \dot{\mathbf{q}}^T \; u_t \; \dot{u}_t]^T \in \mathbb{R}^{10}$, and $\bar{n} = 10$. Now, let us consider as output $\mathbf{y} = [\mathbf{p}_c^T \; \theta_{01}]^T = [\mathbf{y}_1^T \; y_2]^T$. Following the methodology presented in Section 6.2.3 we then make clear the expected functional dependences without the need of explicitly computing the derivatives

$$
\ddot{\mathbf{y}}_1 = \boldsymbol{f}_1(\theta_0, u_t), \quad \dddot{\mathbf{y}}_1 = \boldsymbol{f}_2(\theta_0, \dot{\theta}_0, u_t, \dot{u}_t) \tag{6.17}
$$

and, substituting $\ddot{\theta}_0$ with $f_3$ in (6.16), we have

$$
\ddddot{\mathbf{y}}_1 = \boldsymbol{f}_3(\theta_0, \theta_{01}, \dot{\theta}_0, \dot{\theta}_{01}, u_t, \dot{u}_t, \ddot{u}_t, u_r, \tau), \tag{6.18}
$$

therefore $r_1 = r_2 = 4$. Considering also that, from (6.16), $r_3 = 2$, we have that $r_1 + r_2 + r_3 = 10 = \bar{n}$, which means that the Condition 1 of Definition 7 is satisfied. Therefore it is now worth investigating about the invertibility of $\bar{\mathbf{G}}(\bar{\mathbf{x}})$, which is done in the next proposition.

**Proposition 5.** *The vector $[\mathbf{p}_c^T \; \theta_{01}]^T$ is an exactly linearizing output via dynamic feedback for the generic model with rigid-joint arm (Case RG), as long as $u_t \neq 0$. As a consequence, it is also a flat output.*

*Proof.* See Appendix A.5. ☐

### Derivation of the Algebraic Map from the Flat Output

We shall find now how to explicitly write down the algebraic map that relates $\ddot{\mathbf{p}}_c$, $\dddot{\mathbf{p}}_c$, $\ddddot{\mathbf{p}}_c$, $\theta_{01}, \dot{\theta}_{01}, \ddot{\theta}_{01}$ with $\theta_0$, $\dot{\theta}_0$, and $\mathbf{u}$. Consider the first two equations of (6.16)

$$
\begin{aligned}
m_s\ddot{x}_c &= -\sin\theta_0 u_t \\
m_s\ddot{z}_c &= -\cos\theta_0 u_t + m_s g.
\end{aligned} \tag{6.19}
$$

Define the vector $\mathbf{w} = \ddot{\mathbf{p}}_c - [0 \ g]^T = [w_x \ w_z]^T \in \mathbb{R}^2$, which is a function of $\ddot{\mathbf{p}}_c$. It is clear that $\mathbf{w} = -\frac{u_t}{m_s}[\sin\theta_0 \ \cos\theta_0]^T$. Therefore $\theta_0 = \text{atan2}(-w_x, -w_z)$ and $u_t = m_s||\mathbf{w}||$. Furthermore, differentiating $\theta_0(w_x, w_z)$ we obtain $\dot{\theta}_0(w_x, w_z, \dot{w}_x, \dot{w}_z)$ and $\ddot{\theta}_0(w_x, w_z, \dot{w}_x, \dot{w}_z, \ddot{w}_x, \ddot{w}_z)$, which are all functions of the derivatives of $\mathbf{p}_c$ from the second up to the fourth order. Then we can write

$$u_t = m_s||\mathbf{w}||, \quad \dot{u}_t = \frac{m_s w_1}{||\mathbf{w}||}, \quad \ddot{u}_t = -\frac{m_s \dot{w}_1}{||\mathbf{w}||} - \frac{m_s w_1^2}{||\mathbf{w}||^3},$$

$$\dddot{u}_t = \frac{m_s \ddot{w}_1}{||\mathbf{w}||} - \frac{3m_s \dot{w}_1 w_1}{||\mathbf{w}||^3} + \frac{3m_s w_1^2}{||\mathbf{w}||^5} \quad\quad (6.20)$$

$$\ddddot{u}_t = \frac{m_s \dddot{w}_1}{||\mathbf{w}||} - \frac{4m_s w_1 \ddot{w}_1 + 3m_s \dot{w}_1^2}{||\mathbf{w}||^3} + \frac{9m_s \dot{w}_1 w_1^2 + 6m_s w_1 \dot{w}_1}{||\mathbf{w}||^5} - \frac{15m_s w_1^3}{||\mathbf{w}||^7},$$

where

$$||\mathbf{w}|| = \sqrt{\ddot{p}_{c_x}^2 + \ddot{p}_{c_z}^2 - 2\ddot{p}_{c_z}g + g^2}$$

$$w_1 = \dddot{p}_{c_x}\ddot{p}_{c_x} + \dddot{p}_{c_z}(\ddot{p}_{c_z} - g)$$

$$\dot{w}_1 = \ddddot{p}_{c_x}\ddot{p}_{c_x} + \dddot{p}_{c_x}^2 + \dddot{p}_{c_z}^2 + \ddddot{p}_{c_z}(\ddot{p}_{c_z} - g)$$

$$\ddot{w}_1 = p_{c_x}^{(5)}\ddot{p}_{c_x} + 3\ddddot{p}_{c_x}\dddot{p}_{c_x} + 3\ddddot{p}_{c_z}\dddot{p}_{c_z} + p_{c_z}^{(5)}(\ddot{p}_{c_z} - g)$$

$$\dddot{w}_1 = p_{c_x}^{(6)}\ddot{p}_{c_x} + 4p_{c_x}^{(5)}\dddot{p}_{c_x} + 3\ddddot{p}_{c_x}^2 + 3\ddddot{p}_{c_z}^2 + 4p_{c_z}^{(5)}\dddot{p}_{c_z} + p_{c_z}^{(6)}(\ddot{p}_{c_z} - g),$$

where $\dddot{u}_t$ and $\ddddot{u}_t$ are provided for the convenience of the further analyses. Moreover we can write[7]

$$\theta_0 = \text{atan2}(-w_x, -w_z), \quad w_z = \ddot{p}_{c_z} - g, \quad w_x = \ddot{p}_{c_x}$$

$$\dot{\theta}_0 = \frac{\dot{w}_x w_z - w_x \dot{w}_z}{w_x^2 + w_z^2} = \frac{\dddot{p}_{c_x}(\ddot{p}_{c_z} - g) - \ddot{p}_{c_x}\dddot{p}_{c_z}}{\ddot{p}_{c_x}^2 + (\ddot{p}_{c_z} - g)^2} \quad\quad (6.21)$$

$$\ddot{\theta}_0 = \frac{\ddot{w}_x w_z - w_x \ddot{w}_z}{(w_x^2 + w_z^2)} - \frac{2[(w_z^2 - w_x^2)\dot{w}_x \dot{w}_z + (\dot{w}_x^2 - \dot{w}_z^2)w_x w_z]}{(w_x^2 + w_z^2)^2},$$

where

$$\dot{w}_x = \dddot{p}_{c_x}, \quad \ddot{w}_x = \ddddot{p}_{c_x}, \quad \dot{w}_z = \dddot{p}_{c_z}, \quad \ddot{w}_z = \ddddot{p}_{c_z}.$$

Now considering the last equation of the system dynamics, we can retrieve the motor torque as

$$\tau = \tau(\theta_{01}, \ddot{\theta}_{01}, \theta_0, \dot{\theta}_0, \ddot{\theta}_0, u_t) = m_{ab}(\theta_0, \theta_{01})\ddot{\theta}_0 + m_b\ddot{\theta}_{01} -$$
$$- \frac{m_0 m_1}{m_s}\mathbf{d}_0^T \bar{\mathbf{R}}_1(\theta_0, \theta_{01})\mathbf{d}_1\dot{\theta}_0^2 + \left(\frac{m_1 + m_m}{m_s}d_{0_x} - d_{G_x}\right)u_t \quad\quad (6.22)$$

and using $\theta_0$, $\dot{\theta}_0$, $\ddot{\theta}_0$ from (6.21) and $u_t$ from (6.20), we show that $\tau = \tau(\ddot{\mathbf{y}}_1, \dddot{\mathbf{y}}_1, \ddddot{\mathbf{y}}_1, y_2, \ddot{y}_2)$. Now, replacing $\tau$ from above into the third equation of the system dynamics we have

$$u_r = u_r(\theta_{01}, \dot{\theta}_{01}, \ddot{\theta}_{01}, \theta_0, \dot{\theta}_0, \ddot{\theta}_0, u_t) = m_a\ddot{\theta}_0 + m_b\ddot{\theta}_{01} +$$
$$+ m_{ab}(\theta_0, \theta_{01})(\ddot{\theta}_0 + \ddot{\theta}_{01}) + \frac{m_0 m_1}{m_s}\mathbf{d}_0^T \bar{\mathbf{R}}_1(\theta_0, \theta_{01})\mathbf{d}_1(\dot{\theta}_{01}^2 - \dot{\theta}_0^2) +$$
$$+ \left(\frac{m_1 + m_m}{m_s}d_{0_x} - d_{G_x} + \frac{m_1}{m_s}(d_{1_x}\cos\theta_1 + d_{1_z}\sin\theta_1)\right)u_t, \quad\quad (6.23)$$

---

[7]in the range of $\theta_0$, in which the derivatives of $\text{atan2}(-w_x, -w_z)$ exist.

where by substituting $\theta_0$, $\dot{\theta}_0$, $\ddot{\theta}_0$ from (6.21) and $u_t$ from (6.20), we have
$u_r = u_r(\ddot{\mathbf{y}}_1, \dddot{\mathbf{y}}_1, \ddddot{\mathbf{y}}_1, y_2, \dot{y}_2, \ddot{y}_2)$.

In summary, we obtained $\mathbf{p}_c = \mathbf{y}_1$, $\dot{\mathbf{p}}_c = \dot{\mathbf{y}}_1$, $\ddot{\mathbf{p}}_c = \ddot{\mathbf{y}}_1$ and $\theta_{01} = y_2$, $\dot{\theta}_{01} = \dot{y}_2$, $\ddot{\theta}_{01} = \ddot{y}_2$ from the definition; $u_t = u_t(\ddot{\mathbf{y}}_1)$, $\dot{u}_t = \dot{u}_t(\ddot{\mathbf{y}}_1, \dddot{\mathbf{y}}_1)$, $\ddot{u}_t = \ddot{u}_t(\ddot{\mathbf{y}}_1, \dddot{\mathbf{y}}_1, \ddddot{\mathbf{y}}_1)$ from (6.20); $\theta_0 = \theta_0(\ddot{\mathbf{y}}_1)$, $\dot{\theta}_0 = \dot{\theta}_0(\ddot{\mathbf{y}}_1, \dddot{\mathbf{y}}_1)$, $\ddot{\theta}_0 = \ddot{\theta}_0(\ddot{\mathbf{y}}_1, \dddot{\mathbf{y}}_1, \ddddot{\mathbf{y}}_1)$ from (6.21); and finally $\tau = \tau(\ddot{\mathbf{y}}_1, \dddot{\mathbf{y}}_1, \ddddot{\mathbf{y}}_1, y_2, \ddot{y}_2)$ and $u_r = u_r(\ddot{\mathbf{y}}_1, \dddot{\mathbf{y}}_1, \ddddot{\mathbf{y}}_1, y_2, \dot{y}_2, \ddot{y}_2)$ as shown above[8]. Hence we showed the states and the control inputs of the system as functions of the flat outputs and their finite number of derivatives.

---

**Remark 9.** *Although dependencies of $\ddot{\theta}_0$ and $\ddot{\theta}_{01}$ in (6.16) are the same, $\mathbf{y} = [\mathbf{p}_c^T \; \theta_0]^T$ is not an exactly linearizing output, because in this case it is possible to show that $\det(\bar{\mathbf{G}}) = 0$.*

---

### Impossibility of Exact Tracking of the End-effector Position

In the most interesting cases for aerial manipulation, one needs to control the end-effector position $\mathbf{p}_e$ instead of $\mathbf{p}_c$. Here we introduce a negative result that shows how unfortunately this objective is not feasible for the Case RG.

The expression of $\mathbf{p}_e$ in function of $\mathbf{q}$ is:

$$
\begin{aligned}
\mathbf{p}_e &= f(\mathbf{p}_c, \theta_0, \theta_{01}) \\
&= \mathbf{p}_c + \frac{m_0}{m_s}\mathbf{R}_0\mathbf{d}_0 + \mathbf{R}_{01}\left(\frac{m_0 + m_m}{m_s}\mathbf{d}_1 + \mathbf{d}_e\right),
\end{aligned} \tag{6.24}
$$

which shows that $\mathbf{p}_e$ cannot be computed using only the flat output $[\mathbf{p}_c^T \; \theta_{01}]^T$ since also $\theta_0$ is required in (6.24). Therefore it is impossible to let exactly $\mathbf{p}_e$ track a desired trajectory $\mathbf{p}_e^d(t)$ using control methods based on the flat output $[\mathbf{p}_c^T \; \theta_{01}]^T$.

On the other hand, since

$$
\mathbf{p}_e = \mathbf{p}_m + \mathbf{R}_{01}(\mathbf{d}_1 + \mathbf{d}_e), \tag{6.25}
$$

one can let $\mathbf{p}_e$ exactly track $\mathbf{p}_e^d$, if $[\mathbf{p}_m^T \; \theta_{01}]^T \in \mathbb{R}^3$ is a flat output as well. In order to discover if and under which conditions $[\mathbf{p}_m^T \; \theta_{01}]^T \in \mathbb{R}^3$ is a flat output, let us write the dynamics of the system for the generalized coordinates $\mathbf{q} = [\mathbf{p}_m^T \; \theta_0 \; \theta_{01}]^T \in \mathbb{R}^4$. With respect to these coordinates, the generalized inertia matrix becomes

$$
\mathbf{M} = \begin{bmatrix} m_s\mathbf{I}_2 & * & * \\ \boldsymbol{\alpha}^T(\theta_0) & m_A & * \\ \boldsymbol{\beta}^T(\theta_{01}) & 0 & m_B \end{bmatrix} = \mathbf{M}^T \in \mathbb{R}^{4\times4}, \tag{6.26}
$$

where

$$
\begin{aligned}
m_A &= m_0 \|\mathbf{d}_0\|_2^2 + J_0 & m_B &= m_1 \|\mathbf{d}_1\|_2^2 + J_1 + J_m \\
\boldsymbol{\alpha}(\theta_0) &= -m_0\bar{\mathbf{R}}_0\mathbf{d}_0 \in \mathbb{R}^2 & \boldsymbol{\beta}(\theta_{01}) &= m_1\bar{\mathbf{R}}_{01}\mathbf{d}_1 \in \mathbb{R}^2.
\end{aligned} \tag{6.27}
$$

---

[8]Notice that the high order derivatives of the flat outputs can be computed analytically by taking the time derivatives of the related components of (6.6).

The centrifugal/Coriolis and gravitational forces are

$$\mathbf{c}(\mathbf{q}, \dot{\mathbf{q}}) = \begin{bmatrix} \bar{\alpha}_1(\theta_0)\dot{\theta}_0^2 + \bar{\beta}_1(\theta_{01})\dot{\theta}_{01}^2 \\ \bar{\alpha}_2(\theta_0)\dot{\theta}_0^2 + \bar{\beta}_2(\theta_{01})\dot{\theta}_{01}^2 \\ 0 \\ 0 \end{bmatrix}, \quad \mathbf{g}(\mathbf{q}) = \begin{bmatrix} 0 \\ -m_s g \\ g_3(\theta_0) \\ g_4(\theta_{01}) \end{bmatrix}, \tag{6.28}$$

with $\bar{\boldsymbol{\beta}} = \frac{\partial \boldsymbol{\beta}}{\partial \theta_{01}} = [\bar{\beta}_1 \ \bar{\beta}_2]^T \in \mathbb{R}^2$, and

$$\begin{aligned} \bar{\alpha}_1 &= m_0(d_{0_x}\cos(\theta_0) + d_{0_z}\sin(\theta_0)) \\ \bar{\alpha}_2 &= m_0(d_{0_z}\cos(\theta_0) - d_{0_x}\sin(\theta_0)) \\ \bar{\beta}_1 &= -m_1(d_{1_x}\cos(\theta_{01}) + d_{1_z}\sin(\theta_{01})) \\ \bar{\beta}_2 &= -m_1(d_{1_z}\cos(\theta_{01}) - d_{1_x}\sin(\theta_{01})) \\ g_3 &= -m_0 g(d_{0_x}\cos(\theta_0) + d_{0_z}\sin(\theta_0)) \\ g_4 &= m_1 g(d_{1_x}\cos(\theta_{01}) + d_{1_z}\sin(\theta_{01})). \end{aligned} \tag{6.29}$$

The input matrix is

$$\mathbf{G}(\mathbf{q}) = \begin{bmatrix} -\sin\theta_0 & 0 & 0 \\ -\cos\theta_0 & 0 & 0 \\ -d_{0_x} + d_{G_x} & 1 & -1 \\ 0 & 0 & 1 \end{bmatrix}, \tag{6.30}$$

and, as in the previous case, $\mathbf{f}_E = \mathbf{0}_{4\times 1}$.

Let us now ask ourselves if $\mathbf{y} = [\mathbf{p}_m^T \ \theta_{01}]^T$ is an exactly linearizing output via dynamic feedback (i.e., a flat output). Due to the inertial coupling, $\ddot{\mathbf{y}}$ depends from all the control inputs, therefore the gap between $\bar{n} - r = 2$ will stay for any choice of $s_1$, $s_2$, and $s_3$, which shows that $\ddot{\mathbf{y}}$ is not exactly linearizing and therefore is not flat.

---

**Corollary 1.** *The vectors* $[\mathbf{p}_m^T \ \theta_{01}]^T$, $[\mathbf{p}_1^T \ \theta_{01}]^T$ *and* $[\mathbf{p}_e^T \ \theta_{01}]^T$ *are* not *an exactly linearizing output via dynamic feedback for the generic model with rigid-joint arm (Case RG). As a consequence, they are* not *flat outputs either.*

---

## 6.2.5 Case RC: Rigid-joint Attached to the PVTOL CoM

The negative result of Corollary 1 is a consequence of the strong inertial coupling in (6.26). In this section we show that if we consider a model in which $P_{C_0}$ coincides with $P_M$, i.e., $\mathbf{p}_0 = \mathbf{p}_m$, then the inertial coupling weakens enough to make $[\mathbf{p}_m^T \ \theta_{01}]^T$ an exactly linearizing (i.e., flat) output for the system in exam[9]. In order to prove that, let us choose as generalized coordinates $\mathbf{q} = [\mathbf{p}_m^T \ \theta_0 \ \theta_{01}]^T \in \mathbb{R}^4$. The dynamic model for these coordinates are given in Section 6.2.4, from (6.26) to (6.30). Now assume that $\mathbf{p}_0 = \mathbf{p}_m$. Such case is depicted in Fig. 6.3, where motor and the joint are placed at the CoM of the PVTOL. This is a special case of the generic model, that we call the *coinciding model with rigid joint*[10] (Case RC in Table 6.1).

---

[9]Notice that this inertial coupling disappears as well if the PVTOL mass is small enough, i.e. $m_0 \to 0$ (see (6.27)). However this is not a reasonable assumption.

[10]Notice that a particular case of this one is studied in Thomas et al. (2013), where the three points $P_{C_0}$, $P_G$ and $P_M$ are assumed to be same. In Sec. 6.2.5 we study the more general case in which $P_G$ is not assumed to be coincident. Moreover in that paper, only the differential flatness was studied, while in this section we also prove the exact linearizability and provide the linearizing controller.

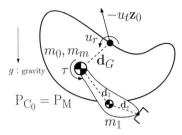

Figure 6.3: A sketch of the PVTOL aerial manipulator for the cases in which the attachment point of the rigid-joint arm is same with the CoM of the PVTOL ($P_{C_0} = P_M$), i.e., Case RC and Case EC. Notice that the point of application of the thrust is still any (i.e., $\mathbf{d}_G \in \mathbb{R}^2$).

Consider the dynamic model given in Section 6.2.4 with $\mathbf{d}_0 = \mathbf{0}_{2\times1}$ (because of the coinciding assumption). We obtain the following simplifications in some of previous expressions:

$$m_A = J_0 \tag{6.31}$$

$$\boldsymbol{\alpha}(\theta_0) = \mathbf{0}_{2\times1} \tag{6.32}$$

$$\bar{\alpha}_1(\theta_0) = \bar{\alpha}_2(\theta_0) = g_3(\theta_0) = 0. \tag{6.33}$$

Moreover, $d_{0_x} = 0$ in (6.30).

The explicit functional dependency of $\ddot{\mathbf{q}}$ then becomes[11]

$$\ddot{x}_m = f_1(\theta_0, \theta_{01}, \dot{\theta}_{01}, u_t, \tau), \quad \ddot{z}_m = f_2(\theta_0, \theta_{01}, \dot{\theta}_{01}, u_t, \tau)$$
$$\ddot{\theta}_0 = f_3(u_t, u_r, \tau), \quad \ddot{\theta}_{01} = f_4(\theta_0, \theta_{01}, \dot{\theta}_{01}, u_t, \tau), \tag{6.34}$$

where we see that now $u_r$ does not appear anymore in neither $\ddot{\mathbf{p}}_m$ nor $\ddot{\theta}_{01}$, as it was instead happening in Case RG. Therefore if we choose $\mathbf{y} = [\mathbf{p}_m^T \; \theta_{01}]^T$ as the output, we obtain from (6.34)

$$\ddot{\mathbf{y}} = \boldsymbol{f}_1(\theta_0, \theta_{01}, \dot{\theta}_{01}, u_t, \tau). \tag{6.35}$$

The fact that the two inputs $u_t$ and $\tau$ are appearing in the three input channels implies that if we choose both $s_1 > 0$ and $s_3 > 0$, it is possible to let $r$ increase more rapidly than $\bar{n}$ when we increase $s_1$ and $s_3$, until the input $u_r$ appears in the higher order derivatives of $\mathbf{y}$ (see Definition 7). Following this intuition, let us consider then $s_1 = 2$, $s_3 = 2$ and $s_2 = 0$. We then obtain as new control inputs $\bar{\mathbf{u}} = [\dddot{u}_t \; u_r \; \ddot{\tau}]^T \in \mathbb{R}^3$, new state $\bar{\mathbf{x}} = [\mathbf{q}^T \; \dot{\mathbf{q}}^T \; u_t \; \dot{u}_t \; \tau \; \dot{\tau}]^T \in \mathbb{R}^{12}$, and $\bar{n} = 12$.

Considering that $\ddot{\theta}_{01}$ is available from $f_4$ of (6.34) we write

$$\dddot{\mathbf{y}} = \boldsymbol{f}_2(\theta_0, \theta_{01}, \dot{\theta}_0, \dot{\theta}_{01}, u_t, \tau, \dot{u}_t, \dot{\tau}) \tag{6.36}$$

and, substituting $\ddot{\theta}_0$ with $f_3$ in (6.34), we have

$$\ddddot{\mathbf{y}} = \boldsymbol{f}_3(\theta_0, \theta_{01}, \dot{\theta}_0, \dot{\theta}_{01}, u_t, \dot{u}_t, \tau, \dot{\tau}, \ddot{u}_t, u_r, \ddot{\tau}). \tag{6.37}$$

---

[11]If one develops the computations, one realizes that $\ddot{\theta}_{01}$ does not depend on $\dot{\theta}_{01}$ since the terms depending on $\dot{\theta}_{01}$ cancel out each other. However this particularity is not necessary to prove the presented result.

Therefore $r_1 = r_2 = r_3 = 4$ and thus $r = 12 = \bar{n}$, which means that the Condition 1 of Definition 7 is satisfied. Therefore it is now worth to analytically search for the invertibility domain of $\bar{\mathbf{G}}(\bar{\mathbf{x}})$, which is stated in the next result.

---

**Proposition 6.** *The vectors* $[\mathbf{p}_c^T\ \theta_{01}]^T$, $[\mathbf{p}_m^T\ \theta_{01}]^T$ *and* $[\mathbf{p}_e^T\ \theta_{01}]^T$ *are all exactly linearizing output via dynamic feedback for the coinciding model with rigid-joint arm (Case RC), as long as* $u_t \neq 0$. *As a consequence, they are also flat outputs.*

---

*Proof.* See Appendix A.6. □

### Derivation of the Algebraic Map from the Flat Output

We shall show now the procedure to explicitly write down the algebraic map that relates $\ddot{\mathbf{p}}_m, \dddot{\mathbf{p}}_m, \ddddot{\mathbf{p}}_m, \theta_{01}, \dot\theta_{01}, \ddot\theta_{01}, \dddot\theta_{01}, \ddddot\theta_{01}$ with $\theta_0, \dot\theta_0$, and $\mathbf{u}$. The position of the CoM of overall system in $\mathcal{F}_W$, and its derivatives are given by

$$\mathbf{p}_c = \mathbf{p}_m + \frac{m_1}{m_s}\mathbf{R}_{01}\mathbf{d}_1 = \mathbf{p}_c(\mathbf{y})$$

$$\begin{bmatrix} x_c \\ z_c \end{bmatrix} = \begin{bmatrix} x_m + \frac{m_1}{m_s}\Big(d_{1_x}c_{01} + d_{1_z}s_{01}\Big) \\ z_m + \frac{m_1}{m_s}\Big(-d_{1_x}s_{01} + d_{1_z}c_{01}\Big) \end{bmatrix},$$

$$\dot{\mathbf{p}}_c = \dot{\mathbf{p}}_m + \frac{m_1}{m_s}\bar{\mathbf{R}}_{01}\mathbf{d}_1\dot\theta_{01} = \mathbf{p}_c(\mathbf{y},\dot{\mathbf{y}})$$

$$\begin{bmatrix} \dot x_c \\ \dot z_c \end{bmatrix} = \begin{bmatrix} \dot x_m + \frac{m_1\dot\theta_{01}}{m_s}\Big(-d_{1_x}s_{01} + d_{1_z}c_{01}\Big) \\ \dot z_m + \frac{m_1\dot\theta_{01}}{m_s}\Big(-d_{1_x}c_{01} - d_{1_z}s_{01}\Big) \end{bmatrix}, \tag{6.38}$$

$$\ddot{\mathbf{p}}_c = \ddot{\mathbf{p}}_m + \frac{m_1}{m_s}\Big(\bar{\mathbf{R}}_{01}\mathbf{d}_1\ddot\theta_{01} - \mathbf{R}_{01}\mathbf{d}_1\dot\theta_{01}^2\Big) = \mathbf{p}_c(\mathbf{y},\dot{\mathbf{y}},\ddot{\mathbf{y}})$$

$$\begin{bmatrix} \ddot x_c \\ \ddot z_c \end{bmatrix} = \begin{bmatrix} \ddot x_m + \frac{m_1}{m_s}\Big(d_{1_x}(-c_{01}\dot\theta_{01}^2 - s_{01}\ddot\theta_{01}) + d_{1_z}(-s_{01}\dot\theta_{01}^2 + c_{01}\ddot\theta_{01})\Big) \\ \ddot z_m + \frac{m_1}{m_s}\Big(d_{1_x}(s_{01}\dot\theta_{01}^2 - c_{01}\ddot\theta_{01}) + d_{1_z}(-c_{01}\dot\theta_{01}^2 - s_{01}\ddot\theta_{01})\Big) \end{bmatrix},$$

where $s_* = \sin(\theta_*)$ and $c_* = \cos(\theta_*)$. The computation of $u_t$ and $\theta_0$ is exactly as in Case RG (see (6.20) and (6.21)). Hence substituting (6.38) in (6.20) and in (6.21), we find $u_t, \dot u_t, \ddot u_t$ and $\theta_0, \dot\theta_0, \ddot\theta_0$ as functions of $\mathbf{y}, \cdots, \ddddot{\mathbf{y}}$. Furthermore, the motor torque can be retrieved from the last equation of the system dynamics as

$$\tau = \tau(\theta_{01}, \ddot{\mathbf{p}}_m, \ddot\theta_{01}) = \boldsymbol{\beta}^T(\theta_{01})\ddot{\mathbf{p}}_m + m_B\ddot\theta_{01} + g_4(\theta_{01}). \tag{6.39}$$

Now, noticing that (from (6.29)) $g_4(\theta_{01}) = -\boldsymbol{\beta}(\theta_{01})g \cdot \mathbf{e}_2$ with $\mathbf{e}_2 = [0\ 1]^T \in \mathbb{R}^2$, and $\cdot$ being the *dot-product*, and recalling that $\bar{\boldsymbol{\beta}} = \frac{\partial\boldsymbol{\beta}}{\partial\theta_{01}}$, we can write

$$\begin{aligned} \dot\tau &= \boldsymbol{\beta}^T\dddot{\mathbf{p}}_m + m_B\dddot\theta_{01} + (\bar{\boldsymbol{\beta}}^T\ddot{\mathbf{p}}_m - \bar{\boldsymbol{\beta}}g\cdot\mathbf{e}_2)\dot\theta_{01} \\ \ddot\tau &= \boldsymbol{\beta}^T\ddddot{\mathbf{p}}_m + m_B\ddddot\theta_{01} + 2\bar{\boldsymbol{\beta}}^T\dddot{\mathbf{p}}_m\dot\theta_{01} + (\bar{\boldsymbol{\beta}}^T\ddot{\mathbf{p}}_m - \bar{\boldsymbol{\beta}}g\cdot\mathbf{e}_2)\ddot\theta_{01} + (\bar{\boldsymbol{\beta}}g\cdot\mathbf{e}_2 - \bar{\boldsymbol{\beta}}^T\ddot{\mathbf{p}}_m)\dot\theta_{01}^2, \end{aligned} \tag{6.40}$$

which means $\dot\tau = \dot\tau(\mathbf{y}, \dot{\mathbf{y}}, \ddot{\mathbf{y}}, \dddot{\mathbf{y}})$, and $\ddot\tau = \ddot\tau(\mathbf{y}, \dot{\mathbf{y}}, \ddot{\mathbf{y}}, \dddot{\mathbf{y}}, \ddddot{\mathbf{y}})$.

Then, using the third row of the system dynamics, we obtain

$$u_r = u_r(\ddot{\theta}_1, u_t, \tau) = J_0\ddot{\theta}_0 + \tau - d_{G_x}u_t, \qquad (6.41)$$

where by knowing $\tau$ from (6.39), and utilizing $\ddot{\theta}_0$ from (6.21) and $u_t$ from (6.20), with taking (6.38) into consideration, we have that $u_r = u_r(\mathbf{y}, \dot{\mathbf{y}}, \ddot{\mathbf{y}}, \dddot{\mathbf{y}}, \ddddot{\mathbf{y}})$.

In summary, we have obtained $\mathbf{p}_m = \mathbf{p}_m(\mathbf{y})$, $\dot{\mathbf{p}}_m = \dot{\mathbf{p}}_m(\dot{\mathbf{y}})$, $\ddot{\mathbf{p}}_c = \ddot{\mathbf{p}}_m(\ddot{\mathbf{y}})$ and $\theta_{01} = \theta_{01}(\mathbf{y})$, $\dot{\theta}_{01} = \dot{\theta}_{01}(\dot{\mathbf{y}})$, $\ddot{\theta}_{01} = \ddot{\theta}_{01}(\ddot{\mathbf{y}})$ from the definition; $u_t = u_t(\mathbf{y}, \dot{\mathbf{y}}, \ddot{\mathbf{y}})$, $\dot{u}_t = \dot{u}_t(\mathbf{y}, \dot{\mathbf{y}}, \ddot{\mathbf{y}}, \dddot{\mathbf{y}})$, $\ddot{u}_t = \ddot{u}_t(\mathbf{y}, \dot{\mathbf{y}}, \ddot{\mathbf{y}}, \dddot{\mathbf{y}}, \ddddot{\mathbf{y}})$ from (6.20) and $\theta_0 = \theta_0(\mathbf{y}, \dot{\mathbf{y}}, \ddot{\mathbf{y}})$, $\dot{\theta}_0 = \dot{\theta}_0(\mathbf{y}, \dot{\mathbf{y}}, \ddot{\mathbf{y}}, \dddot{\mathbf{y}})$, $\ddot{\theta}_0 = \ddot{\theta}_0(\mathbf{y}, \dot{\mathbf{y}}, \ddot{\mathbf{y}}, \dddot{\mathbf{y}}, \ddddot{\mathbf{y}})$ from (6.21) where for both $\mathbf{p}_c$ is computed from (6.38); and finally $\tau = \tau(\mathbf{y}, \ddot{\mathbf{y}})$, $\dot{\tau} = \dot{\tau}(\mathbf{y}, \dot{\mathbf{y}}, \ddot{\mathbf{y}}, \dddot{\mathbf{y}})$, $\ddot{\tau} = \ddot{\tau}(\mathbf{y}, \dot{\mathbf{y}}, \ddot{\mathbf{y}}, \dddot{\mathbf{y}}, \ddddot{\mathbf{y}})$ from (6.39)-(6.40), and $u_r = u_r(\mathbf{y}, \dot{\mathbf{y}}, \ddot{\mathbf{y}}, \dddot{\mathbf{y}}, \ddddot{\mathbf{y}})$ from (6.41). Hence we showed how to compute the states and the control inputs of the system as functions of the flat outputs and their finite number of derivatives.

## 6.2.6 Case EG: Elastic-joint Attached to a Generic Point

In this section we consider the model of the PVTOL given in Fig. 6.2, with an elastic joint between the motor output shaft and the arm link. A sketch of such connection is shown in Fig. 6.4. This case is referred to as Case EG in Table 6.1. The number of generalized coordinates for this case is increased by one with respect to the RG case ($n = 5$) due to the fact that the output shaft of the motor and the link are not rigidly connected and therefore two distinct coordinates are needed to describe the system configuration, namely $\theta_{0m}$ and $\theta_{01}$.

In order to keep the translational dynamics decoupled from the rotational one let us chose as generalized coordinates $\mathbf{q} = [\mathbf{p}_c^T \ \theta_0 \ \theta_{01} \ \theta_{0m}]^T \in \mathbb{R}^5$. Notice that, differently from the rigid-joint cases (Case RG and Case RC), we have $\theta_1 \neq \theta_m$. In fact $\theta_1 = \theta_m + \theta_e$ (see Fig. 6.2 and Fig. 6.4). Whenever $\theta_e = \theta_1 - \theta_m = \theta_{01} - \theta_{0m}$ is nonzero the elastic link is deflected and stores elastic potential energy.

After some algebra it is possible to compute the matrices and vectors in (6.6) for this case. The inertia matrix is

$$\mathbf{M} = \begin{bmatrix} m_s\mathbf{I}_2 & * & * & * \\ \mathbf{0}_{1\times2} & m_a & * & * \\ \mathbf{0}_{1\times2} & m_{ab}(\theta_0, \theta_{01}) & m_b - J_m & * \\ \mathbf{0}_{1\times2} & 0 & 0 & J_m \end{bmatrix} = \mathbf{M}^T \in \mathbb{R}^{5\times5}, \qquad (6.42)$$

where $m_a$ $m_b$ and $m_{ab}$ are given in (6.13). The centrifugal/Coriolis and gravitational forces are

$$\mathbf{c}(\mathbf{q}, \dot{\mathbf{q}}) = \begin{bmatrix} 0 \\ 0 \\ \frac{m_0 m_1}{m_s}\mathbf{d_0}^T\bar{\mathbf{R}}_1\mathbf{d_1}\dot{\theta}_{01}^2 \\ -\frac{m_0 m_1}{m_s}\mathbf{d_0}^T\bar{\mathbf{R}}_1\mathbf{d_1}\dot{\theta}_0^2 \\ 0 \end{bmatrix}, \quad \mathbf{g} = \begin{bmatrix} 0 \\ -m_s g \\ 0 \\ 0 \\ 0 \end{bmatrix}, \qquad (6.43)$$

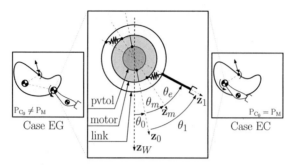

Figure 6.4: An example of elastic-joint between the motor shaft and the link attached to PVTOL. The motor is magnified w.r.t. the PVTOL considering both the EG and the EC cases. The innermost circle is fixed in $\mathcal{F}_0$, thus rigidly attached to the PVTOL. The middle circle is rigidly attached to the motor output shaft, i.e., fixed in $\mathcal{F}_M$. The outermost circle is connected to the middle circle via some elastic components, and it is rigidly connected to the link, thus fixed in $\mathcal{F}_1$.

the control input matrix $\mathbf{G}$ and the elastic forces $\mathbf{f}_E$ due to the elastic potential energy are

$$\mathbf{G}(\mathbf{q}) = \begin{bmatrix} g_{11} & 0 & 0 \\ g_{21} & 0 & 0 \\ g_{31} & 1 & -1 \\ g_{41} & 0 & 0 \\ 0 & 0 & 1 \end{bmatrix}, \quad \mathbf{f}_E(\mathbf{q}) = \begin{bmatrix} 0 \\ 0 \\ 0 \\ f_l(\theta_{0m}, \theta_{01}) \\ f_m(\theta_{0m}, \theta_{01}) \end{bmatrix}, \tag{6.44}$$

where $[g_{11}\ g_{21}\ g_{31}\ g_{41}]^T$ is as the first column of $\mathbf{G}$ in (6.15). Notice that $f_l(\theta_{0m}, \theta_{01})$ is the elastic force acting on the link side, and $f_m(\theta_{0m}, \theta_{01})$ is the elastic force acting on the motor side. These forces can be nonlinear functions of $\theta_{0m}$ and $\theta_{01}$. In the linear spring case, $f_l(\theta_{0m}, \theta_{01}) = k_e(\theta_{0m} - \theta_{01})$ and $f_m(\theta_{0m}, \theta_{01}) = k_e(\theta_{01} - \theta_{0m})$, where $k_e > 0$ is the stiffness of the elastic element.

Writing down the dependences of $\ddot{\mathbf{q}}$ for this case we obtain

$$\begin{aligned} \ddot{x}_c &= f_1(\theta_0, u_t), \quad \ddot{z}_c = f_2(\theta_0, u_t) \\ \ddot{\theta}_0 &= f_3(\theta_0, \theta_{01}, \theta_{0m}, \dot{\theta}_0, \dot{\theta}_{01}, u_t, u_r, \tau) \\ \ddot{\theta}_{01} &= f_4(\theta_0, \theta_{01}, \theta_{0m}, \dot{\theta}_0, \dot{\theta}_{01}, u_t, u_r, \tau) \\ \ddot{\theta}_{0m} &= f_6(\theta_{01}, \theta_{0m}, \tau). \end{aligned} \tag{6.45}$$

As we can see, a part from the introduction of $\theta_{0m}$ the dependency on the other coordinates is the same as the one in (6.16) for Case RG. However, the fact that $n = 5$ (instead of 4) makes the solution adopted for Case RG not immediately applicable. In fact if we set, as in Section 6.2.4, $s_1 = 2$, $s_2 = s_3 = 0$ and we check whether Condition 1 of Definition 7 is satisfied for the output $[\mathbf{p}_c^T\ \theta_{01}]^T$ we fail, since we obtain $\bar{n} = 2n + s_1 = 10 + 2 = 12$ (instead of $\bar{n} = 2n + s_1 = 8 + 2 = 10$) and $r = 10$ (as in Case RG). Therefore it is not straightforward to find the exactly linearizing (flat) output for this case. The reason is that this time we do not gain enough relative degree to reach the new $\bar{n} = 12$. A way to gain more relative degree would be to let $\ddot{\theta}_{01}$ depend on less inputs, since right now is depending on all the inputs. The reason for this dependency is the strong inertial coupling between $\theta_0$ and $\theta_{01}$, see (6.42). Therefore in the following we try whether is possible in some way to loosen this coupling in order to let less inputs appear in $\ddot{\theta}_{01}$.

In order to take a closer look to the rotational coupling let us consider for a moment only the orientation dynamics:

$$\begin{bmatrix} \ddot{\theta}_0 \\ \ddot{\theta}_{01} \end{bmatrix} = \mathbf{B}^{-1} \begin{bmatrix} g_{31}u_t - c_3(\theta_0, \theta_{01}, \dot{\theta}_{01}) + u_r - \tau \\ g_{41}(\theta_0, \theta_{01})u_t + f_l(\theta_{0m}, \theta_{01}) - c_4(\theta_0, \theta_{01}, \dot{\theta}_0) \end{bmatrix}$$

$$\ddot{\theta}_{0m} = J_m^{-1}\left( f_m(\theta_{0m}, \theta_{01}) + \tau \right) \tag{6.46}$$

with

$$\mathbf{B} = \begin{bmatrix} m_a & m_{ab} \\ m_{ab} & m_b - J_m \end{bmatrix}, \begin{bmatrix} c_3 \\ c_4 \end{bmatrix} = \begin{bmatrix} \frac{m_0 m_1}{m_s}\mathbf{d_0}^T\bar{\mathbf{R}}_1\mathbf{d_1}\dot{\theta}_{01}^2 \\ -\frac{m_0 m_1}{m_s}\mathbf{d_0}^T\bar{\mathbf{R}}_1\mathbf{d_1}\dot{\theta}_0^2 \end{bmatrix}. \tag{6.47}$$

This orientation dynamics shares some similarities with the model of a *grounded* planar robot with mixed rigid/elastic joints. It is as if a 'virtual ground base' and the PVTOL are connected through a rigid joint to which it is applied the torque $u_r - \tau$, and the PVTOL and the link are connected through an elastic joint that is actuated by the motor torque $\tau$. However the models are not the same because, e.g., of the the presence of the terms multiplying $u_t$ since in our case the platform is flying.

Mixed rigid/elastic-joints arms for *grounded* manipulators have been studied in De Luca (1998) where the author showed that it is possible to have input-output decoupling and full state linearization for such system, even if there are inertial couplings as in matrix $\mathbf{B}$ given in (6.47), based on dynamic state feedback. Shortly, this is done with a linear dynamic feedback compensator defined for the rigid joint, which let it behave as a *fictitious* elastic joint transmission. To the best of our knowledge this kind of method has never been used for *aerial* manipulators, whose base are (differently from De Luca (1998)) floating and underactuated.

First, let us extend the systems with two new states, $\theta_r$ and $\dot{\theta}_r$ and consider the following dynamic compensator

$$\begin{aligned} u_r - \tau &= k_r(\theta_r - \theta_0) \\ J_r\ddot{\theta}_r + k_r(\theta_r - \theta_0) &= u_n, \end{aligned} \tag{6.48}$$

where $k_r \in \mathbb{R}_{>0}$ and $J_r \in \mathbb{R}_{>0}$ are two additional systems parameters, and $u_n$ is a new control input that replaces $u_r$. Now, replacing $u_r - \tau$ in first equation of (6.46) with the first equation of (6.48), we have

$$\begin{bmatrix} \ddot{\theta}_0 \\ \ddot{\theta}_{01} \end{bmatrix} = \mathbf{B}^{-1} \begin{bmatrix} g_{31}u_t + k_r(\theta_r - \theta_0) - c_3(\theta_0, \theta_{01}, \dot{\theta}_{01}) \\ g_{41}(\theta_0, \theta_{01})u_t + f_l(\theta_{0m}, \theta_{01}) - c_4(\theta_0, \theta_{01}, \dot{\theta}_0) \end{bmatrix}$$

$$\begin{bmatrix} \ddot{\theta}_r \\ \ddot{\theta}_{0m} \end{bmatrix} = \begin{bmatrix} J_r & 0 \\ 0 & J_m \end{bmatrix}^{-1} \begin{bmatrix} k_r(\theta_0 - \theta_r) + u_n \\ f_m(\theta_{0m}, \theta_{01}) + \tau \end{bmatrix}. \tag{6.49}$$

Putting back in place the translational dynamics and writing down the dependences of $\ddot{\mathbf{q}}$ including the new states of the compensator (i.e., considering $\mathbf{q} = [\mathbf{p}_c^T \ \theta_0 \ \theta_{01} \ \theta_r \ \theta_{0m}]^T \in \mathbb{R}^6$)

we obtain

$$\ddot{x}_c = f_1(\theta_0, u_t), \quad \ddot{z}_c = f_2(\theta_0, u_t)$$
$$\ddot{\theta}_0 = f_3(\theta_0, \theta_{01}, \theta_{0m}, \theta_r, \dot{\theta}_0, \dot{\theta}_{01}, u_t)$$
$$\ddot{\theta}_{01} = f_4(\theta_0, \theta_{01}, \theta_{0m}, \theta_r, \dot{\theta}_0, \dot{\theta}_{01}, u_t) \tag{6.50}$$
$$\ddot{\theta}_r = f_5(\theta_0, \theta_r, u_n), \quad \ddot{\theta}_{0m} = f_6(\theta_{01}, \theta_{0m}, \tau).$$

Notice that with the introduction of the compensator the number of states has become $2n = 12$. However, thanks to the compensation applied above, $\ddot{\theta}_{01}$ does not directly depend on $u_r$ and $\tau$ anymore. Therefore there is hope that if we choose as new input a high order derivative of $u_t$ then the the relative degree will be enough this time to let the output $\mathbf{y} = [\mathbf{p}_c^T \; \theta_{01}]^T = [\mathbf{y}_1^T \; y_2]^T$ satisfy Condition 1 of Definition 7.

Let us consider then $s_1 = 4$, $s_2 = s_3 = 0$. With this choice we have $\bar{n} = 2n + 4 = 16$. We then obtain as new control inputs[12] $\bar{\mathbf{u}} = [\dddot{u}_t \; u_n \; \tau]^T \in \mathbb{R}^3$ and the new state $\bar{\mathbf{x}} = [\mathbf{q}^T \; \dot{\mathbf{q}}^T \; u_t \; \dot{u}_t \; \ddot{u}_t \; \dddot{u}_t]^T \in \mathbb{R}^{16}$. The functional dependency of the derivatives of $\mathbf{y}_1$ can be written as follows

$$\ddot{\mathbf{y}}_1 = (f_1 \; f_2)^T = \boldsymbol{\xi}_1(\theta_0, u_t), \quad \dddot{\mathbf{y}}_1 = \boldsymbol{\xi}_2(\theta_0, \dot{\theta}_0, u_t, \dot{u}_t), \tag{6.51}$$

and considering that both $\ddot{\theta}_0$ and $\ddot{\theta}_{01}$ are available from (6.50), we can write

$$\ddddot{\mathbf{y}} = \boldsymbol{\xi}_3(\theta_0, \theta_{01}, \theta_{0m}, \theta_r, \dot{\theta}_0, \dot{\theta}_{01}, u_t, \dot{u}_t, \ddot{u}_t)$$
$$\mathbf{y}_1^{(5)} = \boldsymbol{\xi}_4(\theta_0, \theta_{01}, \theta_{0m}, \theta_r, \dot{\theta}_0, \dot{\theta}_{01}, \dot{\theta}_{0m}, \dot{\theta}_r, u_t, \dot{u}_t, \ddot{u}_t, \dddot{u}_t), \tag{6.52}$$

and using $\ddot{\theta}_r$ and $\ddot{\theta}_{0m}$ from (6.50), we can write

$$\mathbf{y}_1^{(6)} = \boldsymbol{\xi}_5(\theta_0, \theta_{01}, \theta_{0m}, \theta_r, \dot{\theta}_0, \dot{\theta}_{01}, \dot{\theta}_{0m}, \dot{\theta}_r, u_t, \dot{u}_t, \ddot{u}_t, \dddot{u}, \ddddot{u}_t, u_n, \tau),$$

where we stop because $\ddddot{u}$, $u_n$, and $\tau$ appear now linearly in $\mathbf{y}_1^{(6)}$. Therefore we have that $r_1 = r_2 = 6$.

In the same fashion, we can write the derivatives of $y_2$ as

$$\ddot{y}_2 = f_4 = \mu_1(\theta_0, \theta_{01}, \theta_{0m}, \theta_r, \dot{\theta}_0, \dot{\theta}_{01}, u_t) \tag{6.53}$$

and, considering that $\ddot{\theta}_0$ and $\ddot{\theta}_{01}$ are available from (6.50)

$$\dddot{y}_2 = \mu_2(\theta_0, \theta_{01}, \theta_{0m}, \theta_r, \dot{\theta}_0, \dot{\theta}_{01}, \dot{\theta}_{0m}, \dot{\theta}_r, u_t, \dot{u}_t) \tag{6.54}$$

and using $\ddot{\theta}_r$ and $\ddot{\theta}_{0m}$ from (6.50), we can write

$$\ddddot{y}_2 = \mu_3(\theta_0, \theta_{01}, \theta_{0m}, \theta_r, \dot{\theta}_0, \dot{\theta}_{01}, \dot{\theta}_{0m}, \dot{\theta}_r, u_t, \dot{u}_t, \ddot{u}_t, u_n, \tau).$$

Therefore $r_3 = 4$ and $r_1 + r_2 + r_3 = 16 = \bar{n}$, which means that the Condition 1 of Definition 7 is satisfied. Therefore it is now worth to analytically search for the invertibility domain of $\bar{\mathbf{G}}(\bar{\mathbf{x}})$, which is done in the next result.

---

**Proposition 7.** *The vector $[\mathbf{p}_c^T \; \theta_{01}]^T$ is an exactly linearizing output via dynamic feedback for the generic model with an elastic-joint arm (Case EG), as long as $u_t \neq 0$, $k_r \neq 0$ and $k_e \neq 0$ (if the elasticity is linear). As a consequence, it is also a flat output.*

---

*Proof.* See Appendix A.7. □

---

[12]Notice that once $\tau$ and $u_n$ are computed, $u_r$ can be calculated using (6.48).

### Derivation of the Algebraic Map from the Flat Output

We shall now show how to explicitly write down the algebraic map that relates $\ddot{\mathbf{p}}_c$, $\dddot{\mathbf{p}}_c$, $\ddddot{\mathbf{p}}_c$, $\mathbf{p}_c^{(5)}$, $\mathbf{p}_c^{(6)}$, $\theta_{01}$, $\dot{\theta}_{01}$, $\ddot{\theta}_{01}$, $\dddot{\theta}_{01}$, $\ddddot{\theta}_{01}$ with $\theta_0$, $\dot{\theta}_0$, $\theta_{0m}$, $\dot{\theta}_{0m}$, and $\mathbf{u}$.

Similar to the RG case, it is clear from the system dynamics that we can retrieve $u_t$, $\dot{u}_t$, $\ddot{u}_t$ and $\theta_0, \dot{\theta}_0, \ddot{\theta}_0$ from (6.20) and (6.21), respectively. Furthermore, $\theta_{0m}$ can be solved from the fourth equation of the system dynamics as

$$\theta_{0m} = \theta_{0m}(\theta_{01}, \ddot{\theta}_{01}, \theta_0, \dot{\theta}_0, \ddot{\theta}_0, u_t) = \frac{1}{k_e}\Big(m_{ab}(\theta_0, \theta_{01})\ddot{\theta}_0 +$$
$$+ (m_b - J_m)\ddot{\theta}_{01} + c_4(\theta_0, \theta_{01}, \dot{\theta}_0) + k_e\theta_{01} - g_{41}(\theta_0, \theta_{01})u_t\Big), \quad (6.55)$$

where $c_4$ is the fourth row of $\mathbf{c}$ given in (6.43). By introducing $\theta_0 = \theta_0(\ddot{\mathbf{y}}_1)$, $\dot{\theta}_0 = \dot{\theta}_0(\ddot{\mathbf{y}}_1, \dddot{\mathbf{y}}_1)$, $\ddot{\theta}_0 = \ddot{\theta}_0(\ddot{\mathbf{y}}_1, \dddot{\mathbf{y}}_1, \ddddot{\mathbf{y}}_1)$ from (6.21) and $u_t$ from (6.20) we can show that $\theta_{0m} = \theta_{0m}(\ddot{\mathbf{y}}_1, \dddot{\mathbf{y}}_1, \ddddot{\mathbf{y}}_1, y_2, \ddot{y}_2)$, and this implies: $\dot{\theta}_{0m} = \dot{\theta}_{0m}(\ddot{\mathbf{y}}_1, \dddot{\mathbf{y}}_1, \ddddot{\mathbf{y}}_1, \mathbf{y}^{(5)}, y_2, \dot{y}_2, \ddot{y}_2, \dddot{y}_2)$ and $\ddot{\theta}_{0m} = \ddot{\theta}_{0m}(\ddot{\mathbf{y}}_1, \dddot{\mathbf{y}}_1, \ddddot{\mathbf{y}}_1, \mathbf{y}^{(5)}, \mathbf{y}^{(6)}, y_2, \dot{y}_2, \ddot{y}_2, \dddot{y}_2, \ddddot{y}_2)$ with

$$\dot{\theta}_{0m} = \frac{1}{k_e}\Big(\dot{m}_{ab}\ddot{\theta}_0 + m_{ab}\dddot{\theta}_0 + (m_b - J_m)\dddot{\theta}_{01} + \dot{c}_4 +$$
$$+ k_e\dot{\theta}_{01} - \dot{g}_{41}u_t - g_{41}\dot{u}_t\Big)$$
$$\ddot{\theta}_{0m} = \frac{1}{k_e}\Big(\ddot{m}_{ab}\ddot{\theta}_0 + 2\dot{m}_{ab}\dddot{\theta}_0 + m_{ab}\ddddot{\theta}_0 + (m_b - J_m)\ddddot{\theta}_{01} +$$
$$+ \ddot{c}_4 + k_e\ddot{\theta}_{01} - \ddot{g}_{41}u_t - 2\dot{g}_{41}\dot{u}_t - g_{41}\ddot{u}_t\Big), \quad (6.56)$$

where

$$\dot{m}_{ab} = \frac{m_0 m_1}{m_s}\mathbf{d}_0^T\bar{\mathbf{R}}_1\mathbf{d}_1(\dot{\theta}_{01} - \dot{\theta}_0)$$
$$\ddot{m}_{ab} = -\frac{m_0 m_1}{m_s}\mathbf{d}_0^T\mathbf{R}_1\mathbf{d}_1(\dot{\theta}_{01} - \dot{\theta}_0)^2 + \frac{m_0 m_1}{m_s}\mathbf{d}_0^T\bar{\mathbf{R}}_1\mathbf{d}_1(\ddot{\theta}_{01} - \ddot{\theta}_0)$$
$$\dot{c}_4 = -\frac{m_0 m_1}{m_s}\mathbf{d}_0^T\Big(-\mathbf{R}_1(\dot{\theta}_{01} - \dot{\theta}_0)\dot{\theta}_{01}^2 + 2\bar{\mathbf{R}}_1\ddot{\theta}_{01}\Big)\mathbf{d}_1$$
$$\ddot{c}_4 = -\frac{m_0 m_1}{m_s}\mathbf{d}_0^T\Big(-\bar{\mathbf{R}}_1(\dot{\theta}_{01} - \dot{\theta}_0)^2\dot{\theta}_{01}^2 - \mathbf{R}_1(\ddot{\theta}_{01} - \ddot{\theta}_0)\dot{\theta}_{01}^2 - 4\mathbf{R}_1(\dot{\theta}_{01} - \dot{\theta}_0)\ddot{\theta}_{01} + 2\bar{\mathbf{R}}_1\dddot{\theta}_{01}\Big)\mathbf{d}_1$$
$$\dot{g}_{41} = -\frac{m_1}{m_s}(-d_{1_x}\sin\theta_1 + d_{1_z}\cos\theta_1)(\dot{\theta}_{01} - \dot{\theta}_0)$$
$$\ddot{g}_{41} = -\frac{m_1}{m_s}\Big((-d_{1_x}\cos\theta_1 - d_{1_z}\sin\theta_1)(\dot{\theta}_{01} - \dot{\theta}_0)^2 + (-d_{1_x}\sin\theta_1 + d_{1_z}\cos\theta_1)(\ddot{\theta}_{01} - \ddot{\theta}_0)\Big),$$

with $\ddot{\theta}_0$, $\dddot{\theta}_0$, $\ddddot{\theta}_0$ and $u_t, \dot{u}_t, \ddot{u}_t$ an be computed from (6.20) and (6.21). The motor torque is obtained from the fifth equation of the system dynamics, i.e.

$$\tau = J_m\ddot{\theta}_{0m} + k_e\theta_{0m} - k_e\theta_{01}, \quad (6.57)$$

where substituting $\theta_{0m}$ and $\ddot{\theta}_{0m}$ using (6.55), we can show that $\tau = \tau(\ddot{\mathbf{y}}_1, \dddot{\mathbf{y}}_1, \ddddot{\mathbf{y}}_1, \mathbf{y}^{(5)}, \mathbf{y}^{(6)}, y_2, \dot{y}_2, \ddot{y}_2, \dddot{y}_2, \ddddot{y}_2)$.

Finally solving $u_r$ from the third equation of the system dynamics, one obtains

$$u_r = u_r(\theta_{01}, \dot{\theta}_{01}, \ddot{\theta}_{01}, \theta_0, \ddot{\theta}_0, u_t, \tau) = m_a \ddot{\theta}_0 + m_{ab}(\theta_0, \theta_{01}) \ddot{\theta}_{01} + $$
$$+ \frac{m_0 m_1}{m_s} \mathbf{d_0}^T \bar{\mathbf{R}}_1(\theta_0, \theta_{01}) \mathbf{d_1} \dot{\theta}_{01}^2 + \tau - g_{31} u_t, \quad (6.58)$$

where utilizing $\theta_0, \dot{\theta}_0, \ddot{\theta}_0$ from (6.21), $u_t$ from (6.20), and $\tau$ from (6.57) we obtain $u_r = u_r(\ddot{\mathbf{y}}_1, \dddot{\mathbf{y}}_1, \ddddot{\mathbf{y}}_1, \mathbf{y}^{(5)}, \mathbf{y}^{(6)}, y_2, \dot{y}_2, \ddot{y}_2, \dddot{y}_2, \ddddot{y}_2)$.

In summary, we have $\mathbf{p}_c = \mathbf{y}_1$, $\dot{\mathbf{p}}_c = \dot{\mathbf{y}}_1$, $\ddot{\mathbf{p}}_c = \ddot{\mathbf{y}}_1$ and $\theta_{01} = y_2$, $\dot{\theta}_{01} = \dot{y}_2$, $\ddot{\theta}_{01} = \ddot{y}_2$ from the definition; $u_t = u_t(\ddot{\mathbf{y}}_1)$, $\dot{u}_t = \dot{u}_t(\ddot{\mathbf{y}}_1, \dddot{\mathbf{y}}_1)$, $\ddot{u}_t = \ddot{u}_t(\ddot{\mathbf{y}}_1, \dddot{\mathbf{y}}_1, \ddddot{\mathbf{y}}_1)$ from (6.20); $\theta_0 = \theta_0(\ddot{\mathbf{y}}_1)$, $\dot{\theta}_0 = \dot{\theta}_0(\ddot{\mathbf{y}}_1, \dddot{\mathbf{y}}_1)$, $\ddot{\theta}_0 = \ddot{\theta}_0(\ddot{\mathbf{y}}_1, \dddot{\mathbf{y}}_1, \ddddot{\mathbf{y}}_1)$ from (6.21); $\theta_{0m} = \theta_{0m}(\ddot{\mathbf{y}}_1, \dddot{\mathbf{y}}_1, \ddddot{\mathbf{y}}_1, y_2, \ddot{y}_2)$, $\dot{\theta}_{0m} = \dot{\theta}_{0m}(\ddot{\mathbf{y}}_1, \dddot{\mathbf{y}}_1, \ddddot{\mathbf{y}}_1, \mathbf{y}^{(5)}, y_2, \dot{y}_2, \ddot{y}_2, \dddot{y}_2)$, and $\ddot{\theta}_{0m} = \ddot{\theta}_{0m}(\ddot{\mathbf{y}}_1, \dddot{\mathbf{y}}_1, \ddddot{\mathbf{y}}_1, \mathbf{y}^{(5)}, \mathbf{y}^{(6)}, y_2, \dot{y}_2, \ddot{y}_2, \dddot{y}_2, \ddddot{y}_2)$ using (6.55); $\tau = \tau(\ddot{\mathbf{y}}_1, \dddot{\mathbf{y}}_1, \ddddot{\mathbf{y}}_1, \mathbf{y}^{(5)}, \mathbf{y}^{(6)}, y_2, \dot{y}_2, \ddot{y}_2, \dddot{y}_2, \ddddot{y}_2)$ from (6.57) and finally $u_r = u_r(\ddot{\mathbf{y}}_1, \dddot{\mathbf{y}}_1, \ddddot{\mathbf{y}}_1, \mathbf{y}^{(5)}, \mathbf{y}^{(6)}, y_2, \dot{y}_2, \ddot{y}_2, \dddot{y}_2, \ddddot{y}_2)$ from (6.58). Moreover, one can see that $\dddot{u}_t = \dddot{u}_t(\ddot{\mathbf{y}}_1, \dddot{\mathbf{y}}_1, \ddddot{\mathbf{y}}_1, \mathbf{y}_1^{(5)})$, $\ddddot{u}_t = \ddddot{u}_t(\ddot{\mathbf{y}}_1, \dddot{\mathbf{y}}_1, \ddddot{\mathbf{y}}_1, \mathbf{y}_1^{(5)}, \mathbf{y}_1^{(6)})$ using (6.20). Hence we showed the states and the control inputs of the system as functions of the flat outputs and their finite number of derivatives.

## 6.2.7 Case EC: Elastic-joint Attached to the PVTOL CoM

Like for the RG case, the EG case is subject to the same negative result presented in Sec. 6.2.4. Therefore, for the same motivations of the rigid-case (see Secs. 6.2.4 and 6.2.5) let us consider again the model in which $P_{C_0}$ coincides with $P_M$, i.e., $\mathbf{d}_0 = \mathbf{0}_{2\times 1}$, but this time with elastic-joint instead of a rigid one (see Fig. 6.4). This case is referred to as Case EC in Table 6.1. In particular we are interested in finding whether, similarly to the RC case, also in this case the output $\mathbf{y} = [\mathbf{p}_m^T \; \theta_{01}]^T$ is exactly linearizing (i.e., flat).

Let us then consider as generalized coordinates $\mathbf{q} = [\mathbf{p}_m^T \; \theta_0 \; \theta_{01} \; \theta_{0m}]^T \in \mathbb{R}^5$, where, we remind that $\mathbf{p}_m = \mathbf{p}_0$. In this case the inertia matrix is

$$\mathbf{M} = \begin{bmatrix} m_s \mathbf{I}_2 & * & * & * \\ \mathbf{0}_{1\times 2} & J_0 & * & * \\ \boldsymbol{\beta}^T(\theta_{01}) & 0 & m_B - J_m & * \\ \mathbf{0}_{1\times 2} & 0 & 0 & J_m \end{bmatrix} = \mathbf{M}^T \in \mathbb{R}^{5\times 5}, \quad (6.59)$$

where remember that $\boldsymbol{\beta}$ and $m_B$ were defined in (6.27). The centrifugal/Coriolis and gravitational forces are

$$\mathbf{c}(\mathbf{q}, \dot{\mathbf{q}}) = \begin{bmatrix} \bar{\beta}_1(\theta_{01})\dot{\theta}_{01}^2 \\ \bar{\beta}_2(\theta_{01})\dot{\theta}_{01}^2 \\ 0 \\ 0 \\ 0 \end{bmatrix}, \quad \mathbf{g}(\mathbf{q}) = \begin{bmatrix} 0 \\ -m_s g \\ 0 \\ g_4(\theta_{01}) \\ 0 \end{bmatrix}, \quad (6.60)$$

with their components as computed before in (6.29). Notice that the elastic forces $\mathbf{f}_E$ are the same as in (6.44). Finally the control input matrix from generalized forces is

$$\mathbf{G}(\mathbf{q}) = \begin{bmatrix} -\sin(\theta_0) & 0 & 0 \\ -\cos(\theta_0) & 0 & 0 \\ d_{G_x} & 1 & -1 \\ 0 & 0 & 0 \\ 0 & 0 & 1 \end{bmatrix}. \tag{6.61}$$

Replacing $\mathbf{M}$, $\mathbf{c}$, $\mathbf{g}$, $\mathbf{G}$ and $\mathbf{f}_E$ in (6.6) we can derive the explicit dependency of each entry of $\ddot{\mathbf{q}}$, here summarized:[13]

$$\begin{aligned} \ddot{x}_m &= f_1(\theta_0, \theta_{01}, \dot{\theta}_{01}, \theta_{0m}, u_t) \\ \ddot{z}_m &= f_2(\theta_0, \theta_{01}, \dot{\theta}_{01}, \theta_{0m}, u_t) \\ \ddot{\theta}_0 &= f_3(u_t, u_r, \tau) \\ \ddot{\theta}_{01} &= f_4(\theta_0, \theta_{01}, \dot{\theta}_{01}, \theta_{0m}, u_t) \\ \ddot{\theta}_{0m} &= f_5(\theta_{0m}, \theta_{01}, \tau). \end{aligned} \tag{6.62}$$

Let us now consider the output $\mathbf{y} = [\mathbf{p}_m^T, \theta_{01}]^T \in \mathbb{R}^3$ and try to find $s_1$, $s_2$, and $s_3$ that satisfy Condition 1 of Definition 7.

If we compare Case RC with Case EC we have that in the former case $n = 8$ while $n = 10$ in the latter, which implies that a higher total relative degree has to be reached in Case EC to fulfill Condition 1. If we then compare (6.62) to (6.34) we see that the only input appearing in Case EC for $\ddot{\mathbf{y}}$ is $u_t$ while in Case RC both $u_t$ and $\tau$ appear. This is a good sign since in Case RC we had to choose both $s_1 = 2$ and $s_2 = 2$ thus raising $\bar{n}$ to $8 + 4 = 12$ while in Case EC we probably do not need to add two integrators on the $\tau$ channel because $\tau$ it is not appearing already in $\ddot{\mathbf{y}}$.

Let us consider then $s_1 = 2$, and $s_2 = s_3 = 0$. With this choice the new input is $\bar{\mathbf{u}} = [\ddot{u}_t \ u_r \ \tau]^T \in \mathbb{R}^3$, new state $\bar{\mathbf{x}} = [\mathbf{q}^T \ \dot{\mathbf{q}}^T \ u_t \ \dot{u}_t]^T \in \mathbb{R}^{12}$, and $\bar{n} = 12$.

The functional dependency of the derivatives of $\mathbf{y}$ can be written as follows:

$$\ddot{\mathbf{y}} = \boldsymbol{\xi}_1(\theta_0, \theta_{01}, \dot{\theta}_{01}, \theta_{0m}, u_t). \tag{6.63}$$

Let us now further derivate the output until the input appears. Using $\ddot{\theta}_{01}$ from (6.62) we can write

$$\dddot{\mathbf{y}} = \boldsymbol{\xi}_2(\theta_0, \theta_{01}, \theta_{0m}, \dot{\theta}_0, \dot{\theta}_{01}, \dot{\theta}_{0m}, u_t, \dot{u}_t). \tag{6.64}$$

Using $\ddot{\theta}_0$ from (6.62) we can write

$$\ddddot{\mathbf{y}} = \boldsymbol{\xi}_3(\theta_0, \theta_{01}, \theta_{0m}, \dot{\theta}_0, \dot{\theta}_{01}, \dot{\theta}_{0m}, u_t, \dot{u}_t, \ddot{u}_t, u_r, \tau), \tag{6.65}$$

in which the new inputs appear linearly, therefore $r_1 = r_2 = r_3 = 4$ and thus $r = 12 = \bar{n}$, which means that the Condition 1 of Definition 7 is satisfied. Therefore it is now worth to analytically search for the invertibility domain of $\bar{\mathbf{G}}(\bar{\mathbf{x}})$, which is given in the next result.

---

[13]Again, also in this case, if one develops the computations can see that $\ddot{\theta}_{01}$ does not depend on $\dot{\theta}_{01}$ since the terms depending on $\dot{\theta}_{01}$ cancel out each other.

---

**Proposition 8.** *The vectors $[\mathbf{p}_c^T\ \theta_{01}]^T$, $[\mathbf{p}_m^T\ \theta_{01}]^T$ and $[\mathbf{p}_e^T\ \theta_{01}]^T$ are all exactly linearizing output via dynamic feedback for the coinciding model with elastic-joint arm (Case EC), as long as $u_t \neq 0$ and $k_e \neq 0$ (if the elasticity is linear). As a consequence, they are also flat outputs.*

*Proof.* See Appendix A.8. □

### Derivation of the Algebraic Map from the Flat Output

We shall show now the procedure to explicitly derive the algebraic map that relates $\ddot{\mathbf{p}}_m, \dddot{\mathbf{p}}_m, \ddddot{\mathbf{p}}_m, \theta_{01}, \dot{\theta}_{01}, \ddot{\theta}_{01}, \dddot{\theta}_{01}, \ddddot{\theta}_{01}$ with $\theta_0, \dot{\theta}_0, \theta_{0m}, \dot{\theta}_{0m}$, and $\mathbf{u}$.

Consider the position in $\mathcal{F}_W$ of the CoM of the overall system, as in (6.38). By substituting it in (6.20) and in (6.21), we find $u_t, \dot{u}_t, \ddot{u}_t$ and $\theta_0, \dot{\theta}_0, \ddot{\theta}_0$ as functions of $\mathbf{y}, \cdots, \ddddot{\mathbf{y}}$. Furthermore, from the fourth equation of the system dynamics we get

$$\theta_{0m} = \frac{\boldsymbol{\beta}^T \ddot{\mathbf{p}}_m + (m_B - J_m)\ddot{\theta}_{01} + g_4(\theta_{01}) + k_e\theta_{01}}{k_e},\tag{6.66}$$

which is function of solely the flat outputs, i.e., $\theta_{0m} = \theta_{0m}(\mathbf{y}, \ddot{\mathbf{y}})$. Now, recalling that $g_4(\theta_{01}) = -\boldsymbol{\beta}(\theta_{01})g \cdot \mathbf{e}_2$ with $\mathbf{e}_2 = [0\ 1]^T \in \mathbb{R}^2$, we can write

$$\dot{\theta}_{0m} = \frac{\boldsymbol{\beta}^T \dddot{\mathbf{p}}_m + (m_B - J_m)\dddot{\theta}_{01} + (\bar{\boldsymbol{\beta}}^T \ddot{\mathbf{p}}_m - \bar{\boldsymbol{\beta}}g \cdot \mathbf{e}_2 + k_e)\dot{\theta}_{01}}{k_e}$$

$$\ddot{\theta}_{0m} = \frac{\boldsymbol{\beta}^T \ddddot{\mathbf{p}}_m + (m_B - J_m)\ddddot{\theta}_{01} + 2\bar{\boldsymbol{\beta}}^T \dddot{\mathbf{p}}_m\dot{\theta}_{01}}{k_e} +$$

$$+ \frac{(\bar{\boldsymbol{\beta}}^T \ddot{\mathbf{p}}_m - \bar{\boldsymbol{\beta}}g \cdot \mathbf{e}_2 + k_e)\ddot{\theta}_{01} + (\beta g \cdot \mathbf{e}_2 - \boldsymbol{\beta}^T \ddot{\mathbf{p}}_m)\dot{\theta}_{01}^2}{k_e},\tag{6.67}$$

which means $\dot{\theta}_{0m} = \dot{\theta}_{0m}(\mathbf{y}, \dot{\mathbf{y}}, \ddot{\mathbf{y}}, \dddot{\mathbf{y}})$, and $\ddot{\theta}_{0m} = \ddot{\theta}_{0m}(\mathbf{y}, \dot{\mathbf{y}}, \ddot{\mathbf{y}}, \dddot{\mathbf{y}}, \ddddot{\mathbf{y}})$. Moreover, one can rewrite the motor torque using the fifth equation of the system dynamics, namely

$$\tau = \tau(\theta_{01}, \theta_{0m}, \ddot{\theta}_{0m}) = J_m\ddot{\theta}_{0m} + k_e\theta_{0m} - k_e\theta_{01},\tag{6.68}$$

where substituting $\theta_{0m}$ from (6.66) and $\ddot{\theta}_{0m}$ from (6.67), it is $\tau = \tau(\mathbf{y}, \dot{\mathbf{y}}, \ddot{\mathbf{y}}, \dddot{\mathbf{y}}, \ddddot{\mathbf{y}})$.

Finally the PVTOL torque is computed from the third equation of the system dynamics using

$$u_r = u_r(\ddot{\theta}_0, u_t, \tau) = J_0\ddot{\theta}_0 + \tau - d_{G_x}u_t,\tag{6.69}$$

where utilizing $\ddot{\theta}_0$ from (6.21) and $u_t$ from (6.20) by also taking (6.38) into consideration, and $\tau$ from (6.68), we can show that $u_r = u_r(\mathbf{y}, \dot{\mathbf{y}}, \ddot{\mathbf{y}}, \dddot{\mathbf{y}}, \ddddot{\mathbf{y}})$.

In summary, we obtained $\mathbf{p}_m = \mathbf{p}_m(\mathbf{y})$, $\dot{\mathbf{p}}_m = \dot{\mathbf{p}}_m(\dot{\mathbf{y}})$, $\ddot{\mathbf{p}}_c = \ddot{\mathbf{p}}_m(\ddot{\mathbf{y}})$ and $\theta_{01} = \theta_{01}(\mathbf{y})$, $\dot{\theta}_{01} = \dot{\theta}_{01}(\dot{\mathbf{y}})$, $\ddot{\theta}_{01} = \ddot{\theta}_{01}(\ddot{\mathbf{y}})$ from the definition; $u_t = u_t(\mathbf{y}, \dot{\mathbf{y}}, \ddot{\mathbf{y}})$, $\dot{u}_t = \dot{u}_t(\mathbf{y}, \dot{\mathbf{y}}, \ddot{\mathbf{y}}, \dddot{\mathbf{y}})$, $\ddot{u}_t = \ddot{u}_t(\mathbf{y}, \dot{\mathbf{y}}, \ddot{\mathbf{y}}, \dddot{\mathbf{y}}, \ddddot{\mathbf{y}})$ from (6.20) and $\theta_0 = \theta_0(\mathbf{y}, \dot{\mathbf{y}}, \ddot{\mathbf{y}})$, $\dot{\theta}_0 = \dot{\theta}_0(\mathbf{y}, \dot{\mathbf{y}}, \ddot{\mathbf{y}}, \dddot{\mathbf{y}})$, $\ddot{\theta}_0 = \ddot{\theta}_0(\mathbf{y}, \dot{\mathbf{y}}, \ddot{\mathbf{y}}, \dddot{\mathbf{y}}, \ddddot{\mathbf{y}})$ from (6.21) where for both $\mathbf{p}_c$ is obtained from (6.38); $\theta_{0m} = \theta_{0m}(\mathbf{y}, \ddot{\mathbf{y}})$, $\dot{\theta}_{0m} = \dot{\theta}_{0m}(\mathbf{y}, \dot{\mathbf{y}}, \ddot{\mathbf{y}})$, $\ddot{\theta}_{0m} = \ddot{\theta}_{0m}(\mathbf{y}, \dot{\mathbf{y}}, \ddot{\mathbf{y}}, \dddot{\mathbf{y}}, \ddddot{\mathbf{y}})$ from (6.66)-(6.67); and finally $\tau = \tau(\mathbf{y}, \dot{\mathbf{y}}, \ddot{\mathbf{y}}, \dddot{\mathbf{y}}, \ddddot{\mathbf{y}})$ from (6.68) and $u_r = u_r(\mathbf{y}, \dot{\mathbf{y}}, \ddot{\mathbf{y}}, \dddot{\mathbf{y}}, \ddddot{\mathbf{y}})$ from (6.69). Hence we showed how the states and the control inputs of the system can be written as functions of the flat outputs and a finite number of their derivatives.

| Modeling Cases | Linearizing (Flat) Outputs | Relative Degree | New States | New Input |
|---|---|---|---|---|
| *Case RG*: Rigid-Joint Attached to a Generic Point <br> • $P_{C_0} \neq P_M \neq P_G$ <br> • $q = [p_c^T \, \theta_{01}]^T \in \mathbb{R}^4$ | $y = [p_c^T \, \theta_{01}]^T \in \mathbb{R}^3$ | $\dot{\bar{y}} = [\dddot{p}_c^T \, \ddot{\theta}_{01}]^T$ <br> $r = 10$ | $\bar{x} = [q^T \, \dot{q}^T \, u_t \, \dot{u}_t]^T \in \mathbb{R}^{10}$ <br> $\bar{n} = 10$ | $\bar{u} = [\ddot{u}_t \, u_r \, \tau]^T \in \mathbb{R}^3$ |
| *Case RC*: Rigid-Joint Attached to the PVTOL CoM <br> • $P_{C_0} \equiv P_M \neq P_G \implies p_0 = p_m$ <br> • $q = [p_m^T \, \theta_0 \, \theta_{01}]^T \in \mathbb{R}^4$ | also $y = [p_m^T \, \theta_{01}]^T \in \mathbb{R}^3$ <br> and $y = [p_c^T \, \theta_{01}]^T \in \mathbb{R}^3$ | $\dot{\bar{y}} = [\dddot{p}_m^T \, \dddot{\theta}_{01}]^T$ <br> $r = 12$ | $\bar{x} = [q^T \, \dot{q}^T \, u_t \, \dot{u}_t \, \tau \, \dot{\tau}]^T \in \mathbb{R}^{12}$ <br> $\bar{n} = 12$ | $\bar{u} = [\ddot{u}_t \, u_r \, \ddot{\tau}]^T \in \mathbb{R}^3$ |
| *Case EG*: Elastic-Joint Attached to a Generic Point <br> • $P_{C_0} \neq P_M \neq P_G$ <br> • $q = [p_c^T \, \theta_0 \, \theta_{01} \, \theta_r \, \theta_{0m}]^T \in \mathbb{R}^6$ | $y = [p_c^T \, \theta_{01}]^T \in \mathbb{R}^3$ | $\dot{\bar{y}} = [p_c^{(6)T} \, \ddddot{\theta}_{01}]^T$ <br> $r = 16$ | $\bar{x} = [q^T \, \dot{q}^T \, u_t \, \dot{u}_t \, \ddot{u}_t \, \dddot{u}_t]^T \in \mathbb{R}^{16}$ <br> $\bar{n} = 16$ | $\bar{u} = [\ddddot{u}_t \, u_n \, \tau]^T \in \mathbb{R}^3$ |
| *Case EC*: Elastic-Joint Attached to the PVTOL CoM <br> • $P_{C_0} \equiv P_M \neq P_G \implies p_0 = p_m$ <br> • $q = [p_m^T \, \theta_0 \, \theta_{01}, \theta_{0m}]^T \in \mathbb{R}^5$ | also $y = [p_m^T \, \theta_{01}]^T \in \mathbb{R}^3$ <br> and $y = [p_c^T \, \theta_{01}]^T \in \mathbb{R}^3$ | $\dot{\bar{y}} = [\dddot{p}_m^T \, \dddot{\theta}_{01}]^T$ <br> $r = 12$ | $\bar{x} = [q^T \, \dot{q}^T \, u_t \, \dot{u}_t]^T \in \mathbb{R}^{12}$ <br> $\bar{n} = 12$ | $\bar{u} = [\ddot{u}_t \, u_r \, \tau]^T \in \mathbb{R}^3$ |

Table 6.2: A summarizing table of the structural controllability properties for different models of PVTOL aerial manipulators equipped with a rigid-joint or an elastic-joint arm. The first column summarizes the properties of the four different cases, which are deeply studied in Section 6.2. The remaining columns present the corresponding facts discovered in the same section. In every case the total number of states matches with the relative degree, which implies that no destabilizing internal dynamics will arise when an exact feedback linearization controller is applied to the system. This also implies the flatness of the corresponding output.

**Remark 10.** *Notice that in Case RC, both $u_t$ and $\tau$ needed to be delayed twice with a double integrator, while for Case EC this holds only for only $u_t$, in order to match the condition on the relative degree and total number of states $(r = \bar{n})$ . This happens because the spring in Case EC introduces a second order linear system and hence further delaying for $\tau$ is not needed.*

## 6.2.8 Using Flatness for Optimal Trajectory Planning and Control

In this section we formalize an optimal control problem for planning the optimal trajectories of the aerial manipulators which take into account the saturations of the actuators and the bounds of the system state. In order to generate trajectories that satisfy the system dynamics, we show here how to use the differential flatness property of the system in the planning phase. Another advantage of using differential flatness is that one can generate an initial guess of the trajectory by smoothly interpolating the flat output from its initial to final value and analytically compute the all states and control inputs of the system accordingly. In this way, a *warm start* to the optimal solver can be given, which reduces the computation time of the optimal trajectory.

Dynamic feasibility of the optimal trajectories is ensured by the smoothness of the flat output we have. To obtain trajectories that are smooth enough, we use the extension of the system dynamics given in (6.6).

The control of the end-effector positions for the aerial manipulators is an interesting, as well as a relevant problem. Hence here, we focus on the tasks performed by the end-effector of the aerial manipulators. This means, and because of the reasons explained in Sec. 6.2.4, we will use the models described as Case RC and Case EC. After presenting the dynamic extensions for Cases RC and EC, we then use them in the formalization of the optimal control problem together with their exact tracking controllers as presented in Sections 6.2.5 and Sec.6.2.7, respectively.

In Sec. 6.2.5 and in Sec. 6.2.7 we showed the differentially flat outputs of the systems described as Case RC and Case EC, respectively. Now, let us use this knowledge to extend the system dynamics for generating the smooth trajectories.

**Dynamic Extension for Case RC**

Consider the system model in Section 6.2.5. The system dynamics is summarized in (6.34), which can be written in the following form

$$\ddot{\mathbf{q}} = \mathbf{f}(\mathbf{q}, \dot{\mathbf{q}}, \mathbf{u}) \in \mathbb{R}^{4 \times 1}. \tag{6.70}$$

The flat outputs are $\mathbf{y} = [\mathbf{p}_m^T \theta_{01}]^T$ (see Proposition 6) and the implicit functional dependencies of their derivatives are shown in (6.36) and (6.37). Also considering Table 6.2 we know that $\bar{\mathbf{x}} = [\mathbf{q}^T \; \dot{\mathbf{q}}^T \; u_t \; \dot{u}_t \; \tau \; \dot{\tau}]^T \in \mathbb{R}^{12}$ and $\bar{\mathbf{u}} = [\ddot{u}_t \; u_r \; \ddot{\tau}]^T \in \mathbb{R}^3$. Hence, we can write

$$\dot{\bar{\mathbf{x}}} = \begin{bmatrix} \mathbf{0}_4 & \mathbf{I}_4 & \mathbf{0}_4 \\ \mathbf{0}_4 & \mathbf{0}_4 & \mathbf{0}_4 \\ \mathbf{0}_4 & \mathbf{0}_4 & \mathbf{S} \end{bmatrix} \bar{\mathbf{x}} + \begin{bmatrix} \mathbf{0}_{4\times 1} \\ \mathbf{f}(\mathbf{q}, \dot{\mathbf{q}}, \mathbf{u}) \\ \mathbf{s}(\ddot{u}_t, \ddot{\tau}) \end{bmatrix} = \bar{\mathbf{f}}(\bar{\mathbf{x}}, \bar{\mathbf{u}}), \tag{6.71}$$

where $\mathbf{f}$ is available from (6.70) and

$$\mathbf{S} = \begin{bmatrix} 0 & 1 & 0 & 0 \\ 0 & 0 & 0 & 0 \\ 0 & 0 & 0 & 1 \\ 0 & 0 & 0 & 0 \end{bmatrix} \in \mathbb{R}^{4\times 4}, \quad \mathbf{s} = \begin{bmatrix} 0 \\ \ddot{u}_t \\ 0 \\ \ddot{\tau} \end{bmatrix} \in \mathbb{R}^4.$$

Later, we will use the extended system dynamics given in (6.71) for the optimization problem.

**Dynamic Extension for Case EC**

Consider the system model in Section 6.2.7. The system dynamics is summarized in (6.62), which can be written in the following form

$$\ddot{\mathbf{q}} = \mathbf{f}(\mathbf{q}, \dot{\mathbf{q}}, \mathbf{u}) \in \mathbb{R}^{5 \times 1}. \tag{6.72}$$

The flat outputs are $\mathbf{y} = [\mathbf{p}_m^T \theta_{01}]^T$ (see Proposition 8) and the implicit functional dependencies of their derivatives are shown in (6.64) and (6.65). Also considering Table 6.2 we know that $\bar{\mathbf{x}} = [\mathbf{q}^T \; \dot{\mathbf{q}}^T \; u_t \; \dot{u}_t]^T \in \mathbb{R}^{12}$ and $\bar{\mathbf{u}} = [\ddot{u}_t \; u_r \; \tau]^T \in \mathbb{R}^3$. Hence, we can write

$$\dot{\bar{\mathbf{x}}} = \begin{bmatrix} \mathbf{0}_5 & \mathbf{I}_5 & \mathbf{0}_{5\times 2} \\ \mathbf{0}_5 & \mathbf{0}_5 & \mathbf{0}_{5\times 2} \\ \mathbf{0}_{2\times 5} & \mathbf{0}_{2\times 5} & \mathbf{S} \end{bmatrix} \bar{\mathbf{x}} + \begin{bmatrix} \mathbf{0}_{5\times 1} \\ \mathbf{f}(\mathbf{q}, \dot{\mathbf{q}}, \mathbf{u}) \\ \mathbf{s}(\ddot{u}_t) \end{bmatrix} = \bar{\mathbf{f}}(\bar{\mathbf{x}}, \bar{\mathbf{u}}), \tag{6.73}$$

where $\mathbf{f}$ is available from (6.72) and

$$\mathbf{S} = \begin{bmatrix} 0 & 1 \\ 0 & 0 \end{bmatrix} \in \mathbb{R}^{2\times 2}, \quad \mathbf{s} = \begin{bmatrix} 0 \\ \ddot{u}_t \end{bmatrix} \in \mathbb{R}^2.$$

In the following, we will use the extended system dynamics given in (6.73) for the optimization problem.

**Optimal Control Problem**

We consider the following optimization problem

$$
\begin{aligned}
\underset{\bar{\mathbf{x}}(t),\bar{\mathbf{u}}(t)}{\text{minimize}} \quad & J(\bar{\mathbf{x}}(t),\bar{\mathbf{u}}(t),t_L) \\
\text{subject to,} \quad & \forall t \in [t_0, t_L] \\
& \dot{\bar{\mathbf{x}}} = \bar{\mathbf{f}}(\bar{\mathbf{x}}(t),\bar{\mathbf{u}}(t)), \bar{\mathbf{x}}(t_0) = \bar{\mathbf{x}}_0 \\
& \mathbf{q}_m \leq \mathbf{q} \leq \mathbf{q}_M, \quad \mathbf{u}_m \leq \mathbf{u} \leq \mathbf{u}_M \\
& \phi_m \leq \phi \leq \phi_M
\end{aligned}
\tag{6.74}
$$

where $J : \bar{\mathbf{x}}, \bar{\mathbf{u}} \to \mathbb{R}$ is the cost function[14]; $\bar{\mathbf{f}}$ is the system dynamics available from (6.71) for Case RC, and from (6.73) for Case EC; $\mathbf{q}$ and $\mathbf{u}$ are the system coordinates and the inputs[15]; and $\bar{\mathbf{x}}_0$ are the initial conditions. Notice that $\phi$ here defined as the deflection of the elastic element (not the roll of the quadrotor as in Chapters 2 and 4), i.e. $\phi = \theta_e = \theta_1 - \theta_m$, and this condition is added only for Case EC.

Now we have a formal definition of an optimal control problem, which can be used to generate a trajectory for aerial manipulators described as Case RC and RE.

The dimension and hence the complexity of the system at hand is too high to solve it as an optimal control problem in an analytical way. A way for approaching the optimal control problem is to reduce the system complexity, as it is done, e.g., in Lupashin et al. (2010), where authors used the angular velocity of the system as inputs, instead of force/torque inputs. This approach is not viable for aerial manipulators where the dynamical effects cannot be neglected. Therefore, in the following we will consider the full system dynamics and solve the optimization problem using the direct optimization method, such as the one presented in Houska et al. (2011).

### 6.2.9 Numerical Results

In the following, we present the numerical validation of the results found in Section 6.2. Namely, we show various simulation results for controlling particular designs of aerial manipulators: a PVTOL equipped with a joint-arm, which can be rigidly actuated or be compliant. We consider diversified tasks for the aerial manipulator, e.g. *composite trajectory tracking, aerial grasping, link velocity amplification* and *object throwing*.

**Realistic Numerical Tests**

In this section we show the results of extensive simulative tests aimed at validating, in non-ideal conditions, the performances of the feedback controllers and optimal trajectory generators presented in the previous sections. We focus in particular on the algorithms developed for cases RC (Sec. 6.2.5) and EC (Sec. 6.2.7) because they permit to control the end-effector pose, which is the typical task in practical applications. We also test the robustness of those algorithms (shortly denoted as the *RC controller* and the *EC controller* in the following) when applied to the more general RG and EG cases.

---

[14]For example, in Sec. 6.2.9 we consider the *aerial throwing task* in which the cost function is the throwing distance.

[15]Notice that both $\mathbf{q}$ and $\mathbf{u}$ are the part of $\bar{\mathbf{x}}$ in both Case RC and EC. Here by limiting the states and the control inputs of the robots, we also limit $\bar{\mathbf{x}}$.

| Quantity | Symbol | Nom. Value/Range | Unit |
|---|---|---|---|
| PVTOL mass | $m_0$ | 1.00 | [kg] |
| motor mass | $m_m$ | 0.20 | [kg] |
| link mass | $m_1$ | 0.30 | [kg] |
| rotating motor mass | $m_r$ | 0.05 | [kg] |
| object mass | $m_o$ | 0.5 | [kg] |
| PVTOL inertia | $J_0$ | 0.028 | [kgm$^2$] |
| motor solid inertia | $J_{ms}$ | 0.0562e-06 | [kgm$^2$] |
| motor inertia | $J_m$ | 0.4101 | [kgm$^2$] |
| link inertia | $J_1$ | 0.004 | [kgm$^2$] |
| distance vector between $P_{C_0}$ & $P_M$ | $d_0$ | $-[8\ 8]^T \leftrightarrow [8\ 8]^T$ | [cm] |
| distance vector between $P_{C_1}$ & $P_M$ | $d_1$ | $[0\ 0.2]^T$ | [m] |
| distance vector between $P_{C_1}$ & $P_E$ | $d_e$ | $[0\ 0.2]^T$ | [m] |
| distance vector between $P_{C_0}$ & $P_G$ | $d_G$ | $[0.01\ 0.05]^T$ | [m] |
| motor shaft radius | $r_r$ | 0.015 | [m] |
| linear spring stiffness | $k_e$ | $3 \leftrightarrow 30$ | [Nm/rad] |
| motor gear ratio | $g_r$ | 270:1 | - |
| PVTOL thrust range | $T_t$ | $0.1 \leftrightarrow 28$ | [N] |
| PVTOL torque range | $T_r$ | $-3 \leftrightarrow 3$ | [Nm] |
| Motor torque range | $T_m$ | $-5 \leftrightarrow 5$ | [Nm] |
| grasping time | $t_g$ | 2.67 | [s] |
| impact duration | $T_i$ | 0.01 | [s] |

Table 6.3: Nominal parameters of the simulated systems in Sec. 6.2.9.

| Non-idealities | Notation | Value | Unit |
|---|---|---|---|
| deviation in masses | $\delta_m$ | 2 | [% ] |
| deviation in inertias | $\delta_i$ | 10 | [%] |
| deviation in $d_1$ | $\delta_1$ | $[0\ 0.01]^T$ | [m] |
| deviation in $d_G$ | $\delta_G$ | $[0\ 0.01]^T$ | [m] |
| deviation in spring constant $k_e$ | $\delta_s$ | 0,5,10 | [%] |
| 3-sigma Gaussian noise in pos. | $3\sigma_p$ | 0.01 | [m] |
| 3-sigma Gaussian noise in vel. | $3\sigma_v$ | 0.02 | [m/s] |
| 3-sigma Gaussian noise in $\theta_0$ | $3\sigma_0$ | 0.01 | [rad] |
| 3-sigma Gaussian noise in $\dot{\theta}_0$ | $3\sigma_{d0}$ | 0.02 | [rad/s] |
| 3-sigma Gaussian noise in $\theta_1, \theta_m, \theta_e$ | $3\sigma_1$ | 0.001 | [rad] |
| 3-sigma Gaussian noise in $\dot{\theta}_1, \dot{\theta}_m, \dot{\theta}_e$ | $3\sigma_{d1}$ | 0.002 | [rad/s] |

Table 6.4: Deviations from the nominal parameters and standard deviations of the noise used in the simulations. The controllers are not aware of the deviations and use instead the nominal values of Table 6.3.

The controller actions are computed using noisy measurements and nominal (i.e., wrong) values of the system parameters. The system dynamics is integrated using the real parameters values (i.e., nominal + deviations). A summary of the nominal values and the corresponding deviations, as well as of the noise characteristics, can be found in Tables 6.3 and 6.4. Nominal parameters, deviations, and noise are chosen very close to the values available on a real small-size aerial system equipped with standard sensors.

The system is simulated using an Ode8-solver at 1 [kHz] in Matlab Simulink. The noisy positions and velocity measurements are given to the controller at a rate of 30 [Hz], similarly to what a commercial camera+IMU setup would provide. The rate of the noisy orientations and the angular velocities is 500 [Hz], a realistic value for IMU attitude estimation and motor encoder readings.

In the dynamic models, the link attached to PVTOL is considered as a rod, whose inertia is computed using $J_1 = m_1 L^2/12$, where $L = ||\mathbf{d}_1 + \mathbf{d}_e||$. The motor inertia is computed as $J_m = g_r^2 J_{ms}$ where $g_r$ is its gear reduction ratio, and $J_{ms} = m_r r_r^2/2$ is calculated by considering motor as a rotating solid cylinder. The stiffness range of the elastic actuator is chosen similar to the one of *QBMove-VSA*. The physical limits of all the actuators are considered as hard thresholds and provided in Tables 6.3 and 6.5.

In the next plots, for *nominal values* we mean the system behavior in the ideal case, i.e., as if the controllers were fully aware of the real parameters (nominal + deviation) of the system dynamics and there was neiter noise nor under-samplings in the measurements. Notice that this is not the actual case, since there will be always some system or measurement uncertainties in reality. For this reason, we also consider the *actual values*, representing instead the system behavior when the controllers use the nominal parameters, and under-sampled and noisy measurements.

## Pole Placement Strategy

The feedback controllers used in the simulations have been explained in Sec. 6.2.5 and Sec. 6.2.7, in which we have analytically proven that the flat outputs are (in both cases) $\mathbf{y} = [\mathbf{p}_m^T \ \theta_{01}]^T$ (or, equivalently, $[\mathbf{p}_e^T \ \theta_{01}]^T$). Thanks to that results we can apply a nonlinear control loop to bring the system in the form (6.10). Then, given any 3-ple of desired trajectories of class $C^3$, $x_m^d(t)$, $z_m^d(t)$, $\theta_{01}^d(t)$ for $x_m$, $z_m$, and $\theta_{01}$, respectively, the following outer control loop is used

$$
\begin{aligned}
u_{x_m} &= \dddot{x}_m^d + K_{x1}e_x + K_{x2}\dot{e}_x + K_{x3}\ddot{e}_x + K_{x4}\dddot{e}_x \\
u_{z_m} &= \dddot{z}_m^d + K_{z1}e_z + K_{z2}\dot{e}_z + K_{z3}\ddot{e}_z + K_{z4}\dddot{e}_z \\
u_{\theta_{01}} &= \dddot{\theta}_{01}^d + K_{\theta1}e_\theta + K_{\theta2}\dot{e}_\theta + K_{\theta3}\ddot{e}_\theta + K_{\theta4}\dddot{e}_\theta
\end{aligned}
\tag{6.75}
$$

where $\boldsymbol{u}_v = [u_{x_m} \ u_{z_m} \ u_{\theta_{01}}]^T \in \mathbb{R}^3$ as in (6.10), $e_x = x_m^d - x_m$, $e_z = z_m^d - z_m$, $e_\theta = \theta_{01}^d - \theta_{01}$, and $K_{xi}, K_{zi}, K_{\theta i} \in \mathbb{R}_{>0}$, with $i = 1 \ldots 4$, are properly chosen gains. We know that this control law will exponentially steer the three outputs along the desired trajectory, because we have analytically proven that the decoupling matrix $\bar{\mathbf{G}}$ is invertible almost everywhere. To compensate the errors due to uncertainties, an integral term $K_{i*} \int_{t0}^{tf} e_* dt$ is added in the outer loop of each channel, where $* := \{x, z, \theta\}$ and $K_{i*} \in \mathbb{R}_{>0}$.

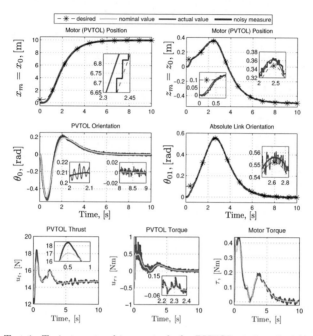

Figure 6.5: Test 1: Trajectory tracking control of a PVTOL equipped with a rigid joint-arm at its CoM (Case RC). A composite trajectory for each flat outputs has been designed and shown with a black-dashed curve. The PVTOL reaches up to 4.3 [m/s] in x-axis with very small tracking error. Two simulations are compared: the ideal case with nominal values, and the non-ideal case with deviations (biases), under-sampling, and noises. The nominal performances in the ideal case are depicted with a yellow solid curve and given only for $\theta_0$ and the control inputs, since the tracking of the flat outputs is in this case perfect. The red solid curves show the actual value of the configurations and inputs in non-ideal case. The purple solid curves depict the noisy, biased, and undersampled measurements used by the controller.

**Remark 11.** *Notice that, as in any dynamic feedback linearization control, the obtained control law is a function of only the measured state. In fact, the derivatives of the output are algebraic function of the state thanks to the system model. Furthermore, the derivative of the actual inputs are internal variables of the state extension. Therefore there is no need to perform any numerical derivation to implement such controller but only to measure the state of the original system, i.e., $(\mathbf{q}, \dot{\mathbf{q}})$.*

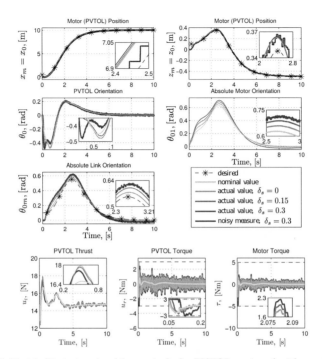

Figure 6.6: Test 1: Trajectory tracking control of a PVTOL equipped with an elastic joint-arm at its CoM (Case EC). The desired trajectory is the same of the simulation in Fig. 6.5 and depicted with black dashed curve. Noises in measurements and deviations in masses, inertias and lengths are included, and the results for three different stiffness deviations are compared: $\delta_s = 0$, $\delta_s = 0.15$ (5%) and $\delta_s = 0.3$ [Nm/rad] (10%). Red color is used for actual values, and its tone gets lighter while $\delta_s$ gets smaller. Purple solid curves represent noisy measurements (used by the controller) for $\delta_s = 0.3$ [Nm/rad]. Yellow solid curves are used for nominal values in the ideal case. Blue dashed lines represent the physical torque limits for the PVTOL and the motor.

## Test 1: Tracking a given Trajectory

### Arm is attached to the PVTOL CoM

Consider the model and the controller presented in Case RC. We compare two simulations: the ideal (nominal) case where there are no parametric deviations in the dynamic model and no noise in measurements; and the non-ideal (actual) case where deviations and noises in Table 6.4 are used. Results are given in Fig. 6.5.

At the beginning of the figures, the initial outputs are different from the desired ones. Nevertheless, the controller lets the system converge to the desired values after a short transient phase. In the non-ideal case, the system requires more control effort to keep the

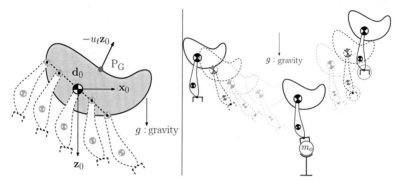

Figure 6.7: – *Left:* the joint is not attached exactly at the PVTOL CoM: $\mathbf{d}_0$ spans the values given in Table 6.3.
– *Right:* a sketch for grasping with PVTOL+arm system.

tracking error small, which, as expected, is never as good as in ideal (nominal) case but is fully acceptable. All the control inputs remain between their thresholds, as defined in Table 6.3. Notice that the PVTOL torque is noisier than the PVTOL thrust and the motor torque. The reason for that is because the control law given in (6.75) is a direct function of the noisy state measurements, and $u_r$ is a direct function of $\boldsymbol{u}_v$. On the other hand, $u_t$ and $\tau$ are smoother since they are computed after two integrations.

Consider the model and the controller presented in Case EC. The linear spring stiffness has been chosen $k_e = 3$ [Nm]. Consider all the non-idealities given in Table 6.4. We aim at comparing the system results for different deviations in $k_e$, namely $\delta_s = 0$, $\delta_s = 0.15$, and $\delta_s = 0.3$ [Nm/rad] (up to 10 %). Results are given in Fig. 6.6. After a transient phase, the system converges to the desired trajectory, similar to the rigid case in Fig. 6.5. The tracking performance degrades nicely for increasing $\delta_s$, for every configuration presented. Moreover, the absolute motor orientation $\theta_{0m}$ increases as well with $\delta_s$, which generates more control effort in torques, as expected.

For small $\delta_s$ the tracking errors are very similar to the rigid case presented in Fig. (6.5). However for a PVTOL with elastic-joint arm, the control efforts in both the PVTOL and the motor torques are much higher than for the rigid controller. Also notice that both PVTOL and motor torques are quite noisy compared to the PVTOL thrust. The reason for that is because $u_r$ and $\tau$ are direct functions of the noisy state-dependent $\boldsymbol{u}_v$ given in (6.75), while $u_t$ is computed after a double integration. Hence we can conclude that the rigid-joint arm is more suitable for trajectory tracking tasks than elastic-joint arm. This can be explained with the need for the controller to 'fight' against the natural tendency of the spring to oscillate at his natural frequency.

### Arm is *not* attached to the PVTOL CoM

The following set of tests validates the capabilities of the proposed controller of tracking a composite trajectory for the desired flat outputs $x_m^d$, $z_m^d$ and $\theta_{01}^d$ in the non-ideal conditions. The plots of the results are shown for the rigid case in Fig. 6.8 and for the elastic case in Fig. 6.9. Notice that in addition to the non-ideal conditions mentioned above, we tested

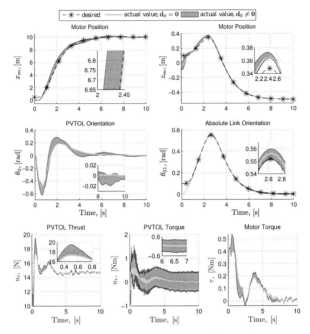

Figure 6.8: Test 1: Tracking of a given trajectory of PVTOL+rigid-joint case in presence of several non-ideal conditions: noise, parameter uncertainty, under-sampling, and attachment point of the joint different from the CoM (see Tables 6.3 and 6.4). Several simulations are run: the maximum and minimum values among the different simulations are plotted with red solid curves, and pink-filled in between.

the two controllers in the case that the arm joint is not perfectly attached to the PVTOL CoM, i.e., $\|\mathbf{d}_0\|$ is not exactly zero (see Fig. 6.7-Left). When doing so, an unstable behavior might appear if $\|\mathbf{d}_0\|$ is too large. However as long as $\|\mathbf{d}_0\|$ is kept in a reasonable bound the behavior remains stable, as illustrated in Hauser et al. (1992).

The main considerations are that: *i)* the controllers do not need a perfect knowledge of the model parameters since the performances degrade smoothly and nicely with the increase of the parameter uncertainty; *ii)* the controllers work well with the typical noise, sampling and quantization that are presents in real systems; *iii)* the control effort in the case of the elastic-joint arm is larger with respect to the rigid-joint case. This happens because the controller needs to suppress the tendency of the spring to oscillate at its natural frequency when steering the system along the desired trajectory.

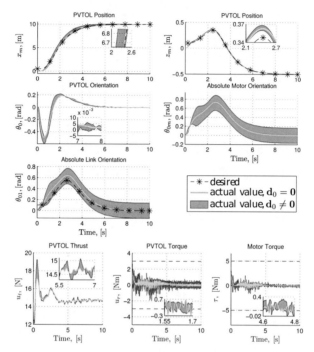

Figure 6.9: Test 1: Tracking of a given trajectory of PVTOL+elastic-joint case in presence of several non-ideal conditions: noise, parameter uncertainty, under-sampling, and attachment point of the joint different from the CoM (see Tables 6.3 and 6.4). Several simulations are run: the maximum and minimum values among the different simulations are plotted with red solid curves, and pink-filled in between.

## Control of Joint Arm with Small Elasticity and Friction

Consider a flying robot equipped with a joint-arm. Now assume a case, where there exist a high-stiffness (small elasticity) and friction between the motor and the link of the arm. In other words the connection is almost rigid but some elasticity is still present. The stiffness and the friction for such case can be modeled using, e.g., existing mechanical engineering techniques as explained in Bathe (1996). Hence, we can use the dynamic model given in Section 6.2.7 (Case EC), with a high value of $k_e$ if the elasticity is linear. For this model, consider a linear friction as well, which generates on the link side a torque $f_{ld} = k_f(\dot{\theta}_{0m} - \dot{\theta}_{01})$, and, on the motor side[16], $f_{md} = k_f(\dot{\theta}_{01} - \dot{\theta}_{0m})$, where $k_f \in \mathbb{R} > 0$.

We compare the results for such model using either the rigid or the elastic controller.

---

[16]Note that $\dot{\theta}_{01} - \dot{\theta}_{0m} = \dot{\theta}_1 - \dot{\theta}_m = \dot{\theta}_e$ is the difference between link and motor velocities, e.g. the signed time derivative of the elastic deflection $\theta_e$.

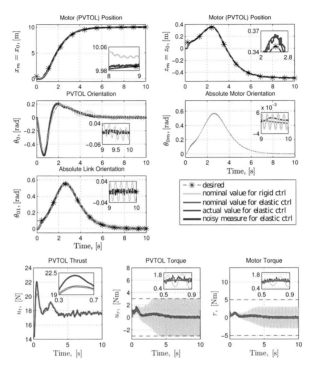

Figure 6.10: Test 1: Trajectory tracking control of a PVTOL equipped with a joint-arm (Case EC) with small elasticity (high stiffness). Same trajectory has been used as before and depicted with black dashed curve. This time, a mass of $m_e = 0.3$ kg is attached to the end-effector. Noises in measurements and deviations in other parameters are included. Results of the elastic controller (of Case EC) and the rigid controller (of Case RC) are compared. Yellow solid curve is for the nominal values of the rigid controller. Red solid curve is used for actual values of the elastic controller, while pink one is for the nominal values. Purple solid curve depicts the noisy, biased and undersampled measurements used by the elastic controller. Blue dashed horizontal lines are for the physical torque limits of the PVTOL and the motor.

Notice that former one neglects the (high) stiffness and friction, while latter one is only unaware of the frictions. We set $k_e = 50$ [Nm/rad] which is much more than the maximum value of the range given in Table 6.3. In addition we set $k_f = 0.01$ [Nms/rad]. Moreover, we connect a mass $m_e = 0.3$ [kg] at the end-effector of the arm, which simulates an aerial transportation task, and increases the challenge of the tracking task. The results are given in Fig. 6.10, where for the rigid controller only the ideal case is considered. It is clear that even in ideal case (with no noises and deviations) and although a truly stiff spring is chosen, the rigid controller is tracking $x_m$ and $\theta_{01}$ worse than the elastic controller (which is tested in non-ideal case unlike the rigid one). Furthermore the rigid controller causes a great effort in the PVTOL and motor torques, which are saturated due to physical motor limits. This is because rigid controller cannot anticipate the motion of the arm in advance. Since rigid controller is unaware of the stiffness, we see steady-state oscillations in $\theta_1$, $\theta_{0m}$ and $\theta_{01}$. Notice that for much greater $k_e$ and $k_f$ values, rigid controller works almost as good as the elastic one.

### Test 2:  Grasping an Object while Flying

In this test we first describe the scenario of grasping a stationary object using both the PVTOL+rigid-joint arm and the PVTOL+elastic-joint arm. A sketch depicting such task is given in the right side of Fig 6.7. The grasped object mass is $m_o > 0$. At time $t_g$ (grasping time instant) the dynamic model of the simulated robot is updated according to the grasping action and a disturbing impact force is also simulated based on the difference between the end-effector and the stationary mass velocities. Consider that at time $t_g$ (grasping time instant) mass $m_o$ is attached to point $P_E$, which is the end-effector. The effect of the grasped mass to the system is accurately and dynamically modeled as in the following:

- $t \geq t_g$ denotes all the times after the object is grasped. It is going to be used as an indicator of the parametric updates in the following.

- Mass of the arm is updated to $m_{1_{t \geq t_g}} = m_1 + m_o$,

- Distances $\mathbf{d}_2$ and $\mathbf{d}_e$ are updated using the formula

$$\mathbf{d}_{1_{t \geq t_g}} = \mathbf{d}_1 + \mathbf{d}_\epsilon$$
$$\mathbf{d}_{e_{t \geq t_g}} = \mathbf{d}_e - \mathbf{d}_\epsilon,$$

  where $\mathbf{d}_\epsilon = \frac{m_o}{m_{1_{t \geq t_g}}} \mathbf{d}_e$,

- Using parallel axis theorem, link inertia is updated to

$$J_{1_{t \geq t_g}} = \frac{m_{1_{t \geq t_g}}}{12} L^2 + m_{1_{t \geq t_g}} ||\mathbf{d}_\epsilon||^2,$$

  where $L = ||\mathbf{d}_1 + \mathbf{d}_e||$ as before.

Moreover, due to differences between the end-effector and the stationary mass velocities, an impact will occur at the moment of grasping. The external force to the system will be then

$$\mathbf{f}_{ext} = \mathbf{J}^T \mathbf{f}_{imp}, \qquad (6.76)$$

where

$$\mathbf{f}_{imp} = -m_o \frac{\dot{\mathbf{p}}_e - \dot{\mathbf{p}}_o}{T_i}$$

with end-effector velocity $\dot{\mathbf{p}}_e$ is computed as

$$\mathbf{p}_e = \dot{\mathbf{p}}_m \bar{\mathbf{R}}_{01}(\mathbf{d}_1 + \mathbf{d}_e)\dot{\theta}_{01},$$

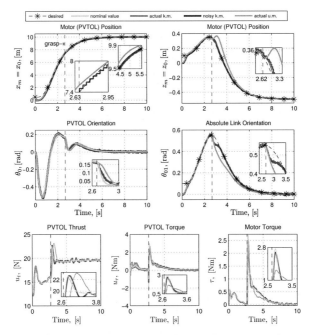

Figure 6.11: Test 2: Aerial grasping with a PVTOL+rigid-joint arm. The grasping instant is shown with a vertical blue dashed line. The nominal values are given with yellow solid curves. In the case of known grasped mass (k.m.) the actual values and noisy measurements are presented with red and purple solid curves, respectively. The pink solid curve shows the values in the case that the grasped mass is unknown (u.m.).

and stationary object velocity is $\dot{\mathbf{p}}_o = \mathbf{0}$. $T_i$ is impact duration and given in Table 6.3. The Jacobian matrix $\mathbf{J}$ is different for rigid-joint arm and elastic-joint arm cases. For former it is

$$\mathbf{J} = \begin{bmatrix} 1 & 0 & 0 & \cos(\theta_{01})(d_{1z} + d_{ez}) - \sin(\theta_{01})(d_{1x} + d_{ex}) \\ 0 & 1 & 0 & -\cos(\theta_{01})(d_{1x} + d_{ex}) - \sin(\theta_{01})(d_{1z} + d_{ez}) \end{bmatrix},$$

and for latter it is

$$\mathbf{J} = \begin{bmatrix} 1 & 0 & 0 & \cos(\theta_{01})(d_{1z} + d_{ez}) - \sin(\theta_{01})(d_{1x} + d_{ex}) & 0 \\ 0 & 1 & 0 & -\cos(\theta_{01})(d_{1x} + d_{ex}) - \sin(\theta_{01})(d_{1z} + d_{ez}) & 0 \end{bmatrix}.$$

Hence, by changing the mass, inertia and the distances of the link as explained above, and applying (6.76) as the external force to the system, we model the aerial grasping scenario.

**Grasping with Rigid-joint Arm**

Consider the model and the controller presented in Section 6.2.5, i.e., Case RC. The composite trajectory used in the previous simulations is suitable for an aerial grasping task, at which an object with $m_o = 0.5$ [kg] is to be grasped by the end-effector at time instant $t_g = 2.67$ [s]. At this second, the joint arm is at its maximum orientation from the initial condition, at a high velocity in $+\mathbf{x}_W$ direction, and at the beginning of its raising up again along the $-\mathbf{z}_W$ axis ($+\mathbf{z}_W$ is facing down because of the NED frame). Results are given in Fig. 6.11. Two cases are compared: known (k.m.) vs unknown (u.m.) grasped mass. After $t_g$, deviations from the desired trajectories are clearly seen for both cases. If the grasped mass is unknown, such deviation is higher for all the flat outputs. In the nominal case, the controller is fully aware of the end-effector velocity and mass, hence it generates high peaks in torques to counterbalance the impact. For the actual cases, controller is aware of the model with some deviations, hence it produces less reaction to the impacts compared to the nominal case, which results as worse tracking performance. Of course, the controller is even less reactive when it does not know about the grasped mass, leading decreasing tracking performance. However, thanks to the robustness of the controller provided by the integral terms of its linear part in (6.75), the outputs always nicely converge to the desired trajectories.

**Grasping with Elastic-joint Arm**

Consider the model presented in Section 6.2.7, i.e., Case EC. The same desired trajectory is used as in Fig. 6.11, where an object with $m_o = 0.5$ [kg] has to be grasped by the end-effector at time instant $t_g = 2.67$[s. Two cases are compared: grasping with low stiffness spring, $k_e = 8$ [Nm] and with high stiffness spring, $k_e = 30$ [Nm]. The results are given in Fig. 6.12. For both the low and the high stiffness cases, the tracking performance of the flat outputs are very close to each other. Moreover it is very similar to the results given in Fig. 6.11, with a clear difference in the absolute link orientation $\theta_{01}$. However, the control effort is much more for Case EC than for Case RC (especially notice the high peaks in the torques for Case EC, which reduces with the increasing stiffness). In fact, for case EC we could not simulate a stable (and reasonable) aerial grasping task when the PVTOL and motor torques were subject to hard saturations. We observed that using very high stiffness joint mitigates this effect and results beneficial for the aerial grasping task.

This is actually a natural result of the system model, since when a elastic-joint arm is exposed to an impact at the grasping moment, it changes the orientation of the arm instantaneously, while the actuation reacts to this change after a second order system dynamics through a spring element. This means, that the controller must generate intensive amount of torques for counter-balancing this instantaneous changes. However, this might not be always feasible, since there are hard control limitations in real implementations. On the other hand, when the grasping is performed by a rigid-joint arm, the impact of the grasping can be accounted by the actuators immediately. If the necessary control inputs are within the physical bounds of the actuators (as in Fig. 6.11), the system can perform a flawless aerial grasping task.

Therefore, one can conclude this simulation set saying that for aerial grasping task and for tracking a generic trajectory Case RC is more advantageous than Case EC in terms of control effort.

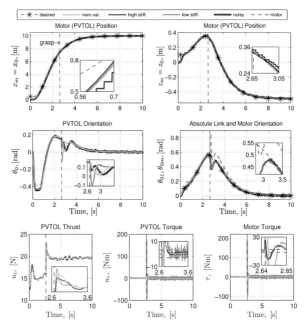

Figure 6.12: Test 2: Aerial grasping with a PVTOL+elastic-joint arm. The grasping instant is shown with a vertical blue dashed line. The nominal values are given with yellow solid curves. In the case in which a high stiffness elastic element is used, the actual values and noisy measurements are presented with red and purple solid curves, respectively. The pink solid curve instead represents the actual values when a low stiffness elastic element is used. A pink dashed curve is used only for actual motor values in case of low stiffness in the second plot on the right column. The horizontal blue dashed lines stand for the physical limits of the motor and PVTOL torques.

### Test 3: High-Speed Swinging: Link velocity amplification

Here we present a case where an elastic-joint arm is more beneficial than a rigid-joint arm attached to the PVTOL. We consider a scenario where the link is swinging at high velocities, which preferably can be used for tasks such as hammering on a surface or throwing an object to far distances. The benefits of amplifying link velocity for ground robots were shown before in the literature (e.g. in Braun et al. (2012) and in Braun et al. (2013)). Actually we have partially studied such designs with an aerial robot in Chapter 5, where a light-weight elastic-joint arm was developed and its link velocity was amplified w.r.t. the motor velocity, and experimental results on board of a flying quadrotor have been provided. However, the controller for the flying robot presented there was not considering the system as a whole but rather as a flying system perturbed by the oscillations of the hanging arm.

In this section, we perform a similar link velocity amplification task, but using the

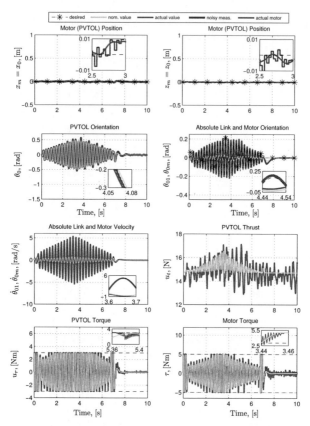

Figure 6.13: Test 3: Link velocity amplification for Case EC. The desired trajectory profile is different from the previous cases, but depicted again with a black-dashed curve. The yellow solid curve stands for the nominal values, and red solid curve for actual values. The noisy measurements of the actual values are given with a purple solid curve. The blue solid curve presents the motor values $(\theta_{0m}, \dot{\theta}_{0m})$, which are shown with the link values $(\theta_{01}, \dot{\theta}_{01})$ in the same subplots. The horizontal blue dashed lines show the physical limits of the PVTOL and motor torques.

controller presented in Sec. 6.2.7 (Case EC). We choose $k_e = 8$ [Nm/rad] for the simulation. The psudo-natural frequency of the system is identified by setting gravity for PVTOL to zero, letting the arm oscillate in free evolution from an initial condition of 60 [deg] and observing its behavior. We found that for the nominal values, the arm swings with period of $T = 0.255$ [s]. We then used this value to generate a desired trajectory for $\theta_{01}$, and set the desired values for the PVTOL positions at zero. Notice that the desired trajectory for $\theta_{01}$ is chosen with an increasing amplitude until some point, and it gradually decreases to

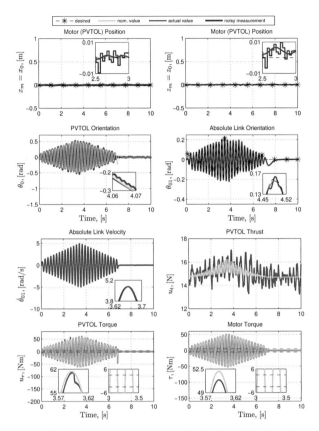

Figure 6.14: Test 3: Oscillating at high link velocity for Case RC. The desired trajectory profile is as in Fig. 6.13, and depicted with a black-dashed curve. The yellow solid curve stands for the nominal values, and the red solid curve for the actual values. The noisy measurements of the actual values are given with a purple solid curve. The horizontal blue dashed lines show the physical limits of PVTOL and motor torques.

zero.

The results are reported in Fig. 6.13, showing a good position tracking performance, with less than 1 [cm] of maximum error. Link and motor velocities are given in third subfigure of the first column, where the link velocity is amplified of more than five times w.r.t. the motor velocity. Notice that the PVTOL and the motor torques are saturated in order to simulate the physical limits, which is the reason of the small tracking errors.

Now, consider the PVOL+rigid-joint arm model, given in Sec. 6.2.5 (Case RC). We asked robot to perform a similar task for reaching a high link velocity. The results are reported in Fig. 6.14. In our simulations we realized that a good tracking performance can be achieved

| Quantities | Notation | Value | Unit |
|---|---|---|---|
| min/max limits of $x_m$ | $\mathbf{q}_m(1) \leftrightarrow \mathbf{q}_M(1)$ | $-0.5 \leftrightarrow 0.5$ | [m] |
| min/max limits of $z_m$ | $\mathbf{q}_m(2) \leftrightarrow \mathbf{q}_M(2)$ | $-0.5$ | [m] |
| min/max limits of $\theta_0$ | $\mathbf{q}_m(3) \leftrightarrow \mathbf{q}_M(3)$ | $-45 \leftrightarrow 45$ | [deg] |
| min/max limits of $\theta_{01}$ | $\mathbf{q}_m(4) \leftrightarrow \mathbf{q}_M(4)$ | $-45 \leftrightarrow 45$ | [deg] |
| min/max limits of $\phi$ | $\phi_m \leftrightarrow \phi_M$ | $-20 \leftrightarrow 20$ | [deg] |
| min. limit of $u_t, u_r, \tau$ | $\mathbf{u}_m$ | $[0.1 \ -1.75 \ -1.5]^T$ | [N] |
| max. limit of $u_t, u_r, \tau$ | $\mathbf{u}_M$ | $[28 \ 1.75 \ 1.5]^T$ | [N] |
| desired $z_m$ | $z_m^*$ | $-1$ | [m] |
| initial values Case RC | $\bar{\mathbf{x}}_0(1) \leftrightarrow \bar{\mathbf{x}}_0(12)$ | $[0 \ -1 \ \mathbf{0}_{1\times6} \ -m_sg \ \mathbf{0}_{1\times3}]^T$ | - |
| initial values Case EC | $\bar{\mathbf{x}}_0(1) \leftrightarrow \bar{\mathbf{x}}_0(12)$ | $[0 \ -1 \ \mathbf{0}_{1\times8} \ -m_sg \ 0]^T$ | - |

Table 6.5: Physical limits used in the aerial throwing tasks (Test:4).

only if the *violation* of the PVTOL and motor torque limits are allowed, and, in fact, when unconstrained, they reach very high values (more than 10 times the limits). In this case, note that there is no amplification of the link velocity w.r.t. the motor velocity; they are the same and very high velocities are achieved in cost of extreme control efforts. In fact, in our simulations, saturating the torques to their physical limits for Case RC has always ended up with an unstable behavior for tracking such high-speed trajectory.

Despite the hard physical limits applied for the Case EC, the requested link velocity is achieved and much less control effort has been used when compared to Case RC. The reason for this is the ability of storing energy in the elastic components. This implies that Case EC (aerial robot with compliant actuator) has more advantages than Case RC (aerial robot with rigid actuator) for the tasks that require high link speed, which can be used, e.g., for throwing or hammering.

**Test 4: Throwing an Object while Flying**

In this test we consider the *aerial throwing* task, in which an object is thrown from the end-effector of the aerial manipulator while the robot is flying. Such task is sketched in Fig. 6.15. Notice that aerial throwing problem is quite different from ground base robots throwing (see Braun et al. (2013)), because in this case, the base of the robot is flying and it needs to compensate the dynamical effects while performing such task. A real scenario of aerial throwing task can be imagined as a situation, where the aerial manipulator is assigned to deliver a package, e.g. a first aid kit, in an hazardous environment, where the arrival point of the package is not suitable for the robot.

We define the following cost function

$$J = -J_d + \int_{t_0}^{t_L} (J_z + J_\tau)dT \quad \text{where}$$

$$J_d = d^2, \quad J_z = (z_m - z_m^*)^2, \quad J_\tau = \tau^2,$$

(6.77)

with upper-script (*) stands for the desired value and $d$ is the thrown distance of the object,

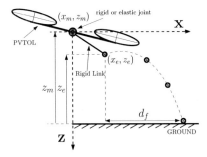

Figure 6.15: Sketch of aerial throwing task using a PVTOL aerial manipulator. The ballistic
trajectory of the thrown object is shown with a dashed curve, and the distance
taken by the object after leaving the aerial manipulator is shown with $d_f$.

computed using the ballistic equation of the flying object, similar to Braun et al. (2013):

$$d(\mathbf{y}, \dot{\mathbf{y}}) = x_e(\mathbf{y}) + \underbrace{\dot{x}_e(\mathbf{y}, \dot{\mathbf{y}}) t_f(\mathbf{y}, \dot{\mathbf{y}})}_{d_f(\mathbf{y}, \dot{\mathbf{y}})}$$

$$t_f(\mathbf{y}, \dot{\mathbf{y}}) = \frac{1}{g}\left(\dot{z}_e + \sqrt{\dot{z}_e^2 + 2g(z_g - z_e)}\right),$$

(6.78)

where recall that $\mathbf{p}_e = [x_e \; z_e]^T$ is the end-effector position and $\dot{\mathbf{p}}_e = [\dot{x}_e \; \dot{z}_e]^T$ is its velocity,
where both can be computed using (6.25)[17]. The height of the ground is $z_g$ in the world
frame, which is the altitude at which the object hits the ground. The total flight time
of the thrown object is $t_f$, and the distance taken by the object after leaving the aerial
manipulator is depicted with $d_f$. The cost function also includes the term $J_z$ for keeping
the aerial robot around its hovering height, and $J_\tau$ for minimizing the actuation costs.

Note that besides the additional saturations on the control inputs, the cost to be
minimized is a direct function of the system's flat outputs. This means, we can directly plan
for the outputs, which we can also exactly control. This demanding example highlights the
importance of the end-effector positions and their being differentially (and exact linearizing)
outputs of the system[18].

Then, by substituting (6.77) in the optimization problem described in (6.74), we compute
the desired trajectories for the aerial manipulators described as Case RC and Case EC,
for achieving aerial throwing task while respecting the system input and state boundaries.
For solving this optimization problem, we used ACADO numerical optimizer presented
in Houska et al. (2011). The parameters for the simulation and the optimization problem
are given in Table 6.5.

The results are given in Fig. 6.16. The optimal trajectories computed using Houska et al.
(2011) are clearly enabling the aerial manipulator to throw the object to a far distance, at

---

[17]Notice that the throwing distance $d$ is a sole function of the flat outputs, i.e. end-effector positions and
velocities. This example directly benefits from the differential flatness property of the end-effectors in
the planning phase.

[18]Unlike controlling the CoM positions of the overall aerial manipulator, e.g. in Acosta et al. (2014),
which are also flat outputs of the system; we directly control the end-effector positions, which are more
practical and reasonable quantities for aerial manipulation tasks.

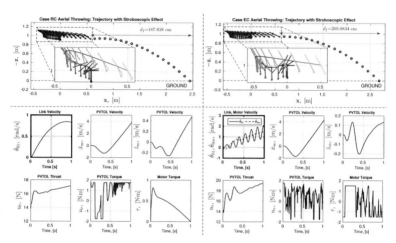

Figure 6.16: Test 4: Throwing object with the aerial manipulators in Case RC (left) and Case EC (right). Throwing is performed at $t_L = 1$ [s].

    – *Left:* Results for Case RC. The first plot on the top shows the trajectory of the aerial robot and the ballistic trajectory of the thrown object using stroboscopic effect. The trajectory of the aerial manipulator is separately emphasized in a zoomed subfigure. The thrown object hits the ground about 188 [cm] away from the end-effector. The link angular velocity, PVTOL linear velocities $(\dot{x}_m, \dot{z}_m)$ and the control inputs are plotted below.

    – *Right:* Results of Case EC. This time, the motor and the rigid link angular velocities are plotted together in the first figure of the second row, where dashed curve depicts the motor velocity and the solid one represents the rigid link velocity. Notice the link velocity amplification w.r.t. the motor velocity using the potential energy stored in the elastic-joint arm. The thrown object hits the ground at about 209 [cm] away from the end-effector.

exactly $t_L = 1$ [s], while keeping the PVTOL at the desired altitude. The results for Case RC (left of Fig. 6.16) and Case EC (right of Fig. 6.16) show that in both cases, the aerial robots are accelerating first backwards and then forwards along the $\mathbf{x}_W$-axis to reach high linear velocities (due to the limits on $x_m$, see Table 6.5). However notice that in Case EC, the aerial manipulator uses the potential energy stored in the elastic-joint for amplifying the link velocity. At the end, the aerial manipulator in Case EC achieves a higher throwing distance than the one of Case RC, by performing an explosive movement. This result is in the line with the high-speed swinging tests via link velocity amplification, presented previously in Fig. 6.13.

We notice that due to the term $J_z$ in (6.77), the aerial manipulator tries to keep itself in the hover condition, which ensures that the system performs a stable flight. However this also limits the system to achieve a better throwing performance. Exploring different definitions of the cost function for aerial throwing is in the scope of our future studies.

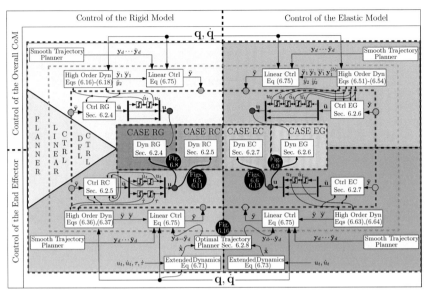

Figure 6.17: Sketch of the trajectory planners, controllers and considered system dynamics presented and used in Sec. 6.2.9 (numerical tests). The following color code is used: upper left yellow block shows the planning/control of Case RG, lower left light blue block shows the same for Case RC, upper right pink block shows the same for Case EG and lower right turquoise block shows the same for Case EC of Section 6.2. Notice that this grid convention matches with Table 6.1. As the triangle on the left indicates; most upper level is dedicated to the trajectory planners (blue-dashed box). Here we use either smooth trajectory planners based on smooth filters, or optimal trajectory planner as described in Sec. 6.2.8. One lower level (covered with orange-dashed box) is where we compute the high order dynamics *analytically*, and implement the linear controllers as explained in Section 6.2.9 (*pole placement*). One more lower level (covered with green-dashed box), is for the controllers based on *Dynamic Feedback Linearization* (DFL) presented in Sec. 6.2. Finally the lowest level (covered with red-dashed box and filled with gray color) represents the four different system models we summarized in Table 6.1. Notice that each components of these blocks are properly cited to their corresponding chapters or equations. Moreover, the dark circles indicates some figures of Sec. 6.2.9, citing numerical examples of the controllers that are placed next to. We note that the overall planning and control of these four models require only the measured system states $(\mathbf{q}, \dot{\mathbf{q}})$. Their numerical derivation is not needed, since we can compute everything else analytically (see also Remark 11).

## 6.2.10 Preliminary Experiments

In this section, we describe our experimental setup, and results of some preliminary flight tests. But before, we note that we tested our controller in a physical simulation considering the full dynamical 3D model of the system, using the CAD model of the experimental setup in SimMechanics, which is known to be a realistic physical simulation toolbox provided by Matlab. The system in 3D consist of a quadrotor equipped with a *Variable Stiffness Actuator* (VSA), as detailed in *QBMove-VSA*. This VSA is also connected to a rigid arm (see Fig. 6.18). Briefly recall that a VSA is indeed a compliant actuator, which has also capacity to change the stiffness parameters of its elastic components. In our experiments, we used VSA with a constant stiffness parameter. It is clear that we have a some type of an elastic-joint arm, where the arm is *not* placed at the CoM of PVTOL ($d_0 \neq 0$ as in Case EG, in Sec. 6.2.6). However, the control we applied is based on the assumption, $d_0 = 0$, as described in Case EC, in Sec. 6.2.7. This resembles the numerical tests shown in Fig. 6.9, where in this case we implement the controller in a real setup.

We first split the 3D model into two planes, Plane-A and Plane-B as shown in Fig. 6.18a. All the motion on Plane-A (including that of the absolute link angle) is controlled using the exact linearizing controller presented in Section 6.2.7 (via thrust, torque around $x_0$ and torque for the VSA). See the lower-left light blue-colored block of Fig. 6.17 for the principle of the controller.

The rest of the quadrotor motion (motion in Plane-B and rotation around the vertical axis $z_0$) is controlled using a *near-hovering controller*, which is explained in Lee et al. (2013). This allows us to test the performance of the controller presented in Sec.6.2.7 in a real experimental scenario. The controllers are tested first with the CAD model of the real setup in SimMechanics, and then later using the experimental setup.

### Preparation of the Qbmove VSA

In our experimental setup we chose to use a variable stiffness actuator, for their wide range of stiffness preset capabilities. This allows the user choosing between high and low stiffness values, depending on the task of the robot[19]. We used Qbmove VSA, an agonistic/antagonistic servo-VSA, whose specifications are available in *QBMove-VSA*. Shortly, it consists of two *PD* controlled servo motors, which allow to regulate independently desired stiffness and output-shaft equilibrium, i.e. in our notations $k_e$ and $\theta_m$, respectively. This VSA provides state measurements at 500 [Hz].

In order for our controller to work with Qbmove VSA, several extra steps need to be taken. First of all, a parametric identification of the QBmove VSA + rigid arm system has been performed, in order to retrieve the parameters of the equivalent motor which was depicted in Fig 6.4. The stiffness (and the damping) parameters of the VSA+arm system are identified by first assuming it as a simple mass-damper system, and then letting the arm swing from an initial condition, without any control action (see Yüksel et al. (2015) for similar a method). Since the VSA features a nonlinear spring, we consider a linear spring for deflection in the range of $\pm 20$ [deg] (this is in line with the range given in Table 6.5). Inertial parameters of the system are found using the system geometry. All the identified and computed parameters are available in Table 6.6.

---

[19]Recall that we had discussed the differences between rigid an elastic joint arms in Sec. 6.2.9, which led us to consider the variable stiffness actuators for the future studies.

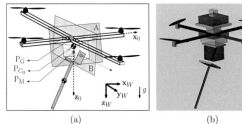
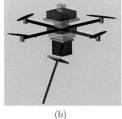
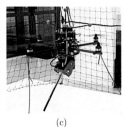

(a)                       (b)                      (c)

Figure 6.18: Evolution from theory to application. *a)* Conceptual sketch of the model in 3D. Motion in Plane-A is controlled using the controller presented in Sec.6.2.7. Motion in Plane-B and the rotation around $z_0$ are controlled using a *near-hovering controller.* *b)* CAD model of the 3D system and a snapshot from *SimMechanics* simulation, where the implemented controllers are tested. Different colors correspond to the different parts of the real system. *c)* Real system on flight. On board of the VTOL, from top to bottom, there are MoCap markers, an Odroid-XU computer, four brushless motor controllers with their power board, a flight controller (incl. IMU), battery pack, Qbmove VSA with its connectors, a rigid arm attached to it. Red ropes are used only for safety reasons, with no tension on them.

Moreover, the control framework we presented requires a torque-controlled motor, while the Qbmove VSA is not allowing this control modality. For this reason we have implemented an outer loop controller around the VSA, which translates the desired torque into a desired position using the estimated parameters and second order system model. This simple *bridge* between the proposed controller (in Sec. 6.2.7) and the Qbmove VSA is directly implemented as a ROS node. The implementation of the outer loop is straight-forward; using the identified parameters (Table 6.6) we implemented the second order mass-spring-damper model (similar to (5.1)), and then invert it for the desired motor accelerations. After integrating it twice, we get both desired motor velocities and the positions to be sent to the VSA. Note that, since the identified parameters are done for a specific range of the states, its performance will be always limited. A better way is to use torque controlled motors, which however in our case were not immediately available.

**Quadrotor Setup**

The experiments are conducted on a flying robot, which is a quadrotor VTOL (see Fig. 2.2 and Fig, 6.18c). The payload of the VTOL is composed of, from top to bottom, MoCap markers, an Odroid-XU[20] computer running Ubuntu 14.04, four brushless motor controllers with their power board, flight controller (incl. IMU), a battery pack, a Qbmove VSA with its connectors, and a rigid arm attached to it (see Fig. 6.18b for the colored items in this order, and Fig. 6.18c for the real setup). Total weight of the system is almost 1.5 [kg] (including safety ropes that are carried by the VTOL), which corresponds to a total hovering thrust of about 14.75 [N]. Each propeller of the VTOL used in these experiments

---

[20]http://www.hardkernel.com/main/products/prdt_info.php?g_code=G137510300620

| Real Parameters | Notation | Value | Unit |
|---|---|---|---|
| mass of the quadrotor | $\tilde{m}_0$ | 1.309 | [kg] |
| mass of the VSA | $\tilde{m}_m$ | 0.06 | [kg] |
| mass of the arm | $\tilde{m}_1$ | 0.098 | [kg] |
| dis. vec. betw. $P_{C_0}$ & $P_G$ | $\tilde{d}_G$ | $[0.0\ 0.0081]^T$ | [m] |
| dis. vec. betw. $P_{C_1}$ & $P_M$ | $\tilde{d}_2$ | $[0\ 0.0979]^T$ | [m] |
| inertia of the PVTOL | $\tilde{J}_0$ | 0.0154 | [kgm$^2$] |
| motor inertia | $\tilde{J}_m$ | 0.4101 | [kgm$^2$] |
| link inertia | $\tilde{J}_1$ | 0.0011 | [kgm$^2$] |
| spring stiffness | $\tilde{k}_e$ | 3.55 | [Nm/rad] |
| spring damping | $\tilde{k}_f$ | 0.07 | [Nms/rad] |

Table 6.6: Measured, computed or identified parameters of the Quadrotor+VSA arm setup. The variable $\tilde{*}$ denotes the quantity $*$ for the experimental setup. Notice that $\tilde{k}_f$ is identified but not used in the controller.

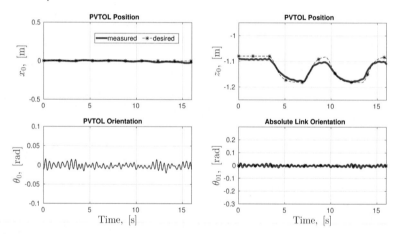

Figure 6.19: First test on controlling the Quadrotor + VSA arm motion along the $\mathbf{z}_W$ direction. A step-like trajectory is followed along the $\mathbf{z}_W$-axis. Notice that negative $\mathbf{z}_W$ is upwards.

can generate lift up to 7 [N], which allows carrying the described payload, and performing the flight.

**Preliminary Experiment of a Quadrotor with a VSA Arm**

Let us now present the results of the first experiments of a Quadrotor VTOL equipped with a Qbmove VSA. Here, we first test the complete system for a trajectory tracking along the $\mathbf{z}_W$ axis while keeping its motion along $\mathbf{x}$ at zero. Results are given in Fig. 6.19, where the

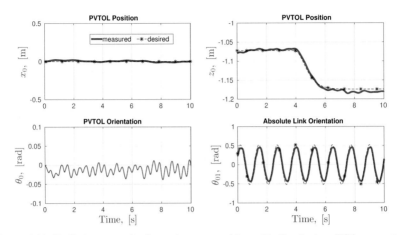

Figure 6.20: Preliminary results for trajectory tracking with Quadrotor+VSA arm setup. The arm attached to the Qbmove VSA is swinging back and forth (see $\theta_{01}$), while Quadrotor VTOL is tracking a stable trajectory along $\mathbf{x}_W$-axis, and a step-like trajectory along the $\mathbf{z}_W$-axis. Oscillations on $\theta_0$ are due to the motion of the arm, against which controller is trying to keep $\mathbf{x}_W$ position constant. Notice that negative $\mathbf{z}_W$ is upwards.

maximum error for both $x_0$ and $z_0$ is around 2 [cm]. In this experiment no desired motion is induced to the VSA arm.

A second experiment is performed, in which the absolute link orientation is following a sinusoidal trajectory, while PVTOL CoM is following another trajectory along the $\mathbf{z}_W$-axis and trying to stay at zero of the $\mathbf{x}_W$-axis. Results are given in Fig. 6.20, where for both $x_0$ and $z_0$ the maximum errors are around 2 [cm].

Notice that for both experiments, steady-state errors are observed, which are due to the unmodeled system dynamics. While choosing better control gains would improve the performance of the tracking, there will always be some tracking errors due to the assumptions used in the controller, e.g. neglecting the damping parameter of the spring. Moreover, there is an obvious difference between the point $P_{C_0}$ and $P_M$ in the real setup (see Fig. 6.18a), while in theory we considered them *coinciding* (see Fig. 6.2).

## 6.2.11 Discussions

In Section 6.2 we studied specific types of aerial manipulators; a PVTOL equipped with either rigid or elastic-joint arm. Such designs are clearly interesting because it can be used for, e.g. aerial grasping, hammering or throwing tasks. In Section 6.2.4 we showed that the attachment point of the arm matters a lot, when one wants to plan for and control the tracking of the end-effector positions. This applies for both, when the arm is actuated rigidly or via some compliant elements. If the arm is attached to the CoM of the PVTOL, i.e. the design is *coinciding*, then the end-effector positions become part of the differentially flat outputs. Although this assumption might not be entirely true for all the aerial manipulators, it is a reasonable one from practical point of view (it makes sense to attach a manipulator to the center of the flying robot). Moreover, even when the attachment point of the arm is not exactly the CoM of the aerial robot, the controllers developed based on the coinciding assumption still works well. Hence making coinciding assumption is quite reasonable, and it allows planning and control directly for the end-effector positions. We will exploit this more in the following sections.

Then, we have observed that both rigid and elastic designs can be beneficial depending on the aerial manipulation task, as explained in Sec. 6.2.9. For a composite trajectory tracking or an aerial grasping task, using rigid-joint arm results better tracking performances and lower control efforts compared to using an elastic-joint arm. This is because the controller for the elastic case needs to fight against the natural tendency of the springs to oscillate. On the other hand, using elastic-joint arm can be very effective when the aerial manipulation task requires explosive movements, and link velocity amplification (e.g. for aerial hammering or throwing).

Clearly, there are different aerial manipulation tasks, and using either rigid- or elastic-joint arm can outcome better or worse results. An obvious trade-off between these two lead us considering the utilization of Variable Stiffness Actuators (VSA) on board of a flying robot. For this reason in Sec. 6.2.10 we performed our experiments for a quadrotor equipped with an arm actuated via a VSA. Furthermore, since the VSA is not torque-controlled, the performance of the proposed controller is limited with the accuracy of the parameter identification. For this reason, it is relevant to seek for a controller, which enjoys the benefits of differential flatness property of the system, and in the same time can be used for the off-the-shelf position/velocity controlled actuators. We will consider this in the next sections, especially in Sec. 6.4.2.

# 6.3 Protocentric Aerial Manipulators (PAMs) with Multiple Redundant Arms

In Section 6.2 we presented the differential flatness property of different kinds of aerial manipulators, when they are equipped with one single-joint arm. Although having 1 DoF arm simplifies most of the control problems (see also Chapter 5), or allows performing tasks e.g. aerial grasping or throwing; increasing the number of DoF of the manipulating arm (or even considering more than one arm) means more dexterity for the aerial manipulator. Clearly, in such case the dynamic coupling between the flying base and the manipulating arm(s) will be more apparent, and the controller needs to account for it when a dynamic aerial manipulation task has to be performed.

Aerial platforms equipped with manipulating arms with more than 1 DoF have been recenlty a point of interest for the researchers. For example, a passive decomposition technique was developed for a generic aerial manipulator in Yang and Lee (2014), an adaptive sliding-mode controller used for a quadrotor equipped with a 2 DoF rigid arm in Kim et al. (2013), and a pure kinematics controller was implemented for a redundant aerial manipulator in Muscio et al. (2016). Common practice in the literature was to develop a controller, considering the CoM dynamics of the aerial manipulator, and then try to extend this for controlling the end-effector motion.

In this section we study the aerial manipulators, having an aerial robot as the flying platform, which are equipped with generic number of manipulating arms, each having generic number of DoFs. Moreover, in our model we consider that any joint of any arm could be connected to its actuated rigidly or via some elastic components. Here, we consider the *coinciding* model of the aerial manipulator, since as also studied in Sec. 6.2.4, in this case, the end-effector positions of the manipulating arm are part of the differentially flat outputs, which clearly simplifies the planning and control of useful aerial manipulation tasks, as discussed in Sec. 6.2.8 and in Sec. 6.2.9. We will call such systems as *Protocentric Aerial Manipulators* (PAMs), meaning that all the manipulating arms are attached to the CoM of the PVTOL.

We note that the study done for this section is published in Yüksel et al. (2016a).

In order to model a generic PAM, we start with the following definitions:

- A PVTOL has $m$ number of manipulating arms (manipulators); an arbitrary arm can be called as $\mu$-th arm, where $\mu \in \{1, 2, 3, \cdots, m\}$.

- The $\mu$-th arm has $n^\mu$ Degrees of Freedom (DoF); an arbitrary joint/motor/link can be called $\nu^\mu$-th joint/motor/link, where $\nu^\mu \in \{1, 2, 3, \cdots, n^\mu\}$.

- The $\mu$-th arm with $n^\mu$-DoF has $k^\mu$-number of elastic joints; an arbitrary elastic joint can be called as $\kappa^\mu$-th elastic joint, where $\kappa^\mu \in \{1, 2, 3, \cdots, k^\mu\}$.

- It is always $k^\mu \leq n^\mu$.

With this convention, we call, e.g., the mass of the $\nu$-th link of the $\mu$-th manipulator as $m_{\nu^\mu}$, or the motor inertia of the $\nu$-th link of the $\mu$-th manipulator is called as $J_{m_{\nu^\mu}}$. This means that the lowest index corresponds to the index of the joint, and the superscript of the lowest index corresponds to the index of the manipulator. Figure 6.21 sketches this numbering convention for the case where $m = 3$.

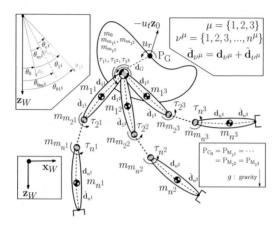

Figure 6.21: Sketch of a protocentric aerial manipulator, i.e., a PVTOL equipped with 3 planar manipulators whose 3 first joints are placed at the VTOL CoM. Each manipulator has different number of degrees of freedom, namely $n^1, n^2$ and $n^3$. This sketch summarizes the assumptions and definitions given in Sec. 6.3 to the case where $m = 3$. On the left up relative and absolute angles of the motor and the link of the first joint of the first manipulator are depicted, where the length of the $\mathbf{z}$ axes are made different just for illustration purposes.

Moreover, in the modeling phase the following assumptions are made, which are essential for developing the control algorithm presented in the next sections:

**A1.** *Only the 2D dynamics of a PVTOL aerial vehicle with m different fully actuated robotic arms is considered.*

**A2.** *All the joints are actuated via a motor, and the rotational center of this motor is the same with the center of the revolute joint that is attached to it.*

**A3.** *[Protocentricity] The first joint of each robotic arm is placed at the Center of Mass (CoM) of the PVTOL, i.e., $P_{C_0} = P_{M_{11}} = P_{M_{12}} = \cdots = P_{M_{1m}}$ (see also Fig. 6.21).*

**A4.** *Each motor is attached to the next link in the chain either rigidly or via some elastic joint.*

As before, we denote with $\mathcal{F}_W : \{P_W, \mathbf{x}_W, \mathbf{z}_W\}$ and $\mathcal{F}_0 : \{P_{C_0}, \mathbf{x}_0, \mathbf{z}_0\}$ the world (inertial) frame and the frame attached to the PVTOL, respectively, where $P_{C_0}$ is the Center of Mass (CoM) of the PVTOL. Define $P_{M_{\nu\mu}}$ as the position of the $\nu$-th motor of the $\mu$-th manipulator in the global frame. Then each joint and motor of the manipulators rotates about an axis parallel to $\mathbf{z}_W \times \mathbf{x}_W$ and passing through its corresponding motor point $P_{M_{\nu\mu}}$. Then the *motor frame* of the $\nu$-th joint of the $\mu$-th manipulator will be defined as $\mathcal{F}_{M_{\nu\mu}} : \{P_{M_{\nu\mu}}, \mathbf{x}_{m_{\nu\mu}}, \mathbf{z}_{m_{\nu\mu}}\}$; this is rigidly attached to the motor output shaft. The joint can be either rigid or elastic, as explained later, therefore for the sake of generality we consider also the *link frame* $\mathcal{F}_{\nu\mu} : \{P_{C_{\nu\mu}}, \mathbf{x}_{\nu\mu}, \mathbf{z}_{\nu\mu}\}$, where $P_{C_{\nu\mu}}$ is the CoM of the $\nu$-th link of the $\mu$-th manipulator. Finally we denote with the point $P_{E\mu}$ the end-effector of the $\mu$-th arm, and $P_C$ the CoM of the whole system (i.e., the VTOL plus all the arms).

Given an angle $\theta_* \in \mathbb{R}$ between the $\mathbf{z}$-axes of two frames (see Fig. 6.21, top left side) the usual rotation matrix definition $\mathbf{R}_* \in SO(2)$ holds, as in (6.5). Therefore, the orientations of, e.g., $\mathcal{F}_0$ in $\mathcal{F}_W$, and $\mathcal{F}_{\nu^\mu}$ in $\mathcal{F}_0$ are expressed by the rotation matrices $\mathbf{R}_0$, and $\mathbf{R}_{\nu^\mu}$, respectively. The absolute angles of the $\nu^\mu$-th motor and link are $\theta_{0m_{\nu^\mu}} = \theta_0 + \sum_{i^\mu=1}^{\nu^\mu} \theta_{m_{i^\mu}}$ and $\theta_{0\nu^\mu} = \theta_0 + \sum_{i^\mu=1}^{\nu^\mu} \theta_{i^\mu}$, respectively (see Fig. 6.21, top left side). Notice that $\theta_{e_{\nu^\mu}} = \theta_{\nu^\mu} - \theta_{m_{\nu^\mu}} = \theta_{0\nu^\mu} - \theta_{0m_{\nu^\mu}}$is constantly zero if the $\nu^\mu$ joint is rigid and can be any if it is elastic.

The constant position of $P_{M_{\nu^\mu}}$ and of $P_{M_{(\nu+1)^\mu}}$ in $\mathcal{F}_{\nu^\mu}$ are denoted with $-\mathbf{d}_{\nu^\mu} = [-d_{\nu^\mu x} \; -d_{\nu^\mu z}]^T \in \mathbb{R}^2$ and with $\tilde{\mathbf{d}}_{\nu^\mu} = [\tilde{d}_{\nu^\mu x} \; \tilde{d}_{\nu^\mu z}]^T \in \mathbb{R}^2$, respectively. The (time-varying) positions of $P_C$, $P_{C_0}$, $P_{C_{\nu^\mu}}$, $P_{M_{\nu^\mu}}$ and $P_{E^\mu}$ in $\mathcal{F}_W$ are denoted with $\mathbf{p}_c = [x_c \; z_c]^T \in \mathbb{R}^2$, $\mathbf{p}_0 = [x_0 \; z_0]^T \in \mathbb{R}^2$, $\mathbf{p}_{\nu^\mu} = [x_{\nu^\mu} \; z_{\nu^\mu}]^T \in \mathbb{R}^2$, $\mathbf{p}_{m_{\nu^\mu}} = [x_{m_{\nu^\mu}} \; z_{m_{\nu^\mu}}]^T \in \mathbb{R}^2$, and $\mathbf{p}_{e^\mu} = [x_{e^\mu} \; z_{e^\mu}]^T \in \mathbb{R}^2$, respectively. The mass and moment of the inertia of the PVTOL and the $\nu^\mu$-th motor and link are denoted with $m_0 \in \mathbb{R}_{>0}$, $J_0 \in \mathbb{R}_{>0}$; $m_{m_{\nu^\mu}} \in \mathbb{R}_{>0}$, $J_{m_{\nu^\mu}} \in \mathbb{R}_{>0}$; $m_{\nu^\mu} \in \mathbb{R}_{>0}$, $J_{\nu^\mu} \in \mathbb{R}_{>0}$, respectively. The gravitational constant is $g \in \mathbb{R}^+$. Also $m_s = m_0 + \sum_{\mu=1}^m \sum_{\nu=1}^{n^\mu} (m_{m_{\nu^\mu}} + m_{\nu^\mu})$ is the total mass of the overall system.

The point $P_G$ is the *center of actuation* of the PVTOL (see Fig 6.21). The constant position of $P_G$ in $\mathcal{F}_0$ is denoted with $\mathbf{d}_G = [d_{G_x} \; d_{G_z}]^T \in \mathbb{R}^2$. The PVTOL is actuated by means of: *i)* a total *thrust force* $-u_t \mathbf{z}_0 \in \mathbb{R}^2$ applied at $P_G$, where $u_t \in \mathbb{R}$ is its magnitude, and *ii)* a *total torque* (moment) $u_r(\mathbf{z}_0 \times \mathbf{x}_0) \in \mathbb{R}^2$ applied also at $P_G$, where $u_r \in \mathbb{R}$ is the torque intensity. Furthermore, an individual motor for each joint applies a torque $\tau_{\nu^\mu}(\mathbf{z}_{\nu^\mu} \times \mathbf{x}_{\nu^\mu})$ at $P_{M_{\nu^\mu}}$ to the joint, where $\tau_{\nu^\mu} \in \mathbb{R}$ is its intensity.

## 6.3.1 Case R: System Dynamics with Only Rigid Joints

Let us first consider the case in which all the joints are rigid, i.e., $k = 0$. There are $m$-number of fully actuated manipulators, and $\mu$-th manipulator has $n^\mu$-DoF. The aerial manipulator has therefore $3+n$ degrees of Freedom (DoFs) corresponding to the generalized coordinates $\mathbf{q} = [\mathbf{q}_p^T \; \mathbf{q}_r^T]^T \in \mathbb{R}^{(3+n)}$, where $\mathbf{q}_p$ is the PVTOL coordinates, and $\mathbf{q}_r$ stands for the arm-side coordinates, which are expressed as

$$\mathbf{q}_p = [\mathbf{p}_0^T \; \theta_0]^T \in \mathbb{R}^3,$$

$$\mathbf{q}_r = [\mathbf{q}_{r1}^T \; \cdots \; \mathbf{q}_{rm}^T]^T \in \mathbb{R}^n, \quad \mathbf{q}_{r\mu}^T = [\theta_{01^\mu} \; \cdots \; \theta_{0n^\mu}]^T \in \mathbb{R}^{n^\mu},$$

$$\mathbf{q}_r = [\underbrace{\theta_{01^1} \; \cdots \; \theta_{0n^1}}_{\mathbf{q}_{r1}^T} \; \underbrace{\theta_{01^2} \; \cdots \; \theta_{0n^2}}_{\mathbf{q}_{r2}^T} \; \cdots \; \underbrace{\theta_{01^m} \; \cdots \; \theta_{0n^m}}_{\mathbf{q}_{rm}^T}]^T \in \mathbb{R}^n,$$

with $n = \sum_{i=1}^m n^i$. Then using Lagrange equation and after some straight-forward algebra we can find the generalized inertia matrix as

$$\mathbf{M} = \begin{bmatrix} \mathbf{M}_p & * \\ \mathbf{M}_{pr} & \mathbf{M}_r \end{bmatrix} = \mathbf{M}^T \in \mathbb{R}^{(3+n)\times(3+n)}$$

$$\mathbf{M}_p = \operatorname{diag}([m_s \; m_s \; J_0]), \quad \mathbf{M}_{pr} = \begin{bmatrix} \mathbf{M}_{pr^1}^T & \cdots & \mathbf{M}_{pr^m}^T \end{bmatrix}^T \qquad (6.79)$$

$$\mathbf{M}_{pr^\mu} = \begin{bmatrix} \mathbf{m}_{01^\mu}(\theta_{01^\mu})^T & 0 \\ \vdots & \vdots \\ \mathbf{m}_{0n^\mu}(\theta_{0n^\mu})^T & 0 \end{bmatrix} \in \mathbb{R}^{n^\mu \times 3},$$

where $m_s$ is the total mass

$$m_s = m_0 + \sum_{j=1}^m \sum_{i=1}^{n^j} \left( m_{ij} + m_{m_{ij}} \right),$$

$\mathbf{M}_p \in \mathbb{R}^{3 \times 3}$ is the PVTOL side inertia matrix, $\mathbf{M}_r(\mathbf{q}_r) \in \mathbb{R}^{n \times n}$ is the manipulator side inertia matrix, and $\mathbf{M}_{pr}(\mathbf{q}_r) \in \mathbb{R}^{n \times 3}$ presents the inertial couplings between the PVTOL and the manipulator arms. We give the detailed computation of the inertia matrix in available in Appendix. B.1.

The gravitational forces are computed as the following

$$\mathbf{g} = \begin{bmatrix} \mathbf{g}_p \\ \mathbf{g}_r \end{bmatrix} \in \mathbb{R}^{(3+n)}, \quad \mathbf{g}_p = \begin{bmatrix} 0 \\ -m_s g \\ 0 \end{bmatrix} \in \mathbb{R}^3$$

$$\mathbf{g}_r = \begin{bmatrix} \mathbf{g}_{r1}^T & \mathbf{g}_{r1}^T & \cdots & \mathbf{g}_{rm}^T \end{bmatrix}^T \in \mathbb{R}^n, \tag{6.80}$$

where for the $\mu$-th manipulator it is

$$\mathbf{g}_{r\mu} = \begin{bmatrix} -g\mathbf{m}_{01\mu}(\theta_{01\mu}) \cdot \mathbf{e}_2 \\ -g\mathbf{m}_{02\mu}(\theta_{02\mu}) \cdot \mathbf{e}_2 \\ \vdots \\ -g\mathbf{m}_{0n\mu}(\theta_{0n\mu}) \cdot \mathbf{e}_2 \end{bmatrix} \in \mathbb{R}^{n^\mu}.$$

with $\mathbf{e}_2 = [0 \ 1]^T$. The Coriolis/Centrifugal forces are found as

$$\mathbf{c} = \begin{bmatrix} \sum_{j=1}^m \sum_{i=1}^{n^j} \bar{\mathbf{m}}_{0ij} \dot{\theta}_{0ij}^2 \\ 0 \\ \mathbf{c}_r(\mathbf{q}_r, \dot{\mathbf{q}}_r) \end{bmatrix} \in \mathbb{R}^{(3+n) \times 1}, \tag{6.81}$$

where $\bar{\mathbf{m}}_{0i\mu} = \frac{\partial \mathbf{m}_{0i\mu}}{\partial \theta_{0i\mu}} \in \mathbb{R}^{2 \times 1}$ and $\mathbf{c}_r(\mathbf{q}_r, \dot{\mathbf{q}}_r) \in \mathbb{R}^n$ is the arm side Coriolis forces. All the explicit steps towards computation of the $\mathbf{g}$ and $\mathbf{c}$ can be found in Appendix B.2.

Finally, the generalized forces are to be

$$\mathbf{f} = \begin{bmatrix} -u_t \sin(\theta_0) \\ -u_t \cos(\theta_0) \\ d_{G_x} u_t + u_r - \sum_{j=1}^m \tau_{1j} \\ \bar{\mathbf{T}} \end{bmatrix} = \mathbf{G}\mathbf{u} \in \mathbb{R}^{(n+3)},$$

$$\bar{\mathbf{T}} = \begin{bmatrix} \bar{\tau}^1 & \cdots & \bar{\tau}^m \end{bmatrix}^T \in \mathbb{R}^n \tag{6.82}$$

$$\bar{\tau}^\mu = \begin{bmatrix} \tau_{1^\mu} - \tau_{2^\mu} & \cdots \tau_{n^\mu-1} - \tau_{n^\mu} & \tau_{n^\mu} \end{bmatrix}^T \in \mathbb{R}^{n^\mu},$$

which leads to the control input matrix in form of

$$\mathbf{G} = \left( \begin{array}{c|c|c} \mathbf{v}(\theta_0) & \mathbf{0} & \mathbf{0} \\ \hline d_{G_x} & 1 & \mathbf{G}_{rp} \\ \hline \mathbf{0} & \mathbf{0} & \mathbf{G}_{rr} \end{array} \right) \in \mathbb{R}^{(n+3) \times (n+2)}, \tag{6.83}$$

where $\mathbf{v} = -\mathbf{z}_0 \in \mathbb{R}^2$, and all other parts of this matrix as well are explicitly given in Appendix B.3. The control input vector is

$$\mathbf{u} = \begin{bmatrix} u_t & u_r & \boldsymbol{\tau}^{1^T} & \boldsymbol{\tau}^{2^T} \cdots & \boldsymbol{\tau}^{m^T} \end{bmatrix}^T \in \mathbb{R}^{(n+2)}, \tag{6.84}$$

where $\boldsymbol{\tau}^\mu = [\tau_{1^\mu} \ \tau_{2^\mu} \ \cdots \ \tau_{n^\mu}]^T \in \mathbb{R}^{n^\mu}$. Then finally the system dynamics can be written in the following form

$$\mathbf{M}\ddot{\mathbf{q}} + \mathbf{c} + \mathbf{g} = \mathbf{G}\mathbf{u}. \tag{6.85}$$

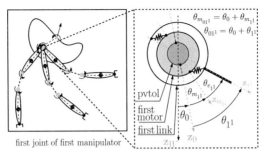

Figure 6.22: An ideal example of elastic joint between the first motor shaft and the first link of the first arm on a PVTOL ($\nu = \mu = 1$). Actuator is magnified w.r.t. the PVTOL. The innermost circle, fixed to $\mathcal{F}_0$, represents the PVTOL. The middle circle, fixed to $\mathcal{F}_{M_{11}}$, represents the actuator (or motor). The outermost circle is connected to the middle circle via elastic components, and it is rigidly connected to the link (fixed to $\mathcal{F}_{1^1}$.)

**Differential Flatness of the Case R**

Choose $\mathbf{y} = [\mathbf{p}_0^T \; \mathbf{q}_r^T]^T \in \mathbb{R}^{(n+2)}$ as an output of the system in (6.85), which has the same size as the control input vector.

---

**Proposition 9.** $\mathbf{y} = [\mathbf{p}_0^T \; \mathbf{q}_r^T]^T \in \mathbb{R}^{(n+2)}$ *is a flat output for the protocentric aerial manipulator with all rigid joints ($k = 0$). The relative degree of each entry of $\mathbf{y}$ is 4, and the total relative degree is $4n + 8$.*

---

*Proof.* See Appendix A.9. □

---

**Corollary 2.** *Since $\exists f_e : \mathbf{p}_0 = f_e(\mathbf{p}_{e^\mu} \; \mathbf{q}_r)$, also $\mathbf{y} = [\mathbf{p}_{e^\mu}^T \; \mathbf{q}_r^T]^T \in \mathbb{R}^{2+n}$ is a flat output of the protocentric aerial manipulator with all rigid joints.*

---

## 6.3.2 Case E: System Dynamics with Rigid and/or Elastic Joints

Now consider the case $k \geq 1$, i.e. when at least one of the joints of the robotic arm is actuated via an elastic component (see Fig. 6.22). Recall the generalized coordinates we presented before in the form of $\mathbf{q} = [\mathbf{q}_p^T \; \mathbf{q}_r^T]^T$. Knowing that the $\mu$-th arm has $n^\mu$ number of links and joints, let us rewrite the arm-side coordinates for the $\mu$-th manipulator as $\mathbf{q}_{r^\mu} = [\mathbf{q}_{r^\mu}^T \; \mathbf{q}_{r_e^\mu}^T]^T \in \mathbb{R}^{n^\mu + k^\mu}$, where $\mathbf{q}_{r^\mu} \in \mathbb{R}^{n^\mu}$ is the absolute orientation of the rigid links (and their actuators if the connection between the link and motor is rigid), and $\mathbf{q}_{r_e^\mu} \in \mathbb{R}^{k^\mu}$ is the absolute orientation of the motors that are connected to their rigid links via elastic components[21]. Such separation is necessary, since in the case when the motor is elastically attached to its link, their orientations are different from each other (see Fig. 6.22), while

---

[21]Notice that the number of elastic connections increase the total number of the generalized coordinates by $k^\mu$.

this was not the case when the connection was rigid. Hence the full generalized coordinates are $\mathbf{q} = [\mathbf{q}_p^T \ \mathbf{q}_r^T]^T = [\mathbf{q}_p^T \ \mathbf{q}_{r_r}^T \ \mathbf{q}_{r_e}^T]^T \in \mathbb{R}^{3+n+k}$, where $\mathbf{q}_{r_r} = [\mathbf{q}_{r_r}^{T^1} \ \mathbf{q}_{r_r}^{T^2} \ \cdots \ \mathbf{q}_{r_r}^{T^m}] \in \mathbb{R}^n$ and $\mathbf{q}_{r_e} = [\mathbf{q}_{r_e}^{T^1} \ \mathbf{q}_{r_e}^{T^2} \ \cdots \ \mathbf{q}_{r_e}^{T^m}] \in \mathbb{R}^k$. Notice that $n = \sum_{i=1}^m n^i$ and $k = \sum_{i=1}^m k^i$. In order to fix the ideas, an idealized elastic connection is sketched in Fig. 6.22, where the first actuator of the first manipulator is considered to be elastic.

Now, to generalize our theory, assume that $k$ number of joints are elastic, where $k \leq n$. Let us define a set, $N := \{N^1, N^2, \cdots, N^m\}$, where $N^\mu$ is another set which contains all the indices of the $\mu$-th robotic arm joints, namely $N^\mu := \{1, 2, \cdots, n^\mu\}$. Then define another set, $K := \{K^1, K^2, \cdots, K^m\}$, where $K^\mu$ is another set that consists of the indices that are belonging to the elastic joints of the $\mu$-th manipulator. Notice that $K^\mu \subset N^\mu$, and $K^\mu = K^\mu \cap N^\mu$. It is obvious that the set $K^\mu$ has $k^\mu$ number of elements, and with $K^\mu\{\kappa^\mu\}$ we denote the $\kappa^\mu$-th element of $K^\mu$, where $\kappa^\mu = \{1, 2, \cdots, k^\mu\}$.

---

**Remark 12.** *For the elements of set $N^\mu$ and $K^\mu$, it is always $N^\mu\{\nu^\mu + 1\} > N^\mu\{\nu^\mu\}$ and $K^\mu\{\kappa^\mu + 1\} > K^\mu\{\kappa^\mu\}$, where $\nu^\mu = \{1, 2, \cdots, n^\mu\}$ and $\kappa^\mu = \{1, 2, \cdots, k^\mu\}$, respectively.*

---

Now for each arm $\mu$, let us define an orthogonal selection matrix, $\mathbf{S}_{N^\mu} \in \mathbb{R}^{n^\mu \times n^\mu}$, in the form of which selects the elements of set $K^\mu$ from the set $N^\mu$, and puts them in set $N^\mu$. Define another selection matrix, $\mathbf{S}_{K^\mu} \in \mathbb{R}^{k^\mu \times n^\mu}$, which selects the elements of set $K^\mu$ from the set $N^\mu$, and puts them in set $K^\mu$. In other words, the diagonal matrix $\mathbf{S}_{N^\mu} \in \mathbb{R}^{n^\mu \times n^\mu}$ whose $\nu^\mu$-th diagonal element is equal to 1 if $\nu^\mu \in K^\mu$ and zero otherwise, and the selection matrix $\mathbf{S}_{K^\mu} \in \mathbb{R}^{k^\mu \times n^\mu}$ obtained from $\mathbf{S}_{N^\mu}$ by removing all the zero row vectors. Let us fix the idea by giving the following examples:

**Example 2.** *Say that $N^\mu = \{1, 2, 3, 4\}$ and $K^\mu = \{2, 4\}$. Define a vector, say $\mathbf{v}^\mu = [v_{1^\mu} \ v_{2^\mu} \ v_{3^\mu} \ v_{4^\mu}]^T \subset N^\mu$. Then the matrix*

$$\mathbf{S}_{N^\mu} = \begin{bmatrix} 0 & 0 & 0 & 0 \\ 0 & 1 & 0 & 0 \\ 0 & 0 & 0 & 0 \\ 0 & 0 & 0 & 1 \end{bmatrix} = \mathbf{S}_{N^\mu}^T \in \mathbb{R}^{n^\mu \times n^\mu}$$

*will select the 2-nd and 4-th components of the vector $\mathbf{v}^\mu$, using $\bar{\mathbf{v}}^\mu = \mathbf{S}_{N^\mu} \mathbf{v} = [0 \ v_{2^\mu} \ 0 \ v_{4^\mu}]^T \subset N^\mu$.*

**Example 3.** *Similar to the same example as above, let's say $N^\mu = \{1, 2, 3, 4\}$ and $K^\mu = \{2, 4\}$. Define a vector, say $\mathbf{v}^\mu = [v_{1^\mu} \ v_{2^\mu} \ v_{3^\mu} \ v_{4^\mu}]^T \subset N^\mu$. Then the matrix*

$$\mathbf{S}_{K^\mu} = \begin{bmatrix} 0 & 1 & 0 & 0 \\ 0 & 0 & 0 & 1 \end{bmatrix} \in \mathbb{R}^{k^\mu \times n^\mu}$$

*can be used to select 2-nd and 4-th elements of $\mathbf{v}^\mu$ and put them in order, using $\bar{\mathbf{v}}^\mu = \mathbf{S}_{K^\mu} \mathbf{v}^\mu = [v_{2^\mu} \ v_{4^\mu}]^T \subset K^\mu$.*

Then define the following block diagonal matrices

$$\mathbf{S}_N = \mathrm{diag}\{\mathbf{S}_{N^1}, \mathbf{S}_{N^2}, \cdots, \mathbf{S}_{N^m}\} \in \mathbb{R}^{n \times n},$$
$$\mathbf{S}_K = \mathrm{diag}\{\mathbf{S}_{K^1}, \mathbf{S}_{K^2}, \cdots, \mathbf{S}_{K^m}\} \in \mathbb{R}^{k \times n}.$$

Notice that considering the generalized coordinates are given in the beginning of the section, it is $\mathbf{q}_{r_e^\mu} = [\theta_{m_{0\{\kappa^\mu=1\}}} \; \theta_{m_{0\{\kappa^\mu=2\}}} \; \cdots \; \theta_{m_{0\{\kappa^\mu=k^\mu\}}}]^T \in \mathbb{R}^{k^\mu}$, where remember that the subscript $*_{\{\kappa^\mu\}}$ stands for the indice/number of the elastic joint of the $\mu$-th manipulator. To have a better understanding of this, let us give the following example.

**Example 4.** *Say, $N^\mu = \{1,2,3,4,5,6,7\}$ and $K^\mu = \{1,5,7\}$. It means that the $\mu$-th manipulator has 7 DoF and three of the motors are connected to their links via some elastic elements, which are the first, fifth and the seventh ones. Then we say $K^\mu\{\kappa^\mu = 3\} = 7$, is the third elastically connected motor, which corresponds to the seventh joint/motor of the $\mu$-th robotic arm. So, we can say e.g. its absolute motor orientation is $\theta_{m_{0\{\kappa^\mu=3\}}} = \theta_{m_{07\mu}}$.*

Let us then first rewrite the generalized inertia matrix as

$$\mathbf{M}_E = \begin{bmatrix} \mathbf{M}_p & * & * \\ \mathbf{M}_{pr} & \mathbf{M}_{rE} & * \\ \mathbf{0} & \mathbf{0} & \mathbf{D}_K \end{bmatrix} = \mathbf{M}^T \in \mathbb{R}^{(3+n+k)\times(3+n+k)}, \tag{6.86}$$

where the inertial terms of the elastically connected motors are summarized in

$$\mathbf{D}_K = \text{diag}\{\mathbf{D}_{K^1}, \mathbf{D}_{K^2}, \cdots, \mathbf{D}_{K^m}\} \in \mathbb{R}^{k\times k},$$
$$\mathbf{D}_{K^\mu} = \text{diag}\{J_{m_{\{\kappa^\mu=1\}}}, J_{m_{\{\kappa^\mu=2\}}}, \cdots, J_{m_{\{\kappa^\mu=k^\mu\}}}\} \in \mathbb{R}^{k\times k}.$$

Similar to Sec. 6.3.1, both $\mathbf{M}_p$ and $\mathbf{M}_{pr}$ are identical to the Case R, as given in (6.3.1). This time we define $\mathbf{M}_{rE}$ for the arm side inertia matrix, which is computed as $\mathbf{M}_{rE} = \mathbf{M}_r - \mathbf{S}_N \mathbf{D}_N$, with $\mathbf{D}_N = \text{diag}\{\mathbf{D}_{N^1}, \mathbf{D}_{N^2}, \cdots, \mathbf{D}_{N^m}\} \in \mathbb{R}^{n\times n}$, where $\mathbf{D}_{N^\mu} = \text{diag}\{J_{m_1\mu}, J_{m_2\mu}, \cdots, J_{m_n\mu}\} \in \mathbb{R}^{n^\mu \times n^\mu}$ is the matrix containing the inertias of all motors of the $\mu$-th manipulator. Following the same order as for the Case R, the gravitational and Coriolis/centrifugal forces are

$$\mathbf{g}_E = \begin{bmatrix} \mathbf{g} \\ \mathbf{0}_{k\times 1} \end{bmatrix} \in \mathbb{R}^{(3+n+k)}, \mathbf{c}_E = \begin{bmatrix} \mathbf{c} \\ \mathbf{0}_{k\times 1} \end{bmatrix} \in \mathbb{R}^{(3+n+k)} \tag{6.87}$$

where $\mathbf{g} \in \mathbb{R}^{(3+n)}$ is given in (6.80) and $\mathbf{c} \in \mathbb{R}^{(3+n)}$ is given in (6.81). The generalized forces will be

$$\mathbf{f}_E = \mathbf{G}_E \mathbf{u} \in \mathbb{R}^{(3+n+k)},$$
$$\mathbf{G}_E = \begin{bmatrix} \mathbf{G} - \bar{\mathbf{S}}_N \\ \bar{\mathbf{S}}_K \end{bmatrix} \in \mathbb{R}^{(3+n+k)\times(n+2)},$$
$$\bar{\mathbf{S}}_N = \begin{bmatrix} \mathbf{0}_{3\times 2} & \mathbf{0}_{3\times n} \\ \mathbf{0}_{n\times 2} & \mathbf{S}_N \in \mathbb{R}^{n\times n} \end{bmatrix} \in \mathbb{R}^{(n+3)\times(n+2)},$$
$$\bar{\mathbf{S}}_K = \begin{bmatrix} \mathbf{0}_{k\times 2} & \mathbf{S}_K \in \mathbb{R}^{k\times n} \end{bmatrix} \in \mathbb{R}^{k\times(n+2)},$$

with $\mathbf{G} \in \mathbb{R}^{(n+3)\times(n+2)}$ and $\mathbf{u} \in \mathbb{R}^{(n+2)}$ available from (6.83)–(6.84).

Remember that, if a joint of a rigid link is connected to its motor via some elastic components, then there will be counteracting torques/forces appearing on both link and motor side of the dynamics, due to the elastic potential energy stored in the elastic elements.

Now, consider the $\nu$-th link of the $\mu$-th manipulator and its joint, connected to its motor via some elastic components. Then we call $f_{l_{\nu\mu}}(\theta_{0\nu\mu}, \theta_{m_{0\nu\mu}})$ as the link-side, and $f_{m_{\nu\mu}}(\theta_{0\nu\mu}, \theta_{m_{0\nu\mu}})$ as the motor-side elastic forces for this joint[22]. Those forces are identically zero if $\nu^\mu \notin K^\mu$. In the case $\nu^\mu \in K^\mu$ they are instead generic functions of $\theta_{0\nu\mu}, \theta_{m_{0\nu\mu}}$. In the linear spring case, $f_{l_{\nu\mu}}(\theta_{0\nu\mu}, \theta_{m_{0\nu\mu}}) = k_{e_{\nu\mu}}(\theta_{m_{0\nu\mu}} - \theta_{0\nu\mu})$ and $f_{m_{\nu\mu}}(\theta_{0\nu\mu}, \theta_{m_{0\nu\mu}}) = k_{e_{\nu\mu}}(\theta_{0\nu\mu} - \theta_{m_{0\nu\mu}})$ where $k_{e_{\nu\mu}} > 0$ is the stiffness of the elastic element.

For the generic representation of the elastic forces in the considered system dynamics, let us first define the vectors $\mathbf{f}_{l^\mu}(\mathbf{q}_{r^\mu}) \in \mathbb{R}^{n^\mu}$ and $\mathbf{f}_{m^\mu}(\mathbf{q}_{r^\mu}) \in \mathbb{R}^n$ for the $\mu$-th manipulator as

$$\mathbf{f}_{l^\mu} = \begin{bmatrix} f_{l_{1\mu}}(\theta_{01\mu}, \theta_{m_{01\mu}}) & \cdots & f_{l_{n\mu}}(\theta_{0n\mu}, \theta_{m_{0n\mu}}) \end{bmatrix}^T \in \mathbb{R}^{n^\mu},$$

$$\mathbf{f}_{m^\mu} = \begin{bmatrix} f_{m_{1\mu}}(\theta_{01\mu}, \theta_{m_{01\mu}}) & \cdots & f_{m_{n\mu}}(\theta_{0n\mu}, \theta_{m_{0n\mu}}) \end{bmatrix}^T \in \mathbb{R}^{n^\mu}.$$

Now notice the fact that if the joint of the $i$-th link is rigidly connected to its motor, then both $f_{m_{\nu\mu}}(\theta_{0\nu\mu}, \theta_{m_{0\nu\mu}}) = f_{l_{\nu\mu}}(\theta_{0\nu\mu}, \theta_{m_{0\nu\mu}}) = 0$, because in rigid connection there are no elastic forces defined[23]. Then we can write

$$\mathbf{f}_L(\mathbf{q}_r) = \mathrm{diag}\{\mathbf{f}_{l^1}(\mathbf{q}_{r^1}), \mathbf{f}_{l^2}(\mathbf{q}_{r^2}), \cdots, \mathbf{f}_{l^m}(\mathbf{q}_{r^m})\} \in \mathbb{R}^n$$

$$\mathbf{f}_M(\mathbf{q}_r) = \mathrm{diag}\{\mathbf{f}_{m^1}(\mathbf{q}_{r^1}), \mathbf{f}_{m^2}(\mathbf{q}_{r^2}), \cdots, \mathbf{f}_{m^m}(\mathbf{q}_{r^m})\} \in \mathbb{R}^n,$$

and then the generalized elastic forces as

$$\mathbf{f}_{El} = \begin{bmatrix} \mathbf{0}_{3\times 1} \\ \mathbf{S}_N \mathbf{f}_L(\mathbf{q}_r) \\ \mathbf{S}_K \mathbf{f}_M(\mathbf{q}_r) \end{bmatrix} \in \mathbb{R}^{(3+n+k)}. \tag{6.88}$$

Hence the system dynamics is in the form of

$$\mathbf{M}_E \ddot{\mathbf{q}} + \mathbf{c}_E + \mathbf{g}_E = \mathbf{G}_E \mathbf{u} + \mathbf{f}_{El}. \tag{6.89}$$

**Differential Flatness of the Case E**

Choose $\mathbf{y} = [\mathbf{p}_0^T \ \mathbf{q}_{r_r}^T]^T \in \mathbb{R}^{(n+2)}$ as output, which has the same size as the control input vector.

---

**Proposition 10.** $\mathbf{y} = [\mathbf{p}_0^T \ \mathbf{q}_r^T]^T \in \mathbb{R}^{(n+2)}$ *is a flat output for the protocentric aerial manipulator with mixed rigid/elastic joints* $(1 \le k \le n)$. *The total relative degree is* $4 + 4 \max_\mu \tilde{k}^\mu + \sum_{\mu=1}^m (2 + 2\tilde{k}^\mu) n^\mu$, *with* $\tilde{k}^\mu = \max(1, k^\mu)$.

---

*Proof.* See Appendix A.10. □

---

**Corollary 3.** *Similar to Corollary 2, since* $\exists f_e : \mathbf{p}_0 = f_e(\mathbf{p}_{e^\mu} \ \mathbf{q}_r)$, *also* $\mathbf{y} = [\mathbf{p}_{e^\mu}^T \ \mathbf{q}_r^T]^T \in \mathbb{R}^{2+n}$ *is a flat output of the protocentric aerial manipulator with mixed rigid/elastic joints.*

---

[22]In this section we do not consider the frictions on the elastic connections.
[23]In the linear spring case setting $k_{\nu\mu} = 0$ is equivalent to saying that the elastic forces are zero, but clearly this cannot be generalized to the nonlinear cases.

---

**Remark 13.** *Notice the different relative degree of the dependencies of $\tau_{\nu^\mu}$ given in (A.58) on the flat outputs for different values of $\nu^\mu$. Assume for instance that both the $(n^\mu - 1)$-th and the $n^\mu$-th link are elastic. Then from bottom to top,*

- *First: from (A.56), we see that $\tau_{n^\mu}$ is a function of $\ddot{\theta}_{m_{0n^\mu}}$; while $\theta_{m_{0n^\mu}}$ is a function of $\ddot{\mathbf{p}}_0$ and $\ddot{\mathbf{q}}_{r^\mu}$, making $\tau_{n^\mu}$ itself a function of $\overset{....}{\mathbf{p}}_0$ and $\overset{....}{\mathbf{q}}_{r^\mu}$.*

- *Second: from (A.56), $\tau_{n^\mu-1}$ is a function of $\ddot{\theta}_{m_{0(n^\mu-1)}}$. But in (A.57), from recursion, $\theta_{m_{0(n^\mu-1)}}$ is a function of $\tau_{n^\mu}$, making $\ddot{\theta}_{m_{0(n^\mu-1)}}$, and thus $\tau_{n^\mu-1}$, a function of $\ddot{\tau}_{n^\mu}$. Knowing from the fist step above $\tau_{n^\mu}$ is a function of $\overset{....}{\mathbf{p}}_0$ and $\overset{....}{\mathbf{q}}_{r^\mu}$, we find $\tau_{n^\mu-1}$ as a function of $\mathbf{p}_0^{(6)}$ and $\mathbf{q}_{r^\mu}^{(6)}$, which are the sixth time derivatives.*

*In general, for a fully elastic manipulator, an increase of two relative degrees per link is to be expected.*

---

**Remark 14.** *We notice that the orientation of the PVTOL, i.e., $\theta_0$, is not part of the flat outputs, conceivably due to the under-actuation of the flying robot. This motivated us to use the absolute representation of the manipulator joint angles, which makes the control torques appear recursively in the manipulator dynamics. Notice from the remark above that while this is not a problem for Case R, for Case E this increases the relative degrees.*

*Hence it is worth noting that using a fully actuated aerial robot might be beneficial if manipulators with compliant actuators are to be used for specific tasks, e.g., safe physical interaction. This does not apply of course if the robotic arm is rigidly actuated. Further study on this remark is in the scope of our future studies.*

### 6.3.3 Control for Case R

In this section we present the exact linearizing controller for the system given in (6.85). We purposely limit our computation to the Case R, since the high relative degree involved in Case E may cause the controller to be unpractical (see Remark 14).

Now, based on our findings in Proposition 9, we take $\mathbf{y} = [\mathbf{p}_0^T \; \mathbf{q}_r^T]^T \in \mathbb{R}^{(n+2)}$ as control variables, leaving out the PVTOL orientation $\theta_0$. We approach the control problem by studying the system with $\theta_0$ removed. We can decompose the inertia matrix $\mathbf{M}$ by defining the following quantities:

$$\mathbf{M} = \left[\begin{array}{c|c|c} \tilde{\mathbf{M}}_p & 0 & \tilde{\mathbf{M}}_{pr}^T \\ \hline 0 & J_0 & 0 \\ \hline \tilde{\mathbf{M}}_{pr} & 0 & \mathbf{M}_r \end{array}\right] \quad \tilde{\mathbf{M}} = \left[\begin{array}{c|c} \tilde{\mathbf{M}}_p & \tilde{\mathbf{M}}_{pr}^T \\ \hline \tilde{\mathbf{M}}_{pr} & \mathbf{M}_r \end{array}\right],$$

where $\mathbf{0}$ is a zero vector or matrix of proper dimensions, $\tilde{\mathbf{M}}_p = \mathrm{diag}(m_s \; m_s)$, and $\tilde{\mathbf{M}}_{pr}$ is simply constituted by the first two columns of $\mathbf{M}_{pr}$. Similarly, for $\mathbf{G}$:

$$
\mathbf{G} = \left[\begin{array}{c|c|c}
\mathbf{v} & \mathbf{0} & \mathbf{0} \\ \hline
d_{G_x} & 1 & \mathbf{G}_{rp} \\ \hline
\mathbf{0} & \mathbf{0} & \mathbf{G}_{rr}
\end{array}\right]
\quad
\tilde{\mathbf{G}} = \left[\begin{array}{c|c}
\mathbf{v} & \mathbf{0} \\ \hline
\mathbf{0} & \mathbf{G}_{rr}
\end{array}\right],
$$

where $\mathbf{v} = [-\sin\theta_0 \ -\cos\theta_0]^T$. This allows us to write:

$$
\tilde{\mathbf{M}}(\mathbf{y})\ddot{\mathbf{y}} + \tilde{\mathbf{n}}(\mathbf{y}, \dot{\mathbf{y}}) = \tilde{\mathbf{G}}(\theta_0)\tilde{\mathbf{u}}, \tag{6.90}
$$

where $\tilde{\mathbf{n}}$ is $\mathbf{n} = \mathbf{c} + \mathbf{g}$ with the 3rd element removed, and $\tilde{\mathbf{u}} = [u_t \ \boldsymbol{\tau}^T]^T$, where $\boldsymbol{\tau} = [\boldsymbol{\tau}^{1T} \ \cdots \ \boldsymbol{\tau}^{mT}]^T$. Now we can differentiate (6.90), yielding (dependencies omitted):

$$
\tilde{\mathbf{M}}\dddot{\mathbf{y}} + \dot{\tilde{\mathbf{M}}}\ddot{\mathbf{y}} + \dot{\tilde{\mathbf{n}}} = \tilde{\mathbf{G}}\dot{\tilde{\mathbf{u}}} + \dot{\tilde{\mathbf{v}}}u_t, \tag{6.91}
$$

where we have evidenced $\tilde{\mathbf{v}} = [\mathbf{v}^T \ \mathbf{0}]^T$. Differentiating further:

$$
\tilde{\mathbf{M}}\ddddot{\mathbf{y}} + 2\dot{\tilde{\mathbf{M}}}\dddot{\mathbf{y}} + \ddot{\tilde{\mathbf{M}}}\ddot{\mathbf{y}} + \ddot{\tilde{\mathbf{n}}} = \tilde{\mathbf{G}}\ddot{\tilde{\mathbf{u}}} + 2\dot{\tilde{\mathbf{v}}}\dot{u}_t + \ddot{\tilde{\mathbf{v}}}u_t, \tag{6.92}
$$

but:

$$
\ddot{\mathbf{v}} = \begin{bmatrix} -\cos(\theta_0)\ddot{\theta}_0 \\ \sin(\theta_0)\ddot{\theta}_0 \end{bmatrix} + \begin{bmatrix} \sin(\theta_0)\dot{\theta}_0^2 \\ \cos(\theta_0)\dot{\theta}_0^2 \end{bmatrix} = \mathbf{h}\ddot{\theta}_0 - \mathbf{v}\dot{\theta}_0^2, \tag{6.93}
$$

where $\mathbf{h} = [-\cos(\theta_0) \ \sin(\theta_0)]^T$. Now, from the 3rd row of the dynamics in (6.85):

$$
\ddot{\theta}_0 = \frac{1}{J_0}\left(d_{G_x}u_t + u_r + \mathbf{G}_{rp}\boldsymbol{\tau}\right). \tag{6.94}
$$

Thus, we can substitute (6.94) in (6.93), and (6.93) in the last term of (6.92):

$$
\ddot{\mathbf{v}}u_t = \frac{u_t}{J_0}\mathbf{h} \cdot (d_{G_x}u_t + u_r + \mathbf{G}_{rp}\boldsymbol{\tau}) - \mathbf{v}\,u_t\dot{\theta}_0^2 = \boldsymbol{\gamma} + \frac{u_t}{J_0}\mathbf{h} \cdot u_r
$$

where we have introduced the new symbol $\boldsymbol{\gamma}$ for compactness. This finally allows us to write:

$$
\tilde{\mathbf{M}}\ddddot{\mathbf{y}} + 2\dot{\tilde{\mathbf{M}}}\dddot{\mathbf{y}} + \ddot{\tilde{\mathbf{M}}}\ddot{\mathbf{y}} + \ddot{\tilde{\mathbf{n}}} - 2\dot{\tilde{\mathbf{v}}}\dot{u}_t - \tilde{\boldsymbol{\gamma}} = \bar{\mathbf{G}}\ddot{\mathbf{u}},
$$

where:

$$
\bar{\mathbf{G}} = \left[\begin{array}{c|c|c}
\mathbf{v} & \frac{u_t}{J_0}\mathbf{h} & \mathbf{0} \\ \hline
\mathbf{0} & \mathbf{0} & \mathbf{G}_{rr}
\end{array}\right]
\quad
\tilde{\boldsymbol{\gamma}} = \begin{bmatrix} \boldsymbol{\gamma} \\ \mathbf{0} \end{bmatrix}
\quad
\bar{\mathbf{u}} = \begin{bmatrix} \ddot{u}_t \\ u_r \\ \ddot{\boldsymbol{\tau}} \end{bmatrix}.
$$

Matrix $\bar{\mathbf{G}} \in \mathbb{R}^{(2+n)\times(2+n)}$ is the decoupling matrix and it is clearly invertible, as long as $u_t \neq 0$, since $|\bar{\mathbf{G}}| = -\frac{u_t}{J_0}$.

The relative degree of the extended system is $r = 4(2 + n) = 8 + 4n$, and the total number of states is $\bar{n} = 2(3 + n) + 2(1 + n) = 8 + 4n = r$; thus, no internal dynamics is left, consistently with the notion that the system is flat. The virtual control input can be computed as

$$
\bar{\mathbf{u}} = \bar{\mathbf{G}}^{-1}\left(\tilde{\mathbf{M}}\ddddot{\mathbf{y}}_r + 2\dot{\tilde{\mathbf{M}}}\dddot{\mathbf{y}} + \ddot{\tilde{\mathbf{M}}}\ddot{\mathbf{y}} + \ddot{\tilde{\mathbf{n}}} - 2\dot{\tilde{\mathbf{v}}}\dot{u}_t - \tilde{\boldsymbol{\gamma}}\right),
$$

where, for a desired $\mathbf{y}$ trajectory denoted as $\mathbf{y}_d$

$$\dddot{\mathbf{y}}_r = \dddot{\mathbf{y}}_d + \mathbf{K}_3(\dddot{\mathbf{y}}_d - \dddot{\mathbf{y}}) + \mathbf{K}_2(\ddot{\mathbf{y}}_d - \ddot{\mathbf{y}})$$
$$+ \mathbf{K}_1(\dot{\mathbf{y}}_d - \dot{\mathbf{y}}) + \mathbf{K}_0(\mathbf{y}_d - \mathbf{y}) + \mathbf{K}_{-1}\int_0^t (\mathbf{y}_d - \mathbf{y})\mathrm{d}t . \tag{6.95}$$

The $\mathbf{K}_i$'s are diagonal positive definite matrices, assigned according to the usual linear pole-placement strategies. Specifically, if $K_{i,j}$ is the $j$-th diagonal element of $\mathbf{K}_i$, then each polynomial

$$p_j(x) = x^5 + K_{3,j}x^4 + K_{2,j}x^3 + K_{1,j}x^2 + K_{0,j}x + K_{-1,j} \tag{6.96}$$

must be Hurwitz, i.e. all its roots must have negative real parts; the introduction of an integral error term provides some ability to reject disturbances, such as carried loads and parameter uncertainty (see Sec. 6.3.4 for its implementation). The inverse of $\bar{\mathbf{G}}$ is easily obtained:

$$\bar{\mathbf{G}}^{-1} = \begin{bmatrix} -\sin\theta_0 & -\cos\theta_0 & \mathbf{0} \\ -\frac{J_0}{u_t}\cos\theta_0 & \frac{J_0}{u_t}\sin\theta_0 & \\ \hline \mathbf{0} & & \mathbf{Grr}^{-1} \end{bmatrix} \tag{6.97a}$$

$$\mathbf{Grr}^{-1} = \begin{bmatrix} \mathbf{Grr}^{1^{-1}} & & \mathbf{0,} \\ & \ddots & \\ \mathbf{0} & & \mathbf{G}_{rr^m}^{-1} \end{bmatrix} \qquad \mathbf{Grr}^{i^{-1}} = \begin{bmatrix} 1 & \cdots & 1 \\ & \ddots & \vdots \\ & & 1 \end{bmatrix} \in \mathbb{R}^{n^i \times n^i} . \tag{6.97b}$$

It should be noticed, the algorithm makes apparent use of higher-order derivatives of the flat outputs, $\ddot{\mathbf{y}}$ and $\dddot{\mathbf{y}}$, which are difficult or impossible to estimate directly. However, these can be computed from $\bar{\mathbf{u}}$ and $\dot{\bar{\mathbf{u}}}$, obtained from integration of appropriate components of the virtual input $\bar{\mathbf{u}}$:

$$\ddot{\mathbf{y}} = \tilde{\mathbf{M}}^{-1}\left(\tilde{\mathbf{G}}\bar{\mathbf{u}} - \tilde{\mathbf{n}}\right)$$
$$\dddot{\mathbf{y}} = \tilde{\mathbf{M}}^{-1}\left(\tilde{\mathbf{G}}\dot{\bar{\mathbf{u}}} + \dot{\tilde{\mathbf{v}}}u_t - \dot{\tilde{\mathbf{M}}}\ddot{\mathbf{y}} - \dot{\tilde{\mathbf{n}}}\right) .$$

## 6.3.4 Numerical Results

In this section we present simulation results for testing the proposed controller in a realistic situation in which measurement noises and sampling errors are considered. We consider a PAM with $m = 2$, $n^1 = 4$ and $n^2 = 3$. Hence the flat output is $\mathbf{y} = [\mathbf{p}_0 \; \theta_{01^1} \; \cdots \; \theta_{04^1} \; \theta_{01^2} \; \cdots \; \theta_{03^2}]^T \in \mathbb{R}^9$. System and simulation parameters are summarized in Table 6.7, which we simulated using Matlab/Simulink. While the system dynamics is simulated at 1 [kHz], the controlled is provided with under-sampled measurements as given in Table 6.7. Moreover, a $\pm 2\%$ of random parametric uncertainty for mass, inertia and distance parameters is used. A pick and place task is chosen for the robot, which is divided in 5 phases: *i)* the robot follows a desired trajectory, *ii)* the two arms grasp two individual point mass objects with unknown mass for the controller (each mass is 0.25 [kg]), *iii)* the objects are carried to another location, where they are unloaded, *iv)* phase (*i*) and (*ii*) are repeated while following a different trajectory, *v)* phase (*iii*) is repeated while following a different trajectory, and then arms return to the initial configuration. The results are given in Fig. 6.23, where the first row shows the CoM and the end-effector positions of the aerial manipulator in the world frame. In the second row the absolute

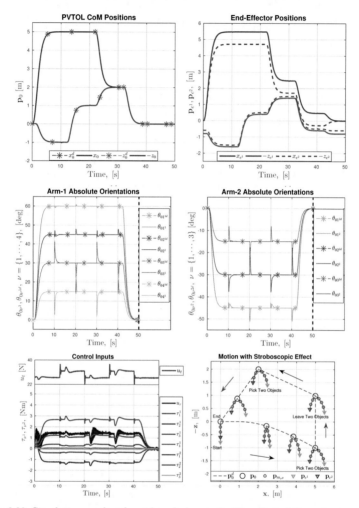

Figure 6.23: Simulation results of a pick and place task. The effect of the unknown grasped masses on the end-effector positions is negligible. In the bottom right, the arrows show the direction of the motion. Red arrows correspond to the mass–carrying phases. In the last row, the thrust input is given on the top, separately.

orientations of the manipulating arms are given. Last row is presenting the control inputs of the system, and the motion of the aerial manipulator with a stroboscopic effect. Notice the effect of the measurement noises in the control inputs. Despite the uncertainties, the tracking performance is almost perfect. At the times of grasping there are small errors on

| Parameters | Notation | Value | Unit |
|---|---|---|---|
| PVTOL mass | $m_0$ | 1 | [kg] |
| PVTOL inertia | $J_0$ | 0.015 | [kgm$^2$] |
| dis. vec. betw. P$_{C_0}$ & P$_G$ | $\mathbf{d}_G$ | $[0\ 0.1]^T$ | [m] |
| partial distance of one link | $\mathbf{d}_{\nu^\mu}$ | $[0\ 0.1]^T$ | [m] |
| partial distance of one link | $\mathbf{d}_{\nu^\mu}$ | $[0\ 0.1]^T$ | [m] |
| mass of one link | $m_{\nu^\mu}$ | 0.2 | [kg] |
| inertia of one link | $J_{\nu^\mu}$ | 0.0007 | [kgm$^2$] |
| mass of one motor | $m_{m_{\nu^\mu}}$ | 0.05 | [kg] |
| inertia of one motor | $J_{m_{\nu^\mu}}$ | 0.0003 | [kgm$^2$] |
| parametric deviations | $\delta$ | 2 | [%] |
| 3-sigma Gauss. noise in pos. | $3\sigma_p$ | 0.001 | [m] |
| 3-sigma Gauss. noise in vel. | $3\sigma_v$ | 0.005 | [m/s] |
| 3-sigma Gauss. noise in $\theta_0$. | $3\sigma_0$ | 0.01 | [rad] |
| 3-sigma Gauss. noise in $\theta_0$. | $3\sigma_{d0}$ | 0.1 | [rad/s] |
| 3-sigma Gauss. noise in $\mathbf{q}_r$. | $3\sigma_r$ | 0.001 | [rad] |
| 3-sigma Gauss. noise in $\dot{\mathbf{q}}_r$. | $3\sigma_{dr}$ | 0.005 | [rad/s] |
| Sampling of lin. pos. and vel. | - | 100 | [Hz] |
| Sampling of angle and ang. vel. | - | 500 | [Hz] |

Table 6.7: Summary of the parameters used in the simulations, whose results were given in Fig. 6.23. The parameters employed by the controller are all subject to a random parametric deviation within ±2%.

the tracking of the flat outputs (see especially the absolute joint angles), which are due to the unknown masses. However these errors goes to zero again thanks to the integral terms defined in the controller (see Sec. 6.3.3), and their effects on both PVTOL CoM and end-effector positions are negligible, which is a particularly interesting result, since they are part of the flat outputs we track (see Proposition 9 and Corollary 2).

## 6.4 Towards Control in 3D

In Section 6.3 we have studied the modeling and control of the protocentric aerial manipulators (PAMs) in a plane. There, we showed the differential flatness property of that kind of system, and a dynamic feedback linearization controller using this property.

Clearly, as many of the robotics implementations, aerial manipulation task have typically to be performed in 3D. So, how can we use what we have found in Section 6.3 for such scenarios? Luckily, there are plenty of tasks in 3D that are actually 2D, and especially for aerial manipulators many of them can be imagined as 2D tasks immersed in a 3D environment (e.g. pick and place, object transportation, etc). This means, although the differential flatness property of PAMs shown in Sec. 6.3 is considering only the planar dynamics, its implementation can be extended into 3D.

In this section, we present a *decentralized flatness-based controller*, which can control the 3D dynamics of a PAM, while in the same time can track its differentially flat outputs. The *decentralized* part of the controller is developed for a generic aerial manipulator, which requires some *feed-forward* quantities, preferably aware of the complex, coupled, nonlinear system dynamics in advance. We provide these terms analytically by using the differential flatness property of the PAMs. Different from a Dynamic Feedback Linearization (DFL) controller (e.g. the one in Sec. 6.3.3), this controller does not perform an exact linearization of the nonlinear system. As also discussed in Section 1.4.2, DFL cancels all the nonlinearities, including the useful ones. Using a *flatness-based* controller as described in this section, we avoid this cancellation. Furthermore, as most of the sophisticated controllers which are aware of the system dynamics, DFL requires torque-controlled motors, and in the case of aerial manipulation small-size light-weight arms with torque controlled actuators are either not available at a low price or not reliable enough in the torque control modality[24]. The controller presented here performs best when the actuators of the manipulating arms are torque-controlled, but it can easily be modified for controlling the off-the-shelf servo motors. In fact, this is exactly what we do for the experiments described in Sec. 6.4.3.

In the following, Sec. 6.4.1 presents the dynamics of a generic aerial manipulator in 3D, using a notation based on the one presented in Sec. 6.3. Then first, a PD-based *decentralized* controller is presented in Sec. 6.4.2, for controlling the motion of a generic aerial manipulator. We are particularly interested in the protocentric designs, i.e. the first joints of each manipulating arm is assumed to be attached to the CoM of the VTOL (see also Assumption 3). Such design removes the concerns put in Remarks 15 and 16. Furthermore, we consider the case when motion of the manipulating arms are constrained on a plane, and all joints are rigidly actuated (due to the concerns raised in Remark 14 we do not consider any elastic actuation here). This allows us using the differential flatness property discovered in Sec. 6.3.1, and implement it in 3D together with the decentralized controller. This is explained in the second part of Sec. 6.4.2. Finally, in Sec. 6.4.3 we present the experimental validation of the results found in Sec. 6.3, using the controller described in this part of the thesis.

We note that this part of the thesis is published online in Tognon et al. (2017).

## 6.4.1 PAM in 3D

Consider a generic aerial manipulator, consisting of a multi-rotor aerial platform (a *Vertical Take-off and Landing* vehicle, VTOL for short) equipped with $m$ robotic arms. Different from Section 6.3, the system is in 3D, but the manipulating arms are moving on a plane. A particular example of the generic aerial manipulator model is shown in Fig. 6.24. This is clearly the extension of the system studied in Sec. 6.3 to 3D, by keeping the arm movements on the $\mathbf{x}_W - \mathbf{z}_W$ plane, which is particularly interesting because of the reasons explained in Sec. 6.4. For now, let us consider a generic aerial manipulator, i.e. the motion of the arms is not necessarily constrained to a plane.

Let us then denote with $\mathcal{F}_W : \{P_W, \mathbf{x}_W, \mathbf{y}_W, \mathbf{z}_W\}$ and $\mathcal{F}_0 : \{P_0, \mathbf{x}_0, \mathbf{y}_0, \mathbf{z}_0\}$ the world (inertial) frame, and the frame attached to the VTOL in 3D, respectively, where $P_0$ is the CoM of the VTOL. The world frame is as always chosen according to the common *North-East-Down* (NED) convention, and the orientation of $\mathcal{F}_0$ in $\mathcal{F}_W$ is described by the rotation matrix $\mathbf{R}_0 = [\mathbf{x}_0 \ \mathbf{y}_0 \ \mathbf{z}_0] \in SO(3)$, same as in (2.1). It is parametrized by the roll-pitch-yaw angles $\boldsymbol{\eta} = [\phi_0 \ \theta_0 \ \psi_0]^T \in \mathbb{R}^3$, this time a subscript $*_0$ is appearing for denoting the flying base orientations.

---

[24]In Sec 6.2.10 we needed to perform some parametric identifications and use an inverse linear model of the motor for turning a servo motor into a torque-controlled one. Note that identification was done for a certain parametric range of the actuator, which limits the performance of such inversion.

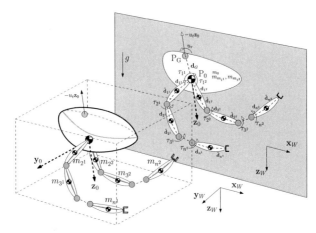

Figure 6.24: Sketch of a PAM with two manipulating arms ($m = 2$) in 3D. Notice that in the 3D model of the PAM (the one on the left down), the axis of rotation for each individual joint of each arm is parallel to $\mathbf{y}_0$. Hence its projection on $\mathbf{x}_W - \mathbf{z}_W$ plane looks like the one in the grayed area (right up). Notice the possibility that the geometric center of base actuation, $P_G$, and the CoM of the flying base, $P_0$ are not coinciding in the same point.

The joint configurations of the manipulating arms are described by the vector $\boldsymbol{\theta}_\mu = [\theta_{1\mu} \; \cdots \; \theta_{n\mu}]^T \in \mathbb{R}^{n\mu}$, which contains the relative orientations of the joints (notice that this vector is different from $\mathbf{q}_r$ described in Sec. 6.3, which contains the absolute joint orientations). The total number of links of the generic aerial manipulator is again $n = \sum_{\mu=1}^m n^\mu$. A possible set of generalized coordinates of the whole platform is

$$\mathbf{q} = [\mathbf{p}_0^T \; \boldsymbol{\eta}^T \; \boldsymbol{\theta}^T]^T \in \mathbb{R}^{6+n}, \tag{6.98}$$

where $\mathbf{p}_0 = [x_0 \; y_0 \; z_0]^T \in \mathbb{R}^3$ is the position of $P_0$ expressed in $\mathcal{F}_W$ and $\boldsymbol{\theta} = [\boldsymbol{\theta}_1^T \; \cdots \; \boldsymbol{\theta}_m^T]^T \in \mathbb{R}^n$.

We can collectively refer to all joint torques as $\boldsymbol{\tau} = [\boldsymbol{\tau}_1^T \; \cdots \; \boldsymbol{\tau}_m^T]^T \in \mathbb{R}^n$. Recall that the flying base (VTOL) is actuated by the thrust $u_t$, which is a scalar force value acting perpendicularly to the platform (in the direction of $-\mathbf{z}_0$), and by the torque $\mathbf{u}_r \in \mathbb{R}^3$. Also remember that with $P_G$ we denote the *center of actuation* of the PVTOL, whose constant position in $\mathcal{F}_0$ is denoted with $\mathbf{d}_G = [d_{G_x} \; d_{G_y} \; d_{G_z}]^T \in \mathbb{R}^3$. The *thrust vector* $\mathbf{f}_0 = -u_t \mathbf{z}_0 \in \mathbb{R}^3$ and the base torque $\mathbf{u}_r$ are applied at and around $P_G$, respectively. The overall control input of the whole generic aerial manipulator is

$$\mathbf{u} = [u_t \; \mathbf{u}_r^T \; \boldsymbol{\tau}^T]^T \in \mathbb{R}^{4+n}. \tag{6.99}$$

The system can be modeled dynamically using the classical Lagrangian notation

$$\mathbf{M}(\mathbf{q})\ddot{\mathbf{q}} + \mathbf{c}(\mathbf{q}, \dot{\mathbf{q}}) + \mathbf{g} = \mathbf{G}(\boldsymbol{\eta})\mathbf{u}, \tag{6.100}$$

where $\mathbf{M}$ is the inertia matrix, $\mathbf{c}$ is the vector of Coriolis and centrifugal forces, $\mathbf{g}$ is the vector of gravity forces, and $\mathbf{G}$ is the input matrix

$$\mathbf{G}(\boldsymbol{\eta}) = \begin{bmatrix} -\mathbf{R}_0 \mathbf{e}_3 & \mathbf{0} \\ \mathbf{0} & \mathbf{I}_{3+n} \end{bmatrix} \in \mathbb{R}^{(6+n) \times (4+n)}, \tag{6.101}$$

where $\mathbf{I}_k$ is the $k \times k$ identity matrix, $\mathbf{e}_3$ is the third column of $\mathbf{I}_3$, and $\mathbf{0}$ is the zero matrix, of appropriate dimension.

Since the control input has less elements $(4 + n)$ than the configuration variables $(6 + n)$, the system is underactuated.

Further, the inertia matrix has the following structure

$$\mathbf{M}(\mathbf{q}) = \begin{bmatrix} m_s\mathbf{I}_3 & \mathbf{M}_{pr} \\ \mathbf{M}_{pr}^T & \mathbf{M}_r \end{bmatrix} \in \mathbf{R}^{(6+n)\times(6+n)}, \qquad (6.102)$$

where $m_s$ is the total mass of the system.

---

**Remark 15.** *Because of the underactuation, commonly in multi-rotor platforms the position* $\mathbf{p}_0$ *and the yaw* $\psi_0$ *are controlled, while the roll* $\phi_0$ *and the pitch* $\theta_0$ *are used as virtual inputs or they are left uncontrolled. In the case of aerial manipulation, the position of the end-effectors does not only depend on* $\mathbf{p}_0$, $\psi_0$ *and* $\boldsymbol{\theta}$, *but also on* $\phi_0$ *and* $\theta_0$. *Consequently, it is not possible to plan exclusively for* $\mathbf{p}_0$, $\psi_0$ *and* $\boldsymbol{\theta}$ *if the position of the end-effectors is to be controlled.*

---

**Remark 16.** *The inertia matrix* $\mathbf{M}$ *exhibits dynamic couplings between all elements of the state. This considerably complicates the control problem.*

---

## 6.4.2 Decentralized Flatness-Based Control

In this section we first present a *decentralized* controller for a generic aerial manipulator in 3D. By decentralization, we mean that the controller does not consider the dynamic coupling of the complex system, *explicitly*. However, it does take the system dynamics *implicitly* into account, by using some *feed-forward* terms. Moreover, it uses *feed-back* terms for steering the system to a desired behavior while providing some robustness to the closed-loop system.

Then we consider a particular design of the aerial manipulators, namely protocentric aerial manipulators (PAMs), with all its joints are rigidly actuated and the joint motion is constrained on a plane. A sketch of such system is given in Fig. 6.24. Since from Sec. 6.3 we are aware of the differential flatness property of such systems, we us it for computing the feed-forward terms of the decentralized controller, which makes it aware of the complex system dynamics in advance.

### Decentralized Controller

Now, say $\mathbf{y}^d(t)$ stands for the desired output of the system given in (6.100), and our objective is to track this output. If the desired output trajectory is consistent with the underactuation, it is in theory possible to find some corresponding *desired* states and inputs as

$$\mathbf{q}^d(t) = [\mathbf{p}_0^{d^T} \ \boldsymbol{\eta}^{d^T} \ \boldsymbol{\theta}^{d^T}]^T, \quad \mathbf{u}^d = [u_t^d \ \mathbf{u}_r^{d^T} \ \boldsymbol{\tau}^{d^T}]^T,$$

$$\dot{\mathbf{q}}^d(t) = [\dot{\mathbf{p}}_0^{d^T} \ \dot{\boldsymbol{\eta}}^{d^T} \ \dot{\boldsymbol{\theta}}^{d^T}]^T, \qquad\qquad\qquad (6.103)$$

where we assume that these desired values are given; hence we will call them *feed-forward* terms. Notice that these terms can be computed as the *nominal* states and the inputs using the differential flatness property of the aerial manipulator (as shown in Sec. 6.3). In fact, doing so, we will be using the knowledge of the system dynamics in a decentralized controller.

Now, let us first address the control of the aerial platform, in this case a VTOL. We develop a hierarchical approach based on the separation of the translational and rotational dynamics, which eventually tracks the position $\mathbf{p}_0^d$. Firstly let us define the controlled thrust vector as:

$$\mathbf{f}_0 = \mathbf{f}_0^d + \mathbf{f}_0^\star = \mathbf{f}_0^d + \mathbf{K}_{\mathbf{p}_0}^P(\mathbf{p}_0^d - \mathbf{p}_0) + \mathbf{K}_{\mathbf{p}_0}^D(\dot{\mathbf{p}}_0^d - \dot{\mathbf{p}}_0), \tag{6.104}$$

where $\mathbf{K}_{\mathbf{p}_0}^P, \mathbf{K}_{\mathbf{p}_0}^D \in \mathbb{R}_{\geq 0}^{3\times3}$. Notice that $\mathbf{f}_0$ is computed as a combination of the feed-forward terms $(\cdot^d)$, and the feedback term $(\cdot^\star)$ proportional to the state error of the system with respect to the nominal one. From the controlled thrust vector we can retrieve the commanded thrust as

$$u_t = -(\mathbf{R}_0\mathbf{e}_3)^T\mathbf{f}_0, \tag{6.105}$$

and the commanded attitude as

$$\begin{aligned} \mathbf{z}_0^c &= \mathbf{f}_0/\left\|\mathbf{f}_0\right\|, & \mathbf{y}_0^c &= \mathbf{z}_0^c \times \mathbf{e}_1, \\ \mathbf{x}_0^c &= \mathbf{y}_0^c \times \mathbf{z}_0^c, & \mathbf{R}_0^c &= [\mathbf{x}_0^c\ \mathbf{y}_0^c\ \mathbf{z}_0^c]. \end{aligned} \tag{6.106}$$

This closes the *outer-loop control*. The controlled attitude is then passed to the *inner-loop control* as the desired attitude, to compute the controller torque as:

$$\begin{aligned} [\mathbf{e}_{\mathbf{R}_0}]_\wedge &= 1/2(\mathbf{R}_0^{cT}\mathbf{R}_0 - \mathbf{R}_0^T\mathbf{R}_0^c) \\ \mathbf{e}_{\boldsymbol{\omega}_0} &= \boldsymbol{\omega}_0^d - \boldsymbol{\omega}_0 \\ \mathbf{u}_r &= \mathbf{u}_r^d + \mathbf{u}_r^\star = \mathbf{u}_r^d + \mathbf{K}_{\mathbf{R}_0}^P\mathbf{e}_{\mathbf{R}_0} + \mathbf{K}_{\mathbf{R}_0}^D\mathbf{e}_{\boldsymbol{\omega}_0}, \end{aligned} \tag{6.107}$$

where, $[*]_\wedge$ represents the skew operation, $\boldsymbol{\omega}_0 \in \mathbb{R}^3$ and $\boldsymbol{\omega}_0^d \in \mathbb{R}^3$ are the current and desired angular velocities of the VTOL body in body-fixed frame[25], and $\mathbf{K}_{\mathbf{R}_0}^P, \mathbf{K}_{\mathbf{R}_0}^D \in \mathbb{R}_{\geq 0}^{3\times3}$.

Now, let us give the control of the generic $\nu^\mu$-th joint, in order to track the relative desired angle. For a torque-controlled motor, we design the control law based on a PD strategy as

$$\tau_{\nu^\mu} = \tau_{\nu^\mu}^d + \tau_{\nu^\mu}^\star = \tau_{\nu^\mu}^d + k_{\nu^\mu}^P(\theta_{\nu^\mu}^d - \theta_{\nu^\mu}) + k_{\nu^\mu}^D(\dot{\theta}_{\nu^\mu}^d - \dot{\theta}_{\nu^\mu}), \tag{6.108}$$

where $k_{\nu^\mu}^P, k_{\nu^\mu}^D \in \mathbb{R}_{\geq 0}$. This controller ensures the best performances. Nonetheless, for kinematically controlled motors, it is possible to adapt the controller for achieving good results. For instance, for a velocity-controlled motor, the commanded velocity can be given as

$$\dot{\theta}_{\nu^\mu} = \dot{\theta}_{\nu^\mu}^d + k_{\nu^\mu}^P(\theta_{\nu^\mu}^d - \theta_{\nu^\mu}). \tag{6.109}$$

See also Sec. 6.4.3 for its implementation.

In summary, the VTOL thrust, $u_t$, is computed in (6.105); its torque is given in (6.107); and the control input of the individual motors of the manipulators, $\tau_{\nu^\mu}$ in (6.108) which

---

[25]Notice that $\boldsymbol{\omega}_0$ can be easily computed using $\boldsymbol{\eta}$ and $\dot{\boldsymbol{\eta}}$. This also applies to $\boldsymbol{\omega}_0^d$ using $\boldsymbol{\eta}^d$ and $\dot{\boldsymbol{\eta}}^d$ available from (6.103).

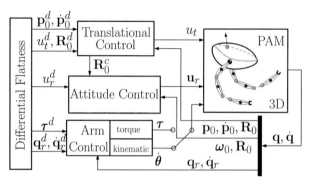

Figure 6.25: Control of the PAM depicted in Fig. 6.24, using the decentralized controller explained in Sec. 6.4.2 and its differential flatness property found before in Sec. 6.3.1.

collectively builds the torque input $\tau$. Hence, we have all the control inputs $\mathbf{u}$ of the system in (6.100). A simple variant, as in (6.109), allows the use of this controller for kinematically-controlled motors. A schematic representation of the controller is shown in Fig. 6.25. Now let us show how to use the differential flatness of a specific type of aerial manipulator for computing the feed-forward terms of this controller.

**Using Differential Flatness of PAMs**

For the decentralized controller presented above to track a desired output properly, an algorithm computing all the nominal states and inputs (feed-forward terms) is required. We can do this by using the differential flatness property of the PAMs, as found in Sec. 6.3 of this thesis. There, a PAM (with any number of manipulator arms, each having any number of DoFs, with rigid or compliant transmission) is characterized by all manipulator arms being attached to the CoM of the flying base. In Sec. 6.3, we studied the properties of such systems in the 2D vertical plane, and we found that they are differentially flat w.r.t. a set of flat outputs given by the CoM position of the flying base and the absolute rotations of the manipulator links. The choice of the absolute joint angles as system coordinates, together with the protocentric design, overcomes both difficulties highlighted in Remarks 15 and 16. In particular, also the position of the end-effector and the absolute rotations of the manipulator links are flat outputs, which makes such platforms of particular interest.

Now, consider a PAM in 3D, where the motion of all manipulators are constrained to a plane, i.e. $y_0 = 0$ and $\mathbf{y}_0^T \mathbf{z}_W = \mathbf{y}_0^T \mathbf{x}_W = 0$. A sketch of such design is depicted in Fig. 6.24, where each joint of all manipulators rotate around an axis parallel to $\mathbf{x}_0 \times \mathbf{z}_0$. Now, notice the similarity between the projection of the considered PAM on the $\mathbf{x}_W - \mathbf{z}_W$ plane, and the system discussed in Sec. 6.3; they are the same for the case when all joints are rigid.

Given the above, the generalized coordinates of a PAM in 2D can be re-written as

$$\mathbf{q}_2 = \begin{bmatrix} \mathbf{p}_0^T & \theta_0 & \mathbf{q}_r^T \end{bmatrix}^T \in \mathbb{R}^{3+n}, \tag{6.110}$$

where this time $\mathbf{p}_{0_{xz}} = [x_0 \ z_0]^T \in \mathbb{R}^2$ is the position of the CoM of the flying base in the $\mathbf{x}_W - \mathbf{z}_W$ plane, $\theta_0$ is the pitch as before, and $\mathbf{q}_r = [\mathbf{q}_{r1}^T \ \cdots \ \mathbf{q}_{rm}^T]^T \in \mathbb{R}^n$ is the vector combining the *absolute* orientations of each joint of every arm as, with $\mathbf{q}_{r\mu}^T = [\theta_{01\mu} \cdots \theta_{0n\mu}]^T \in \mathbb{R}^{n\mu}$ written for the $\mu$-th manipulating arm, as before. Remember from Sec. 6.3 that $\theta_{0\nu\mu}$ is the absolute orientation of the $\nu$-th joint of the $\mu$-th arm, and that $\theta_{0k\mu} = \theta_0 + \sum_{\nu=1}^{k} \theta_{\nu\mu}$. Let us re-write the set of inputs as

$$\mathbf{u}_2 = [u_t \quad u_r \quad \boldsymbol{\tau}^T]^T \in \mathbb{R}^{2+n}. \tag{6.111}$$

Notice that the PAM in 2D is also underactuated, as the one in 3D. Finally, we define this time $\mathbf{p}_{e_{xz}^\mu}$ as the position of the $\mu$-th end-effector in the considered plane.

We can now turn the Proposition 9 and Corollary 2 into the following fact:

---

**Fact 2.** $\mathbf{y} = [\mathbf{p}_{0_{xz}}^T \ \mathbf{q}_r^T]^T \in \mathbb{R}^{(n+2)}$ *is a flat output of a PAM modeled in 2D. Hence, clearly,* $\mathbf{y}_e = [\mathbf{p}_{e_{xz}^\mu}^T \ \mathbf{q}_r^T]^T \in \mathbb{R}^{(n+2)}$ *is a flat output, for any* $\mu$.

---

This means we can control the motion of a PAM as shown in Fig. 6.24 in the $\mathbf{x}_W - \mathbf{z}_W$ plane, by combining the decentralized controller presented in above and the flatness property proven in Sec. 6.3. By setting the desired motions of all the other DoFs to *zero*, we can control the overall system in 3D, pawing the way for the experimental validation of the results presented in Sec. 6.3.

This requires the computation of the nominal states and the control inputs of the PAM in the $\mathbf{x}_W - \mathbf{z}_W$ plane, as functions of $\mathbf{y}, \dot{\mathbf{y}}, \ddot{\mathbf{y}}, \dddot{\mathbf{y}}, \ddddot{\mathbf{y}}$. Considering Fact 2, the nominal states and the control inputs are to be computed as sole functions of the flat ouputs are $\theta_0, \dot{\theta}_0$ (flying base pitch and its time derivatives), and $u_t, u_r, \tau_{\nu\mu}$.

The thrust, and pitch with its derivatives can be computed using the translational dynamics of the CoM position of the PAM given in (A.51). Remember there, we defined the vector

$$\mathbf{w} = \mathbf{w}(\mathbf{y}, \dot{\mathbf{y}}, \ddot{\mathbf{y}}) = \ddot{\mathbf{p}}_c - [0 \ g]^T = [w_x \ w_z]^T \in \mathbb{R}^2, \tag{6.112}$$

which is direct function of only the flat outputs. Notice that $\mathbf{w} = -\frac{u_t}{m_s}[\sin(\theta_0) \ \cos(\theta_0)]^T$. Hence,

$$\theta_0 = \theta_0(\ddot{\mathbf{p}}_c) = \operatorname{atan2}(-w_x, -w_z), \quad \dot{\theta}_0 = \dot{\theta}_0(\ddot{\mathbf{p}}_c, \dddot{\mathbf{p}}_c) = \frac{w_z \dot{w}_x - w_x \dot{w}_z}{w_x^2 + w_z^2}$$

$$\ddot{\theta}_0 = \ddot{\theta}_0(\ddot{\mathbf{p}}_c, \dddot{\mathbf{p}}_c, \ddddot{\mathbf{p}}_c) = \frac{\ddot{w}_x w_z - w_x \ddot{w}_z}{w_x^2 + w_z^2} - \frac{2[(w_z^2 - w_x^2)\dot{w}_x \dot{w}_z + (\dot{w}_x^2 - \dot{w}_z^2)w_x w_z]}{(w_x^2 + w_z^2)^2} \tag{6.113}$$

$$u_t = u_t(\ddot{\mathbf{p}}_c) = m_s ||\mathbf{w}||.$$

Therefore, we need to compute the time derivatives of $\mathbf{p}_c$ from second up to the fourth order, as sole functions of the flat outputs. This is done in Tognon et al. (2017) and for brevity we do not report them here. Then, by placing those terms in (6.113) appropriately, one can compute the nominal values of $\theta_0(\mathbf{y}, \dot{\mathbf{y}}, \ddot{\mathbf{y}})$, $\dot{\theta}_0(\mathbf{y}, \dot{\mathbf{y}}, \ddot{\mathbf{y}}, \dddot{\mathbf{y}})$, and $u_t(\mathbf{y}, \dot{\mathbf{y}}, \ddot{\mathbf{y}})$ as sole functions of the flat outputs and their derivatives up to the fourth order.

The nominal torque of the $\nu^\mu$-th motor is computed as in (A.53), which brings it in the form of $\tau_{\nu\mu} = \tau_{\nu\mu}(\mathbf{y}, \dot{\mathbf{y}}, \ddot{\mathbf{y}})$. The PVTOL torque is computed in (A.54) with $\ddot{\theta}_0$ is available from (6.113), which means it can be represented as $u_r = u_r(\mathbf{y}, \dot{\mathbf{y}}, \ddot{\mathbf{y}}, \dddot{\mathbf{y}}, \ddddot{\mathbf{y}})$.

Finally, we have showed how to use the flat outputs $\mathbf{y}$ given in Fact 2 and their derivatives up to the fourth order, for computing the nominal values of $\theta_0, \dot{\theta}_0, u_t, u_r, \boldsymbol{\tau}$. It is clear

| Flat outputs | From Fact 2, $\mathbf{y} = [\mathbf{p}_{0_{zz}}^T, \mathbf{q}_r^T]^T \in \mathbb{R}^{2+n}$. Since $\exists f_e : \mathbf{p}_{0_{zz}} = f_e(\mathbf{p}_{e_{zz}^\mu}, \mathbf{q}_r)$, it is also $\mathbf{y}_e = [\mathbf{p}_{e_{zz}^\mu}^T, \mathbf{q}_r^T]^T \in \mathbb{R}^{2+n}$. |
|---|---|
| Nominal States | Considering (6.110); $\mathbf{p}_{0_{zz}}$, $\dot{\mathbf{p}}_{0_{zz}}$, $\mathbf{q}_r$ and $\dot{\mathbf{q}}_r$ are direct functions of $\mathbf{y}$ and $\dot{\mathbf{y}}$. This leaves $\theta_0$ and $\dot{\theta}_0$ to be computed. |
| | From (6.113), $\theta_0 = \theta_0(\ddot{\mathbf{p}}_c)$. It is shown in Tognon et al. (2017): $\ddot{\mathbf{p}}_c = \ddot{\mathbf{p}}_c(\ddot{\mathbf{p}}_{0_{zz}}, \ddot{\mathbf{p}}_{m_{\nu\mu}}, \ddot{\mathbf{p}}_{\nu\mu})$, $\ddot{\mathbf{p}}_{m_{\nu\mu}} = \ddot{\mathbf{p}}_{m_{\nu\mu}}(\mathbf{y}, \dot{\mathbf{y}}, \ddot{\mathbf{y}})$, |
| | $\ddot{\mathbf{p}}_{\nu\mu} = \ddot{\mathbf{p}}_{\nu\mu}(\mathbf{y}, \dot{\mathbf{y}}, \ddot{\mathbf{y}}) \implies \ddot{\mathbf{p}}_c = \ddot{\mathbf{p}}_c(\mathbf{y}, \dot{\mathbf{y}}, \ddot{\mathbf{y}})$. Then $\underline{\theta_0 = \theta_0(\mathbf{y}, \dot{\mathbf{y}}, \ddot{\mathbf{y}})}$. |
| | From (6.113), $\dot{\theta}_0 = \dot{\theta}_0(\ddot{\mathbf{p}}_c, \dddot{\mathbf{p}}_c)$. It is shown in Tognon et al. (2017): $\dddot{\mathbf{p}}_c = \dddot{\mathbf{p}}_c(\dddot{\mathbf{p}}_{0_{zz}}, \dddot{\mathbf{p}}_{m_{\nu\mu}}, \dddot{\mathbf{p}}_{\nu\mu})$, $\dddot{\mathbf{p}}_{m_{\nu\mu}} = \dddot{\mathbf{p}}_{m_{\nu\mu}}(\mathbf{y}, \dot{\mathbf{y}}, \ddot{\mathbf{y}}, \dddot{\mathbf{y}})$, |
| | $\dddot{\mathbf{p}}_{\nu\mu} = \dddot{\mathbf{p}}_{\nu\mu}(\mathbf{y}, \dot{\mathbf{y}}, \ddot{\mathbf{y}}, \dddot{\mathbf{y}}) \implies \dddot{\mathbf{p}}_c = \dddot{\mathbf{p}}_c(\mathbf{y}, \dot{\mathbf{y}}, \ddot{\mathbf{y}}, \dddot{\mathbf{y}})$. Then $\underline{\dot{\theta}_0 = \dot{\theta}_0(\mathbf{y}, \dot{\mathbf{y}}, \ddot{\mathbf{y}}, \dddot{\mathbf{y}})}$. |
| Nominal Inputs | Considering (6.113) and $\ddot{\mathbf{p}}_c$ from the above; first $u_t = u_t(\ddot{\mathbf{p}}_c) \implies \underline{u_t = u_t(\mathbf{y}, \dot{\mathbf{y}}, \ddot{\mathbf{y}})}$. |
| | From (A.53), $\exists f_\tau : \tau_{\nu\mu} = \tau_{\nu^\mu+1} + f_\tau(\ddot{\mathbf{p}}_{0_{zz}}, \mathbf{q}_r, \dot{\mathbf{q}}_r, \ddot{\mathbf{q}}_r) \implies \underline{\tau_{\nu\mu} = \tau_{\nu\mu}(\mathbf{y}, \dot{\mathbf{y}}, \ddot{\mathbf{y}})}$ where $\tau_{\nu^\mu+1} = 0$ for $\nu^\mu = n^\mu$. |
| | From (A.54), $u_r = J_0\ddot{\theta}_0 + \sum_{j=1}^m \tau_{1j} - d_{G_z}u_t$. Above it is show that $u_t = u_t(\mathbf{y}, \dot{\mathbf{y}}, \ddot{\mathbf{y}})$, $\tau_{1^\mu} = \tau_{1^\mu}(\mathbf{y}, \dot{\mathbf{y}}, \ddot{\mathbf{y}})$, |
| | and from (6.113) it is $\ddot{\theta}_0 = \ddot{\theta}_0(\mathbf{y}, \dot{\mathbf{y}}, \ddot{\mathbf{y}}, \dddot{\mathbf{y}}, \ddddot{\mathbf{y}})$. Then, $\underline{u_r = u_r(\mathbf{y}, \dot{\mathbf{y}}, \ddot{\mathbf{y}}, \dddot{\mathbf{y}}, \ddddot{\mathbf{y}})}$. |

Table 6.8: A summarizing table of the differential flatness of PAMs in 2D when all the joints are rigid. Different outputs of the system are given on the top. The nominal states and inputs as implicit functions of the flat outputs and their derivatives up to the fourth order are provided. Remember that for the $\nu^\mu$-th element of the system; $\mathbf{p}_{\nu^\mu}$ and $\mathbf{p}_{m_{\nu^\mu}}$ are the (time varying) individual link and motor CoM positions presented in $\mathcal{F}_W$, respectively.

that all the other states are actually the flat outputs themselves. This implies that, $\exists h : \mathbf{b} = h(\mathbf{y}, \dot{\mathbf{y}}, \ddot{\mathbf{y}}, \dddot{\mathbf{y}}, \ddddot{\mathbf{y}})$, where $\mathbf{b} = [\mathbf{q}_2^T \; \dot{\mathbf{q}}_2^T \; \mathbf{u}_2^T]^T$ is a vector combining all the states and inputs of the PAM in 2D. Finally, the relative joint angles can be easily obtained as $\theta_{1\mu} = \theta_{01\mu} - \theta_0$ and $\theta_{\nu\mu} = \theta_{0\nu\mu} - \theta_{0(\nu-1)\mu}$ for $\nu > 1$, which constructs the vector $\boldsymbol{\theta}$. The computations of the nominal states and the inputs is summarized in Table 6.8 for the convenience of the reader.

Now, for tracking a desired $\mathbf{y}^d(t)$, where $\mathbf{y}$ is as in Fact 2, using the differential flatness property in 2D, but together with the controller developed in 3D, we can say

$$\mathbf{R}_0^d = \begin{bmatrix} c_{\theta_0^d} & 0 & s_{\theta_0^d} \\ 0 & 1 & 0 \\ -s_{\theta_0^d} & 0 & c_{\theta_0^d} \end{bmatrix} \tag{6.114}$$

$$\mathbf{f}_0^d = -u_t^d \mathbf{R}_0^d \mathbf{e}_3, \qquad \boldsymbol{\omega}_0^d = \dot{\theta}_0^d \mathbf{e}_2, \tag{6.115}$$

$$\mathbf{u}_r^d = u_r^d \mathbf{e}_2, \qquad y_0^d \equiv 0 \tag{6.116}$$

will impose the necessary constraints. Notice that $\theta_0^d, \dot{\theta}_0^d, u_t^d$ are computed as in (6.113) and $u_r^d$ as in (A.54) for $\mathbf{y}^d$. Clearly, $\tau_{\nu\mu}^d$ will be computed in the same way using (A.53). Then we can use these values as the feed-forward terms of the controller as shown in Fig. 6.25.

Notice that the flatness of $\mathbf{y}_e$ in Fact. 2 is quite obvious, thanks to the protocentric design and the absolute joint coordinates. Let us give the following remark:

---

**Remark 17.** *The flat outputs of a PAM are:*

- $\mathbf{y} = [\mathbf{p}_{0_{xz}}^T \; \mathbf{q}_r^T]^T \in \mathbb{R}^{(2+n)}$ *from Sec. 6.3,*
- $\mathbf{y}_e = [\mathbf{p}_{e_{xz}^\mu}^T \; \mathbf{q}_r^T]^T \in \mathbb{R}^{(2+n)}$ *for any $\mu$, since $\exists f_e : \mathbf{p}_{0_{xz}} = f_e(\mathbf{p}_{e_{xz}^\mu}, \mathbf{q}_r)$.*

---

| Physical param. | VTOL | $1^{st}$-Link | $2^{nd}$-Link |
|---|---|---|---|
| Mass [kg] | 1.3 | 0.145 | 0.123 |
| Rot. inertia [kgm$^2$] | 0.03 | $1.2 \cdot 10^{-3}$ | $0.9 \cdot 10^{-3}$ |
| Length [m] | 0.4 (diam.) | 0.29 | 0.25 |

| Controller | $\mathbf{K}^P_{\mathbf{p}_0}$ | $\mathbf{K}^D_{\mathbf{p}_0}$ | $\mathbf{K}^P_{\mathbf{R}_0}$ | $\mathbf{K}^D_{\mathbf{R}_0}$ | $k^P_1$ | $k^P_2$ |
|---|---|---|---|---|---|---|
| Gain | $12\mathbf{I}_3$ | $7\mathbf{I}_3$ | $3\mathbf{I}_3$ | $0.3\mathbf{I}_3$ | 1.8 | 1.6 |

| Trajectory Parameters | $a^x_{\mathbf{p}_0}$ | $a^1_{\mathbf{q}_r}$ | $a^2_{\mathbf{q}_r}$ | $\omega_t$ |
|---|---|---|---|---|
| | [m] | [deg] | [deg] | [deg/s] |
| (a) | 0 | 30 | 60 | $2\pi/3$ |
| (b) | 0.5 | -40 | -70 | $2\pi/3$ |
| (c) | 0.5 | 40 | 70 | $2\pi/3$ |

Table 6.9: Starting from the top: physical parameters of the real system; controller gains; and the parameters of the three trajectories. Length and the inertia are the one on the 2D vertical plane need to compute nominal state and inputs by the flatness. We note that since both motors of the arm are rigidly placed on the quadrotor body, we compute their weights as part of the VTOL.

### 6.4.3 Experimental Results

In this section we show the results of some preliminary experiments for validating the controller proposed in Sec. 6.4.2. Furthermore, we analyze its performances by comparing it with other standard control techniques. there we show that the controller developed based on the differential flatness property of the PAMs (see Sec. 6.3) outperforms the other standard techniques for tracking various trajectories.

We note that the experiments described here are performed at LAAS-CNRS, Toulouse, France, by Marco Tognon.

Let us first describe briefly the testbed used for the experiments (see Fig. 6.26). The aerial manipulator consists of a quadrotor VTOL and a 2 DoF manipulating arm. The quadrotor VTOL is identical to the one presented in Sec. 2.3.1 and depicted in Fig. 2.2. The arm is developed in home, but its design and construction is not part of this thesis. However let us briefly describe it here. The arm structure is based on carbon fiber bars and printed plastic parts, whose design was inspired by the work in Cano et al. (2013). A big difference of this design is that the actuators of both joints are placed at the base of the arm, rigidly attached to the VTOL. The first joint is connected to its actuator[26] (a dynamixel MX-64 motor) directly, while the second one is connected to its motor (a dynamixel MX-28) via a metal-reinforced plastic belt (with very low elasticity).

A detailed description of the setup is given in Fig. 6.26. Such design allows us to have a

---

[26]http://en.robotis.com/index/

149

light-weight arm reducing the mass of each joint and in particular their inertia. This in turn allows to use a relative small quadrotor (diameter 0.4 [m], maximum thrust per propeller of about 5.26 [N] with respect to the ones normally used in the literature for arms of similar length, as, e.g., in Jimenez-Cano et al. (2013). Since the motors are rigidly attached to the aerial vehicle, their mass can be seen as part of the total VTOL mass. For the physical parameters of the system, please refer to Tab. 6.9. Since the motors cannot be controlled in torque but at best in velocity (as almost all the affordable motors suitable for aerial manipulation) to control the arm we use (6.109), except for a slight modification needed to cope with the fact that the second link is not directly attached to its motor.

The control law presented in Sec. 6.4.2, implemented in Matlab–Simulink, runs on a desktop PC sending the commanded propeller velocities at 500 [Hz] and the commanded arm motor velocities at 250 [Hz] through a serial communication. The gains used for the controller are given in Tab. 6.9. The control loop is then closed based on the measurements of: i) position, attitude, linear and angular velocities of the quadrotor at 1 [KHz] using the hardware and software explained in Section 2.3; ii) the position and velocity of the arm motors provided by their internal absolute encoders at 250 [Hz]. In order to read the motor values corresponding to zero joint angles, a calibration procedure is implemented *once*, using the Mo-Cap markers on the manipulator arm (see Fig. 6.26).

We tested the proposed controller with a parametric and multi-DoF sinusoidal-like trajectory, i.e.:

$$
\mathbf{y}^d =
\begin{bmatrix}
\mathbf{p}^d_{0_{xz}} \\
\mathbf{q}^d_r
\end{bmatrix}
=
\begin{bmatrix}
a^x_{\mathbf{p}_0} & 0 & a^1_{\mathbf{q}_r} & a^2_{\mathbf{q}_r}
\end{bmatrix}^T
\sin(\omega_t t)
\tag{6.117}
$$

for three different sets of parameters corresponding to three qualitatively different task trajectories:
(a) the arm is oscillating and quadrotor is fixed,
(b) the arm and quadrotor are oscillating with opposite phases,
(c) the arm and quadrotor are oscillating with the same phase.
These task trajectories are understandable from Fig. 6.27, and the parameters of the trajectories are given in Tab. 6.9.

For each of the three task trajectory, we compared the performance of the proposed controller using three different types of *feedforward methods*:
1) *minimal compensation*: on the quadrotor side only the total mass is compensated, i.e., $u^d_t = -m_s g \mathbf{e}^T_3 \mathbf{R}_0 \mathbf{e}_3$. In this way the VTOL and the arm virtually are considered as two independent systems (even if in practice they are not).
2) *static compensation*: only the static effects due to the gravity are compensated, i.e., the nominal state and the inputs are computed considering all the derivatives of the desired trajectory are equal to zero, i.e., $\mathbf{y}^{d^{(l)}} = \mathbf{0}$ for $l = 1, \ldots, 4$, $(\mathbf{y}^d \neq \mathbf{0})$. This method is often used for the control of the aerial manipulators, for so called *quasi-static* operations in order to partially compensate the effects of the manipulator on the aerial vehicle.
3) *dynamic compensation*: this corresponds to our proposed method where we exploit the flatness of the system. We compute the nominal states and inputs as functions of the desired trajectory to be tracked, and provide them to the controller as explain in Sec. 6.4.2.

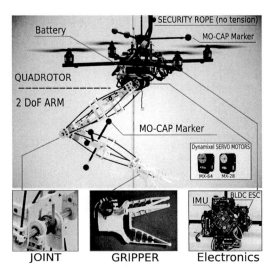

Figure 6.26: Experimental setup of the aerial manipulator. A quadrotor VTOL is equipped with a 2 DoF manipulating arm. The quadrotor setup is the one presented in Section 2.3.1, and the manipulating arm is built in home, as part of a project outside of this thesis.

The performances of these three methods are shown in Figs. 6.28, 6.29 and 6.30, using red, green and blue curves, respectively. In particular the plots show the evolution of the position of the VTOL CoM and the end-effector[27] in the first two rows, the remaining configuration variables in 3D (third and fourth row), and the inputs on the vertical plane, as well as the nominal relative quantities (with a dashed black line). Note that superscript $*^i$ in the labels showing the number of the method tested, which should not be mixed with the notation of number of manipulating arms, $\nu$, introduced in Sec. 6.3 (it was placed as the superscript of the lowest subscript as in $*_{\nu^\mu}$).

Looking at the tracking of the desired VTOL CoM and end-effector position one can see that the minimal compensation (method 1) shows good tracking performances (similar to the one with our method) only for trajectory (b). On the other hand, for trajectories (a) and (c) the tracking error is considerably larger than the one with dynamic compensation.

For the static compensation (method 2), the tracking performances result to be good (similar to the one with our proposed method 3) only for trajectory (a). Indeed, since trajectory (a) is the less dynamic one (quadrotor not moving), the static compensation is enough to obtain good performances. However, for more dynamical trajectories as (b) and (c) the performances rapidly get worse.

On the contrary, our proposed method 3 shows good tracking performances for all the types of trajectories validating the fact the dynamic compensation based on the flatness is a good control strategy for both static and dynamic trajectories. Moreover, thanks to the

---

[27]Notice that for this particular case; i.e a PAM with $m = 1$, $n = 2$ and all joints are rigidly attached, the joint angles or the position of the end-effector is equivalent since they are both flat outputs.

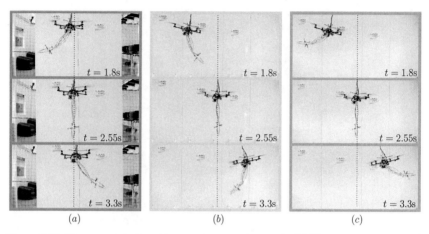

Figure 6.27: Nine moments from the experiments using method 3 (dynamic compensation). From the left to the right column the configurations of the trajectories (a), (b) and (c) are shown, respectively. From the top to the bottom row the configurations at the start (top), intermediate (middle) and end (bottom) moments of half period of each the trajectory are shown, respectively.

feedback terms, the controller is robust enough to the non perfect protocentricity of the real system. Indeed in the testbed used during the experiments, along the $z_0$-axis of $\mathcal{F}_0$ there is a non zero offset of about 6 [cm] between the position of CoM of the VTOL and the first joint. Nevertheless the controller is able to keep the tracking error small even for dynamic trajectories.

In addition to the good results obtained with our method (explained in Sec. 6.4.2) (the decentralized flatness-based controller presented in Sec. 6.4.2), it is also very interesting to notice that for trajectory (b), the method 1 based on the minimal compensation is better than the method 2 based on the static compensation in terms of tracking error. This brings us two interesting results.

The first is due to the the the fact that for some dynamic trajectories it is more suitable to just compensate the effect of the total mass rather than try to compensate the static configuration only. This error in the compensation leads to undesired effects and in turn to a large tracking error, as seen in Fig. 6.29.

The second interesting aspect is that for some particular dynamic trajectories, as for trajectory (b), the arm could help the aerial vehicle to move toward the desired direction, implying the need of smaller compensations and in turn of smaller control efforts. Indeed, looking at Fig. 6.27.b one could notice the similarity between: i) the motion of the robotic arm and the one of the legs of a person sitting on a swing when trying to enhance the angular motion of the swing; ii) the thrust force and the tension along the cables attached to the swing to win the gravity and the centrifugal terms. This is why for trajectory (b) the minimal compensation shows similar results to the one obtained with our method. Based on this consideration we believe that the studies on optimal trajectory generation become even more fundamental to achieve aerial manipulation tasks exploiting the dynamic properties (such as the flatness) of the systems.

## 6.4.4 Discussions

In summary, we have presented in Sec. 6.3 a control method for certain type of aerial manipulators, namely PAMs, and discovered their differential flatness property. An important fact is that, when the manipulating arms are attached to the CoM of the flying platform, their end-effectors become part of the differentially flat outputs. In other words, instead of looking for a different output of the system, we assume a certain dynamic model and for that model showed the *useful* outputs. This allowed us developing high-performance tracking controllers, e.g. DFL controller. Noticing few disadvantages of DFL as explained in Sec. 1.4.2, as many other dynamics-aware controller it also requires torque-controlled actuators. This can be a problem, since most of the actuators available for the aerial manipulation are not torque-controlled. However, the decentralized flatness-based controller proposed in Sec. 6.4.2 can overcome this issue.

An important fact is that, the number of total relative degree of the aerial manipulator system increases quadratically, with the number of elastic joints (see Proposition 10). Clearly, this means that if the aerial manipulator has high number of compliant actuators, only much smoother and slower trajectories can be tracked using a differential flatness based controller. As also noted in Remark 14, this is because the robotic arm configuration is represented using the *absolute joint angles*. The reason for that was the underactaion of the aerial platform. This might imply that if the flying platform is fully actuated, this drawbacks of having elastic actuator might disappear.

We have put the differential flatness property of PAMs (when all joints are rigid) in test, both numerically (in Sec. 6.3.4) and experimentally (in Sec. 6.4.3). Especially in the experiments, we show how well our (decentralized differential-based) controller performs w.r.t. the other conventional controllers. This is because it is aware of the complex nonlinear system dynamics, and this awareness is induced to the controller thanks to the differential flatness property of the PAMs.

In overall Chapter 6, we showed the differential flatness property of the aerial manipulators, when their motion is constrained on a plane. Of course, the results presented here are reasonably implementable in 3D (e.g. in Sec. 6.4.1), since many aerial manipulation tasks are 2D, but immersed in 3D. However, showing the differential flatness of the end-effector positions of the 3D system is not yet done, as well as its proper control. This is in the scope of our future works.

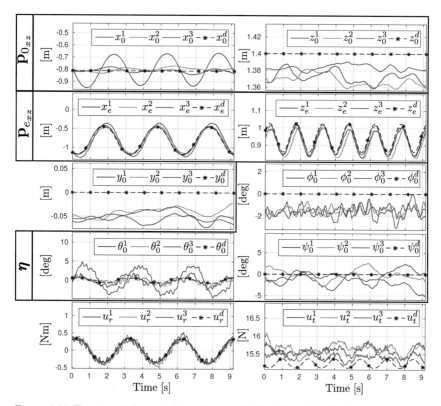

Figure 6.28: Experimental results for trajectory (a) (see Fig. 6.27.a for the snapshots from the experiment). In all plots, the flat outputs and the nominal states/inputs are depicted with black dashed lines (and stars). First two rows show the CoM and end-effector positions in $\mathbf{x}_W - \mathbf{z}_W$ plane, and they are highlighted together with the quadrotor orientation $\boldsymbol{\eta}$. Again in all plots, red, green and blue curves stands for the results of the controller with *minimal compensaton, static compensation,* and *dynamic compensation,* respectively. While the first controller perform worse for tracking of the all outputs, third one (proposed controller) is always performing good especially when tracking the end-effector positions.

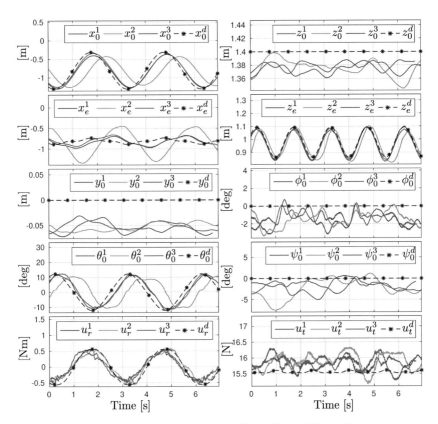

Figure 6.29: Experimental results for trajectory (b) (see Fig. 6.27.b for the snapshots from the experiment). Same color and line code is used as in Fig. 6.28, and the order of plots are the same. The proposed controller (blue) achieves always a better tracking performance. An interesting result is that for such a dynamic tracking task, the controller with the static compensation (green) performs worse than the one with minimum compensation (red).

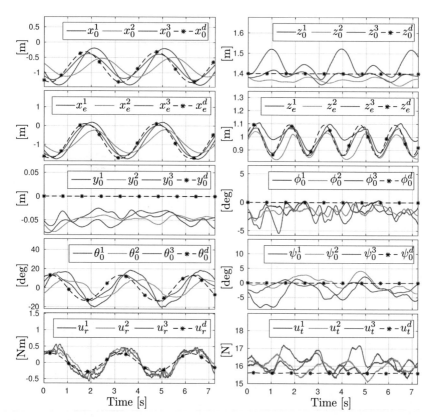

Figure 6.30: Experimental results for trajectory (c) (see Fig. 6.27.c for the snapshots from the experiment). Same color and line code is used as in Fig. 6.28, and the order of plots are the same. The proposed controller (blue) again outperforms the other methods. Note that here and in Figs. 6.28 and 6.29 superscript $*^i$ in the labels showing the number of the method tested, which should not be mixed with the notation of number of manipulating arms, $\nu$, introduced in Sec. 6.3 (it was placed as the superscript of the lowest subscript as in $*_{\nu^\mu}$).

# Chapter 7

# Conclusions

In this thesis, we studied the design, modeling and control for the APhI and aerial manipulation. For both to be performed, an appropriate aerial platform (or flying robot) had to be chosen. Our choice was a quadrotor, because of the reasons explained in Section 1.1.2. Note that a detailed literature research is provided for aerial robots (in Sec. 1.1), VTOLs (in Sec. 1.1.1), APhI (in Sec. 1.2), and aerial manipulation (in Sec. 1.3).

The aerial platform, a quadrotor VTOL, is described in detail (from its mathematical model to its realization) in Chapter 2. This chapter does not only introduce the flying robot carefully, but also exposes all its electronics, sensors with their hardware and software, in detail.

For achieving a proper APhI, we had to find a way to acquire the external wrench information. In Chapter 3, two methods are presented; one is the (indirect) estimation and the other one is the (direct) measurement of the external wrenches. Each method outperforms another one for different reasons, but at the end, we chose to use a light-weight cheap F/T sensor onbard of the quadrotor. Briefly, the estimation of the wrenches is a cheap method in terms of weight, computational power (most of the time) and money. On the other hand an F/T sensor provides precise and reliable measurements (sometimes used even as ground truths), however it brings additional weight to the flying robot. As it is explained in Sec. 3.2 we utilized a low cost, light weight F/T sensor and compared its result with a model-based external wrench estimator presented in Sec. 3.3. Based on the results, we decided to use the F/T sensor for the upcoming APhI tasks. We believe that, when especially the APhI task requires high precision and/or it is performed outdoor, usage of the F/T sensors would be even more essential.

In the light of our observations from Chapter 3, we showed how to perform APhI in Chapter 4 with a quadrotor and an F/T sensor on board of it. The most important contribution of this chapter is the improved version of the Interconnection and Damping Assignment-Passivity Based Controller (IDA-PBC) method for the quadrotors. Briefly, IDA-PBC allows controlling an APhI task for the quadrotors, by reshaping its physical properties. Moreover, it renders the controlled system to a passive one (or cyclo-passive in our case), ensuring the stability of the interaction. This method is considered as a *low-level* controller, responsible for controlling the APhI, but not tracking of e.g. position or velocity trajectories (see also Remark 6). However, it accepts high-level inputs, which can be sent by an outer control loop developed for, e.g. force/torque tracking. In simulations (Sec. 4.3), we tested the IDA-PBC controller for different interaction tasks, e.g. sliding on an uneven surface. Later in the experiments (Sec. 4.3.2) we realized this sliding task using IDA-PBC together with a position tracker.

Later, we approached to the problem of APhI and aerial manipulation from the *design* point of view, and in Chapter 5 we presented a new light-weight elastic joint arm design to be used on board of the miniature flying robots. After producing it in-house and identifying its physical parameters, we used it on-board of a quadrotor VTOL for two different tasks; link velocity amplification and stable physical interaction. The former task is achieved thanks to the elastic components of the arm, and by oscillating it at its natural frequency. This simple but explosive movement task can be achieved only thanks to the elastic potential energy stored in the springs, and be interpreted as hammering or throwing. Then the latter task is achieved due to the intrinsic safe behavior of the compliant elements, absorbing the impacts from the environment and blocking them before they are transmitted to the flying robot. What it was missing in this chapter, was that the controller for both quadrotor and the elastic arm were not aware of each other's dynamics, so both were considering each other as *disturbances*. However for performing a meaningful and dynamic aerial manipulation task, the controller needs to account for the dynamic couplings between these two systems. This leads us to the following chapter.

In Chapter 6 we studied the control of the aerial manipulators when they are equipped with manipulating arms driven by rigid or compliant actuators. In this chapter we considered aerial robots with various manipulation types; with single joint-arm or multiple manipulating arms, with one or generic number of DoF, with rigid or elastic actuation, and when the arm is attached to any point of the flying robot or specifically to its CoM. All these choices effect the controllability of the system. We showed the differentially flat (and so the exactly linearizing) outputs of all these systems, and how to use them for the control of the aerial manipulator. Particularly, we noticed that the usage of rigid or elastic actuation dramatically effects the task capability of the aerial manipulator. In example, for simple trajectory tracking and aerial grasping tasks, rigid actuators are more preferable. On the other hand, when an explosive dynamic motion is required, e.g. aerial hammering or throwing, then the elastically actuated joints perform better. Furthermore, placing the manipulating arms at the CoM of the flying platform has a striking effect; it makes the end-effector positions as part of the differential flat outputs. This is especially interesting, because it allows us generating dynamic trajectories directly for the end-effector positions and develop advanced controllers for them. A common approach in the literature was to control the CoM position of the overall system (which is always a flat output of the system). However, this is not a practical implementation, since CoM of the overall system is a time-varying position w.r.t. the body of the flying robot where all the sensors are usually placed. Also, controlling the CoM positions is impractical, because typically in useful tasks one wants to control the end-effector positions. Another approach was trying to control the end-effector position directly, but then one has to deal with the internal dynamics, and this is not an easy and solved problem. In short words again, if one designs the platform in a smart way (we call it PAM), which is also not cumbersome at all, but it rather makes sense, then the end-effector position becomes also an additional flat output (together with the CoM position of the overall aerial manipulator) and this simplifies dramatically the control problem because, among other things, the internal dynamics disappears. So rather than finding another output, we have found a mechanical design condition for which the output that we want to control is always nicely controllable. In this chapter we presented both numerical and experimental results, highlighting the importance of the theoretical contributions.

The work done in this thesis can be extended in many directions:

- IDA-PBC method (Chapter 4) can be improved by studying its robustness, proposing new high-level controllers (e.g. wrench tracking), and tuning its parameters better (maybe in an adaptive manner) for achieving more precise interaction tasks.

- Control of the Variable Stiffness Actuator (VSA) on a quadrotor (see Sec. 6.2.10) clearly can be improved, by actively controlling the variable stiffness as well. Being able to change the stiffness of the VSA, in real time, would increase the range of aerial manipulation tasks that can be performed by one robot. A demanding scenario could be a case, where the aerial robot is grasping an object with a high-stiff arm, and later throwing it away (or if the grasped object is a hammer than using it for hammering) by amplifying the velocity of the low-stiff arm. We are already working in this direction.

- Controlling the end-effector positions of an aerial manipulator directly in 3D is another step forward. So far, we have controlled them in 3D, but the dynamic motions are constrained on a plane. Performing dynamic aerial manipulation tasks in 3D, and differential flatness analysis for the end-effectors of the aerial manipulators is something we are currently working on as well.

- With the increasing number of elastic actuation, the relative degree of the flat outputs increases w.r.t. the system inputs. This is due to the underactuation of the flying platform. Hence, aerial manipulators with fully actuated flying base deserve to be studied when they are equipped with compliant manipulating arms.

- Last, but not least, more emphasis on planning in the Cartesian space of the end-effectors are needed, considering the recently discovered differential flatness property of the PAMs. In fact, this can be further improved for the scenarios, where multiple aerial manipulators are performing the manipulation task.

# Appendix A

# Technical Proofs

## A.1 Proof of Proposition 1

We will prove that the estimation error defined in (3.9) will asymptotically vanish by showing that the error dynamics (3.11) is asymptotically stable. Let

$$V(\mathbf{e}_o, \mathbf{q}_q) = \mathbf{e}_o^T \mathbf{B}(\mathbf{q}_q) \mathbf{e}_o \qquad (A.1)$$

be a positive definite candidate Lyapunov function. Considering (3.11) and (3.15), we can write:

$$\frac{dV(\mathbf{e}_o, \mathbf{q}_q)}{dt} = 2\mathbf{e}_o^T \mathbf{B} \dot{\mathbf{e}}_o + \mathbf{e}_o^T \dot{\mathbf{B}} \mathbf{e}_o = -2\mathbf{e}_o^T \mathbf{B} \mathbf{L} \mathbf{e}_o + \mathbf{e}_o^T \dot{\mathbf{B}} \mathbf{e}_o = -2c_o \mathbf{e}_o^T \mathbf{e}_o + \mathbf{e}_o^T \dot{\mathbf{B}} \mathbf{e}_o. \qquad (A.2)$$

The first component of the right hand side of (A.2) is negative definite for $c_o \in \mathbb{R}^+$. The second component has an indefinite sign. Nevertheless, since $\mathbf{B}(\mathbf{q}_q) = \mathbf{B}^T(\mathbf{q}_q)$, $\dot{\mathbf{B}}(\mathbf{q}_q, \dot{\mathbf{q}}_q)$ is symmetric and, therefore, its eigenvalues are real. From (3.6) we can compute:

$$\dot{\mathbf{B}}(\mathbf{q}_q, \dot{\mathbf{q}}_q) = \begin{pmatrix} \mathbf{0}_3 & \mathbf{0}_3 \\ \mathbf{0}_3 & {}^W\dot{\mathbf{M}}_{qr}(\boldsymbol{\eta}, \dot{\boldsymbol{\eta}}) \end{pmatrix}. \qquad (A.3)$$

Considering that ${}^W\mathbf{M}_{qr}(\boldsymbol{\eta}) = \mathbf{T}(\boldsymbol{\eta})^T \mathbf{M}_{qr} \mathbf{T}(\boldsymbol{\eta}) \in \mathbb{R}^{3 \times 3}$ as in Sec. 2.1 with $\mathbf{T}(\boldsymbol{\eta})$ as in (2.2), it is possible to write

$$ {}^W\dot{\mathbf{M}}_{qr}(\boldsymbol{\eta}, \dot{\boldsymbol{\eta}}) = \boldsymbol{\Phi}(\boldsymbol{\eta})\dot{\phi} + \boldsymbol{\Theta}(\boldsymbol{\eta})\dot{\theta}, \qquad (A.4)$$

where

$$\boldsymbol{\Phi} = \begin{bmatrix} 0 & 0 & 0 \\ 0 & \Phi_{22} & \Phi_{23} \\ 0 & \Phi_{32} & \Phi_{33} \end{bmatrix}, \quad \boldsymbol{\Theta} = \begin{bmatrix} 0 & 0 & \Theta_{13} \\ 0 & \Theta_{22} & \Theta_{23} \\ \Theta_{31} & \Theta_{32} & \Theta_{33} \end{bmatrix} \qquad (A.5)$$

with

$$\begin{aligned} \Phi_{22} &= 2J_{zz}c_\phi s_\phi \\ \Phi_{23} = \Phi_{32} &= c_\theta(-J_{zz}c_\phi^2 + J_{yy}c_\phi c_\theta + J_{zz}s_\phi^2) \\ \Phi_{33} &= 2c_\theta^2 c_\phi s_\phi(J_{yy} - J_{zz}) \\ \Theta_{13} = \Theta_{31} &= -J_{xx}c_\theta \\ \Theta_{22} &= -2J_{yy}c_\theta s_\theta \\ \Theta_{23} = \Theta_{32} &= (J_{zz}c_\phi - 2J_{yy}c_\theta)s_\phi s_\theta \\ \Theta_{33} &= (J_{xx} - J_{zz}c_\phi^2 - J_{yy}s_\phi^2)s_\theta, \end{aligned} \qquad (A.6)$$

where $s_* = \sin(*)$, $c_* = \cos(*)$. Since $|\dot{\phi}| < \tilde{\phi}$ and $|\dot{\theta}| < \tilde{\theta}$, from (A.6) it is easy to find two finite numbers $\alpha, \beta \in \mathbb{R}$ such that $\alpha < \dot{B}_{ij} < \beta$, $i, j \in \{1, \ldots, 6\}$, where $\dot{B}_{ij}$ is the $ij$-th element of $\dot{\mathbf{B}}$. Thus, as shown in Zhan (2006), it is always possible to find a finite upper bound $\lambda_B$ for all the possible eigenvalues of $\dot{\mathbf{B}}(\mathbf{q}_q, \dot{\mathbf{q}}_q)$:

$$\max_{\mathbf{q}_q, \dot{\mathbf{q}}_q} \lambda_M \left\{ \dot{\mathbf{B}}(\mathbf{q}_q, \dot{\mathbf{q}}_q) \right\} \leq \lambda_B < \infty \tag{A.7}$$

where $\lambda_M \left\{ \dot{\mathbf{B}}(\mathbf{q}_q, \dot{\mathbf{q}}_q) \right\}$ is the maximum eigenvalue of $\dot{\mathbf{B}}(\mathbf{q}_q, \dot{\mathbf{q}}_q)$. Thus, we have that

$$\mathbf{e}_o^T \dot{\mathbf{B}} \mathbf{e}_o \leq \lambda_B \mathbf{e}_o^T \mathbf{e}_o. \tag{A.8}$$

It is therefore possible to choose a $c_o > \frac{\lambda_B}{2}$ which implies $\dot{V}$ is negative definite and that, therefore, $\mathbf{e}_o(t) \to \mathbf{0}$ which proves the statement. $\square$

---

**Remark A.1.1.** *We note that the finite upper bound in (A.7) is found for some fixed* $(\mathbf{q}_q, \dot{\mathbf{q}}_q)$*. On the other hand, $\dot{\mathbf{B}}$ in (A.3) is a time-varying matrix, i.e. it is actually* $\dot{\mathbf{B}}(\mathbf{q}, \dot{\mathbf{q}}, t)$ *with t for the time, meaning that the eigenvalues of $\dot{\mathbf{B}}$ are changing in time. Finding a maximum for the eigenvalues of this matrix over the time might not be feasible; this maximum might not exist.*

*We see two possible ways for addressing this problem:*

- *By establishing an upper bound for the eigenvalues of $\dot{\mathbf{B}}$ independent of time. In this way, the choice of a constant $c_o > \frac{\lambda_B}{2}$ will ensure $\mathbf{e}_o(t) \to \mathbf{0}$. In the scope of a future work, this would be done by considering the components of $\dot{\mathbf{B}}$ more in detail and analyzing the stability of the time-varying (non-autonomous) systems under the light of Theorem 4.8 of Khalil (2001).*

- *Or, by computing the upper bound $\lambda_B(t)$ for every time step, and making sure that it is always $c_o(t) > \frac{\lambda_B(t)}{2}$, by actively computing $c_o$ over the time instead of using a preset one.*

## A.2 Proof of Proposition 2

Consider the energy function defined in (4.12). Using (4.11) we obtain:

$$\begin{aligned} \dot{H} &= \begin{bmatrix} \frac{\partial^T H}{\partial \mathbf{q}_q} & \frac{\partial^T H}{\partial \mathbf{p}} \end{bmatrix} \begin{bmatrix} \dot{\mathbf{q}}_q \\ \dot{\mathbf{p}} \end{bmatrix} \\ &= -\frac{\partial^T H}{\partial \mathbf{p}} \mathcal{R} \frac{\partial H}{\partial \mathbf{p}} + \frac{\partial^T H}{\partial \mathbf{p}} \mathbf{G} \mathbf{u}_i + \frac{\partial^T H}{\partial \mathbf{p}} \mathbf{w}_{ext}. \end{aligned} \tag{A.9}$$

Considering that $\mathcal{R} \geq 0$ we obtain that

$$\dot{H} \leq \frac{\partial^T H}{\partial \mathbf{p}} \mathbf{G} \mathbf{u}_i + \frac{\partial^T H}{\partial \mathbf{p}} \mathbf{w}_{ext}, \tag{A.10}$$

which proofs the statement after considering Def. 2, and Remark 1. $\square$

**Corollary A.2.1.** *Notice that the considered outputs of the system in* (4.11) *are relative degree one with respect to its inputs. Moreover with the inequality in* (A.10) *we show that the system is* weakly minimum phase *(equilibrium of the zero dynamics is stable, i.e. for the unforced system it is $\dot{H} \leq 0$). Then considering Def. 3 and Remark 1, we can say that the system in* (4.11) *is* cyclo-passive *w.r.t. the input-output pair given in Prop. 2.*

## A.3 Proof of Proposition 3

After straight-forward algebra, the desired dissipation matrix in (4.31) can be written in the following form

$$
\mathcal{R}_d = \begin{bmatrix} \left(\frac{m_d}{m}\right)^2 k_T \mathbf{g}_1^T \mathbf{g}_1 & \mathbf{0} \\ \mathbf{0} & k_d \mathbf{N} + \mathbf{N}\mathbf{K}_R\mathbf{N} \end{bmatrix} \in \mathbb{R}^{4\times 4}. \tag{A.11}
$$

The first matrix on the diagonal is trivially positive definite. The second matrix on the diagonal is positive definite because it is the sum of two positive definite matrices. In fact, $k_d \in \mathbb{R}^+$ and therefore $k_d\mathbf{N} > \mathbf{0}$. Furthermore, since $\mathbf{N}$ and $\mathbf{K}_R$ are positive definite, $\mathbf{N}\mathbf{K}_R\mathbf{N}$ is positive definite[1]. This concludes that $\mathcal{R}_d \geq \mathbf{0}$.

The structure of the desired dissipation matrix in (A.11) is influenced both by the under-actuation of the quadrotor and by the change of momentum. Because of the underactuation, the damping in the Cartesian space is influenced only by the parameter $k_T$ and, therefore, it is not possible to set arbitrary damping factors along the three Cartesian directions. On the other hand, it is possible to achieve any desired damping for the rotational dynamics by properly tuning the matrix $\mathbf{K}_R$. The damping force is an external force and, because of the change of momentum in the target dynamics, the desired inertia affects the achievable damping. Nevertheless, setting (4.32) we have

$$
\mathcal{R}_d = \begin{bmatrix} \bar{k}_T \|\mathbf{g}_1\|^2 & \mathbf{0} \\ \mathbf{0} & \bar{\mathbf{K}}_R \end{bmatrix} \in \mathbb{R}^{4\times 4}, \tag{A.12}
$$

where $\| * \|$ stands for *2-norm* of $*$. Hence, it is possible to achieve any desired damping $\bar{k}_T \in \mathbb{R}^+$ along the actuated Cartesian direction and any rotational damping matrix $\mathbb{R}^{3\times 3} \ni \bar{\mathbf{K}}_R > \mathbf{0}$. □

## A.4 Proof of Proposition 4

Consider the energy function defined in (4.16). Using (4.37) we obtain:

$$
\begin{aligned}
\dot{H}_d &= \begin{bmatrix} \frac{\partial^T H_d}{\partial \mathbf{q}_q} & \frac{\partial^T H_d}{\partial \bar{\mathbf{p}}} \end{bmatrix} \begin{bmatrix} \dot{\mathbf{q}}_q \\ \dot{\bar{\mathbf{p}}} \end{bmatrix} \\
&= -\frac{\partial^T H_d}{\partial \bar{\mathbf{p}}} \mathcal{R}_d \frac{\partial H_d}{\partial \bar{\mathbf{p}}} + \frac{\partial^T H_d}{\partial \bar{\mathbf{p}}} \mathbf{M}_d \mathbf{M}^{-1} \mathbf{G}\mathbf{u}_o + \frac{\partial^T H_d}{\partial \bar{\mathbf{p}}} \tilde{\mathbf{w}}_{ext}
\end{aligned} \tag{A.13}
$$

---

[1] $\forall \mathbf{z} \in \mathbb{R}^{3\times 1} \neq \mathbf{0}$ we have $\mathbf{N}\mathbf{z} = \mathbf{z}' \neq \mathbf{0}$ and $\mathbf{z}^T(\mathbf{N}\mathbf{K}_R\mathbf{N})\mathbf{z} = \mathbf{z}^T\mathbf{N}^T\mathbf{K}_R\mathbf{N}\mathbf{z} = \mathbf{z}'^T\mathbf{K}_R\mathbf{z}' > 0$.

Considering that $\mathcal{R}_d \geq \mathbf{0}$ from Proposition 3 we obtain that

$$\dot{H}_d \leq \frac{\partial^T H_d}{\partial \bar{\mathbf{p}}} \mathbf{M}_d \mathbf{M}^{-1} \mathbf{G} u_o + \frac{\partial^T H_d}{\partial \bar{\mathbf{p}}} \tilde{\mathbf{w}}_{ext} \tag{A.14}$$

which proofs the statement after considering Def. 2, and Remark 1. □

---

**Corollary A.4.1.** *Notice that the considered outputs of the system in (4.37) are relative degree one with respect to its inputs. Moreover with the inequality in (A.14) we show that the system is* weakly minimum phase *(equilibrium of the zero dynamics is stable, i.e. for the unforced system it is $\dot{H}_d \leq 0$). Then considering Def. 3 and Remark 1, we can say that the system in (4.37) is cyclo-passive w.r.t. the input-output pair given in Prop. 4.*

---

## A.5  Proof of Proposition 5

Let us divide first the generalized coordinates into two parts; $\mathbf{q} = [\mathbf{p}_c^T \; \mathbf{q}_r^T]^T \in \mathbb{R}^4$, where $\mathbf{q}_r = [\theta_0 \; \theta_{01}]^T \in \mathbb{R}^2$. Then,

$$m_s \ddot{\mathbf{p}}_c = \mathbf{v} u_t + \begin{bmatrix} 0 \\ m_s g \end{bmatrix}, \quad \mathbf{v} = \begin{bmatrix} -\sin\theta_0 \\ -\cos\theta_0 \end{bmatrix} \in \mathbb{R}^2. \tag{A.15}$$

Differentiating twice with respect to time we obtain

$$m_s \dddot{\mathbf{p}}_c = \ddot{\mathbf{v}} u_t + 2\dot{\mathbf{v}} \dot{u}_t + \mathbf{v} \ddot{u}_t, \quad \dot{\mathbf{v}} = \bar{\mathbf{v}} \dot{\theta}_0$$

$$\ddot{\mathbf{v}} = \bar{\mathbf{v}} \ddot{\theta}_0 - \mathbf{v} \dot{\theta}_0^2, \quad \bar{\mathbf{v}} = \frac{\partial \mathbf{v}}{\partial \theta_0} = \begin{bmatrix} -\cos\theta_0 \\ \sin\theta_0 \end{bmatrix}. \tag{A.16}$$

Now let us write the rotational dynamics of the system

$$\ddot{\mathbf{q}}_r = \begin{bmatrix} \ddot{\theta}_0 \\ \ddot{\theta}_{01} \end{bmatrix} = \mathbf{W} \begin{bmatrix} -c_3(\theta_0, \theta_{01}, \dot{\theta}_{01}) + g_{31} u_t + u_r + \tau \\ -c_4(\theta_0, \theta_{01}, \dot{\theta}_0) + g_{41}(\theta_0, \theta_{01}) u_t + \tau \end{bmatrix}, \tag{A.17}$$

where $c_3$ and $c_4$ are the 3-rd and 4-th elements of $\mathbf{c}$ in (6.14), and $g_{31}$ and $g_{41}$ are the 3-rd and 4-th elements of the first row of $\mathbf{G}$ in (6.15). Moreover, $\mathbf{W} = \mathbf{M}_r^{-1}$, where $\mathbf{M}_r$ is given in (6.13).

Now, use $\ddot{\theta}_0$ from the first column of (A.17) in (A.16) and $\ddot{\theta}_{01}$ from the last column of (A.17); and stack them together as

$$\begin{bmatrix} \dddot{\mathbf{p}}_c \\ \ddot{\theta}_{01} \end{bmatrix} = \mathbf{h}(\mathbf{q}_r, \dot{\mathbf{q}}_r, u_t, \dot{u}_t) + \bar{\mathbf{G}} \bar{\mathbf{u}}, \tag{A.18}$$

where

$$\underbrace{\begin{bmatrix} \frac{\mathbf{v}}{m_s} & \frac{\bar{\mathbf{v}}}{m_s} W_{11} u_t & \frac{\bar{\mathbf{v}}}{m_s}(W_{12} - W_{11}) u_t \\ \hline 0 & W_{12} & W_{22} - W_{11} \end{bmatrix}}_{\bar{\mathbf{G}}} \underbrace{\begin{bmatrix} \ddot{u}_t \\ u_r \\ \tau \end{bmatrix}}_{\bar{\mathbf{u}}}. \tag{A.19}$$

The matrix $\bar{\mathbf{G}}$ is the new input matrix, with $W_{ij}$ being the $ij$-th component of $\mathbf{W}$. After some algebra we can express the determinant of $\bar{\mathbf{G}}$ as

$$|\bar{\mathbf{G}}| = -\frac{u_t(W_{11}W_{22} - W_{12}^2)}{m_s^2} = -\frac{u_t|\mathbf{W}|}{m_s^2} = -\frac{u_t}{m_s^2|\mathbf{M}_r|}, \tag{A.20}$$

meaning that $\bar{\mathbf{G}}$ is invertible[2] as long as $u_t \neq 0$. Furthermore it must hold $|\mathbf{M}_r| \neq 0$ for $\bar{\mathbf{G}}$ to be well-defined.

Let us re-write the components of $\mathbf{M}_r$ in the following form

$$\begin{aligned} m_a &= m_\alpha \mathbf{d}_\alpha^T \mathbf{d}_\alpha + J_0 \\ m_b &= m_\beta \mathbf{d}_\beta^T \mathbf{d}_\beta + J_1 + J_m \\ m_{ab} &= m_\gamma \mathbf{d}_\alpha^T \mathbf{d}_\beta, \end{aligned} \tag{A.21}$$

with $\mathbf{d}_\alpha = \bar{\mathbf{R}}_0 \mathbf{d}_0$ and $\mathbf{d}_\beta = \bar{\mathbf{R}}_1 \mathbf{d}_1$, and $m_\alpha = \frac{m_0(m_1+m_m)}{m_s}$, $m_\beta = \frac{m_1(m_0+m_m)}{m_s}$ and $m_\gamma = \frac{m_0 m_1}{m_s}$. Then we can write

$$\begin{aligned} |\mathbf{M}_r| &= m_a m_b - m_{ab}^2 \\ &= m_{pos} + m_\alpha m_\beta (\mathbf{d}_\alpha^T \mathbf{d}_\alpha)(\mathbf{d}_\beta^T \mathbf{d}_\beta) - m_\gamma^2 (\mathbf{d}_\alpha^T \mathbf{d}_\beta)(\mathbf{d}_\alpha^T \mathbf{d}_\beta), \end{aligned} \tag{A.22}$$

where[3] $m_{pos} = m_\alpha(J_1 + J_m)\mathbf{d}_\alpha^T \mathbf{d}_\alpha + m_\beta J_0 \mathbf{d}_\beta^T \mathbf{d}_\beta + J_0(J_1 + J_m) > 0$. Notice that it is always $m_\alpha m_\beta > m_\gamma^2$. Moreover, $(\mathbf{d}_\alpha^T \mathbf{d}_\alpha)(\mathbf{d}_\beta^T \mathbf{d}_\beta) - (\mathbf{d}_\alpha^T \mathbf{d}_\beta)(\mathbf{d}_\alpha^T \mathbf{d}_\beta) = (d_{\alpha_1} d_{\beta_2} - d_{\alpha_2} d_{\beta_1})^2 > 0$ and it is always $(\mathbf{d}_\alpha^T \mathbf{d}_\alpha)(\mathbf{d}_\beta^T \mathbf{d}_\beta) \geq 0$, where $d_{\alpha_i}$ and $d_{\beta i}$ are the $i$-th components of $\mathbf{d}_\alpha$ and $\mathbf{d}_\beta$, respectively. Hence it is always $|\mathbf{M}_r| > 0$. This proofs that $[\mathbf{p}_c^T \ \theta_{01}]^T$ is exact linearizing output for Case RG. From Fact 1 it is differentially flat output as well. $\qquad \square$

# A.6 Proof of Proposition 6

For $[\mathbf{p}_c^T \ \theta_{01}]^T$ this descends from Proposition 5 since Case RC is a special case of Case RG. Concerning $[\mathbf{p}_e^T \ \theta_{01}]^T$, this descends from (6.25) and from the flatness of $[\mathbf{p}_m^T \ \theta_{01}]^T$, which we shall prove in the following.

First let us write the system dynamics in the following form

$$\ddot{\mathbf{q}} = \mathbf{W} \left[ \begin{array}{c} \mathbf{v} u_t - \bar{\beta}\dot{\theta}_{01}^2 + \begin{bmatrix} 0 \\ m_s g \end{bmatrix} \\ \hline \begin{bmatrix} d_{G_x} u_t + u_r - \tau \\ \tau \end{bmatrix} - \begin{bmatrix} 0 \\ g_4(\theta_{01}) \end{bmatrix} \end{array} \right], \tag{A.23}$$

where $\bar{\beta} = \frac{\partial \beta}{\partial \theta_{01}} = [\bar{\beta}_1 \ \bar{\beta}_2]^T \in \mathbb{R}^2$, $\beta$ as in (6.27), $g_4$ as in (6.29) and $\mathbf{v}$ is in (A.15). In this case we decompose the inverse of the inertia matrix as

$$\mathbf{W} = \mathbf{M}^{-1} = \left[ \begin{array}{c|c} \mathbf{W}_{11} & \mathbf{W}_{21}^T \\ \hline \mathbf{W}_{21} & \mathbf{W}_{22} \end{array} \right], \ \mathbf{W}_{21} = \begin{bmatrix} \mathbf{0}_{1\times 2} \\ \tilde{\mathbf{W}}_{21} \in \mathbb{R}^{1\times 2} \end{bmatrix}, \tag{A.24}$$

---

[2]Notice that all masses and inertias are positive.
[3]Recall that $\mathbf{d}_\alpha^T \mathbf{d}_\alpha = \|\mathbf{d}_0\|_2^2 \geq 0$ and $\mathbf{d}_\beta^T \mathbf{d}_\beta = \|\mathbf{d}_1\|_2^2 \geq 0$, as also stated in (6.13). Notice that even in case of $\mathbf{d}_0 = \mathbf{d}_1 = \mathbf{0}$, it is still $m_{pos} > 0$ and $|\mathbf{M}_r| > 0$, implying that $\bar{\mathbf{G}}$ is invertible.

where $\mathbf{W}_{11} \in \mathbb{R}^{2 \times 2}$ and $\mathbf{W}_{22} = \text{diag}([W_{22^1}, W_{22^2}]) \in \mathbb{R}^{2 \times 2}$. Then, we can write

$$\ddot{\mathbf{p}}_m = \mathbf{W}_{11}\mathbf{v}u_t + \tilde{\mathbf{W}}_{21}^T\tau - \mathbf{W}_{11}\bar{\beta}\dot{\theta}_{01}^2 + \mathbf{W}_{11}\begin{bmatrix} 0 \\ m_sg \end{bmatrix} - \tilde{\mathbf{W}}_{21}^T\mathbf{g}_4(\theta_{01}). \tag{A.25}$$

Moreover from the third equation of (A.23) it is

$$\ddot{\theta}_1 = W_{22^1}u_r + W_{22^1}(d_{G_x}u_t - \tau) \tag{A.26}$$

and from the fourth one it is

$$\ddot{\theta}_{01} = \tilde{\mathbf{W}}_{21}\mathbf{v}u_t + W_{22^2}\tau - \tilde{\mathbf{W}}_{21}\bar{\beta}\dot{\theta}_{01}^2 + \tilde{\mathbf{W}}_{21}\begin{bmatrix} 0 \\ m_sg \end{bmatrix} - W_{22^2}\mathbf{g}_4(\theta_{01}). \tag{A.27}$$

Differentiating (A.25) and (A.27) twice w.r.t. time, and utilizing $\ddot{\theta}_0$ from (A.26), we get

$$\begin{bmatrix} \dddot{\mathbf{p}}_m \\ \dddot{\theta}_{01} \end{bmatrix} = \mathbf{h}(\mathbf{q}_r, \dot{\mathbf{q}}_r, u_t, \tau, \dot{u}_t, \dot{\tau}) + \bar{\mathbf{G}}\bar{\mathbf{u}}, \tag{A.28}$$

where $\mathbf{q}_r$ is defined as in Appendix A.5 and

$$\bar{\mathbf{G}}\bar{\mathbf{u}} = \underbrace{\begin{bmatrix} \mathbf{W}_{11}\mathbf{v} & \mathbf{W}_{11}\bar{\mathbf{v}}W_{22^1}u_t & \tilde{\mathbf{W}}_{21}^T \\ \hline \tilde{\mathbf{W}}_{21}\mathbf{v} & \tilde{\mathbf{W}}_{21}\bar{\mathbf{v}}W_{22^1}u_t & W_{22^2} \end{bmatrix}}_{\bar{\mathbf{G}}} \underbrace{\begin{bmatrix} \dot{u}_t \\ u_r \\ \dot{\tau} \end{bmatrix}}_{\bar{\mathbf{u}}}, \tag{A.29}$$

where $\bar{\mathbf{v}}$ is as in (A.16), and $\bar{\mathbf{G}}$ is the new input matrix, whose determinant can easily be computed, after some algebra, as

$$|\bar{\mathbf{G}}| = -\frac{u_t}{J_0m_s\Big((J_1 + J_m)m_s + m_1(m_0 + m_m)\|\mathbf{d}_1\|_2^2\Big)}, \tag{A.30}$$

which is always invertible as long as $u_t \neq 0$. This proofs that $\mathbf{y} = [\mathbf{p}_m^T \; \theta_{01}]^T$ is exact linearizing output for Case RC. From Fact 1 it is differentially flat output as well. $\qquad\square$

## A.7  Proof of Proposition 7

Let us re-formalize the system dynamics given in (6.50) (assuming linear spring case) as

$$m_s\ddot{\mathbf{p}}_c = \mathbf{v}u_t + \begin{bmatrix} 0 \\ m_sg \end{bmatrix}, \quad \mathbf{v} = \begin{bmatrix} -\sin(\theta_0) \\ -\cos(\theta_0) \end{bmatrix} \in \mathbb{R}^2 \tag{A.31a}$$

$$\begin{bmatrix} \ddot{\theta}_0 \\ \ddot{\theta}_{01} \end{bmatrix} = \mathbf{W}\begin{bmatrix} g_{31}u_t + k_r(\theta_r - \theta_0) - c_3(\theta_0, \theta_{01}, \dot{\theta}_{01}) \\ g_{41}(\theta_0, \theta_{01})u_t + k_e(\theta_{0m} - \theta_{01}) - c_4(\theta_0, \theta_{01}, \dot{\theta}_0) \end{bmatrix} \tag{A.31b}$$

$$\begin{bmatrix} \ddot{\theta}_r \\ \ddot{\theta}_{0m} \end{bmatrix} = \begin{bmatrix} J_r & 0 \\ 0 & J_m \end{bmatrix}^{-1} \begin{bmatrix} k_r(\theta_0 - \theta_r) + u_n \\ k_e(\theta_{01} - \theta_{0m}) + \tau \end{bmatrix}, \tag{A.31c}$$

where this time $\mathbf{q}_r = [\theta_0 \ \theta_{01} \ \theta_r \ \theta_{0m}]^T \in \mathbb{R}^4$ and $\mathbf{W} = \mathbf{B}^{-1}$ with $\mathbf{B}$ as in (6.47). In the following, by $W_{ij}$, we denote the $ij$-th component of $\mathbf{W} = \mathbf{W}^T \in \mathbb{R}^{2\times2}$.

Then differentiating (A.31a) four times w.r.t. the time, we get

$$m_s \mathbf{p}_c^{(6)} = \mathbf{v}\,\dddot{\ddot{u}}_t + \bar{\mathbf{v}}u_t\,\dddot{\theta}_0 + \mathbf{h}_1(\mathbf{q}_r, \dot{\mathbf{q}}_r, u_t, \dot{u}_t, \ddot{u}_t, \dddot{u}_t). \tag{A.32}$$

The quantity $\dddot{\theta}_0$ can be analytically expressed by differentiating twice $\ddot{\theta}_0$, whose analytical expression is available from the first equation of (A.31b), substituting: $\ddot{\theta}_{01}$ from the second equation of (A.31b), $\ddot{\theta}_r$, and $\ddot{\theta}_{0m}$ from (A.31c). In this way we obtain

$$\dddot{\theta}_0 = \frac{W_{11}k_r}{J_r}u_n + \frac{W_{12}k_e}{J_m}\tau + h_2(\mathbf{q}_r, \dot{\mathbf{q}}_r, u_t, \dot{u}_t, \ddot{u}_t) \tag{A.33}$$

and utilizing it in (A.32), we have

$$\mathbf{p}_c^{(6)} = \frac{\mathbf{v}}{m_s}\dddot{\ddot{u}}_t + \bar{\mathbf{v}}u_t\frac{W_{11}k_r}{m_sJ_r}u_n + \bar{\mathbf{v}}u_t\frac{W_{12}k_e}{m_sJ_m}\tau + \mathbf{h}_A(\mathbf{q}_r, \dot{\mathbf{q}}_r, u_t, \dot{u}_t, \ddot{u}_t, \dddot{u}_t), \tag{A.34}$$

where $\mathbf{h}_A = \mathbf{h}_1 + \bar{\mathbf{v}}u_t h_2 \in \mathbb{R}^{2\times1}$ and $\bar{\mathbf{v}}$ is as in (A.16).

Similarly, we express analytically $\dddot{\theta}_{01}$ by differentiating twice $\ddot{\theta}_{01}$ and substituting $\ddot{\theta}_r$ and $\ddot{\theta}_{0m}$ using (A.31), thus getting

$$\dddot{\theta}_{01} = \frac{W_{12}k_r}{J_r}u_n + \frac{W_{22}k_e}{J_m}\tau + h_b(\mathbf{q}_r, \dot{\mathbf{q}}_r, u_t, \dot{u}_t, \ddot{u}_t). \tag{A.35}$$

Then, using (A.34) and (A.35) we can write

$$\begin{bmatrix} \mathbf{p}_c^{(6)} \\ \dddot{\theta}_{01} \end{bmatrix} = \mathbf{h}(\mathbf{q}_r, \dot{\mathbf{q}}_r, u_t, \dot{u}_t, \ddot{u}_t, \dddot{u}_t) + \bar{\mathbf{G}}\bar{\mathbf{u}}, \tag{A.36}$$

where $\mathbf{h} = [\mathbf{h}_A \ h_b]^T \in \mathbb{R}^{3\times1}$ and

$$\bar{\mathbf{G}}\bar{\mathbf{u}} = \underbrace{\begin{bmatrix} \dfrac{\mathbf{v}}{m_s} & u_t\bar{\mathbf{v}}\dfrac{W_{11}k_r}{m_sJ_r} & u_t\bar{\mathbf{v}}\dfrac{W_{12}k_e}{m_sJ_m} \\ 0 & \dfrac{W_{12}k_r}{J_r} & \dfrac{W_{22}k_e}{J_m} \end{bmatrix}}_{\bar{\mathbf{G}}} \underbrace{\begin{bmatrix} \dddot{\ddot{u}}_t \\ u_n \\ \tau \end{bmatrix}}_{\bar{\mathbf{u}}}, \tag{A.37}$$

where $\bar{\mathbf{G}}$ is the sought input matrix (remember that $u_r$ can be computed using (6.48)). The determinant of $\bar{\mathbf{G}}$ is

$$|\bar{\mathbf{G}}| = -\frac{u_t k_r k_e(W_{11}W_{22} - W_{12}^2)}{J_m J_r m_s^2} = -\frac{u_t k_r k_e|\mathbf{W}|}{J_m J_r m_s^2}$$

$$= -\frac{u_t k_r k_e}{J_m J_r m_s^2 |\mathbf{B}|}. \tag{A.38}$$

By construction it is always $J_m J_r m_s^2 > 0$. In order to show that the determinant is well defined we now show that it is also $|\mathbf{B}| \neq 0$. In fact we have:

$$\begin{aligned} |\mathbf{B}| &= m_a m_b - m_{ab}^2 - m_a J_m \\ &= m_{pos2} + m_\alpha m_\beta(\mathbf{d}_\alpha^T\mathbf{d}_\alpha)(\mathbf{d}_\beta^T\mathbf{d}_\beta) - m_\gamma^2(\mathbf{d}_\alpha^T\mathbf{d}_\beta)(\mathbf{d}_\alpha^T\mathbf{d}_\beta), \end{aligned} \tag{A.39}$$

where $m_{pos2} = m_\alpha J_1 \mathbf{d}_\alpha^T \mathbf{d}_\alpha + m_\beta J_0 \mathbf{d}_\beta^T \mathbf{d}_\beta + J_1 J_2 > 0$. Moreover, $(\mathbf{d}_\alpha^T \mathbf{d}_\alpha)(\mathbf{d}_\beta^T \mathbf{d}_\beta) - (\mathbf{d}_\alpha^T \mathbf{d}_\beta)(\mathbf{d}_\alpha^T \mathbf{d}_\beta) = (d_{\alpha_1} d_{\beta_2} - d_{\alpha_2} d_{\beta_1})^2 > 0$ and it is always $(\mathbf{d}_\alpha^T \mathbf{d}_\alpha)(\mathbf{d}_\beta^T \mathbf{d}_\beta) \geq 0$, where $d_{\alpha_i}$ and $d_{\beta_i}$ are the $i$-th components of $\mathbf{d}_\alpha$ and $\mathbf{d}_\beta$, respectively (see also Appendix A.5 for similar results). Hence we have that $|\mathbf{B}| > 0$.

Since the denominator in (A.38) is always positive the matrix $\bar{\mathbf{G}}$ is invertible as long as $u_t \neq 0$, $k_r \neq 0$, $J_r \neq 0$ and $k_e \neq 0$ (if the elasticity is linear). This proofs that $[\mathbf{p}_c^T \ \theta_{01}]^T$ is exact linearizing output for Case EG. From Fact 1 it is differentially flat output as well. □

## A.8  Proof of Proposition 8

For $[\mathbf{p}_c^T \ \theta_{01}]^T$ this descends from Proposition 7 since Case EC is a special case of Case EG. Concerning $[\mathbf{p}_e^T \ \theta_{01}]^T$, it is enough to prove the flatness of $[\mathbf{p}_m^T \ \theta_{01}]^T$ and apply (6.25). In the following we then prove only the flatness of $[\mathbf{p}_m^T \ \theta_{01}]^T$.

First, notice that the inertia matrix cannot be decoupled as nicely as in Case EG. However, we can re-concatenate the generalized coordinates in the form of; $\tilde{\mathbf{q}} = \mathbf{S}\mathbf{q} = [\mathbf{p}_m^T \ \theta_{01} \ \theta_0 \ \theta_{0m}]^T \in \mathbb{R}^5$, where $\mathbf{S}$ is an orthogonal selection matrix in form of

$$\mathbf{S} = \begin{bmatrix} 1 & 0 & 0 & 0 & 0 \\ 0 & 1 & 0 & 0 & 0 \\ 0 & 0 & 0 & 1 & 0 \\ 0 & 0 & 1 & 0 & 0 \\ 0 & 0 & 0 & 0 & 1 \end{bmatrix} = \mathbf{S}^T \in \mathbb{R}^{5\times5}. \tag{A.40}$$

The new inertia matrix becomes $\tilde{\mathbf{M}} = \mathbf{S}^T \mathbf{M} \mathbf{S}$, where

$$\tilde{\mathbf{M}} = \begin{bmatrix} \tilde{\mathbf{M}}_{11} & \mathbf{0}_{3\times2} \\ \hline \mathbf{0}_{2\times3} & \tilde{\mathbf{M}}_{22} \end{bmatrix}, \tilde{\mathbf{M}}_{11} = \begin{bmatrix} m_s \mathbf{I}_2 & \beta \\ \hline \beta^T & m_B - J_m \end{bmatrix} \in \mathbb{R}^{3\times3}$$

$$\tilde{\mathbf{M}}_{22} = \begin{bmatrix} J_1 & 0 \\ \hline 0 & J_m \end{bmatrix} \in \mathbb{R}^{2\times2}; \tag{A.41}$$

the Coriolis/centrifugal forces become $\tilde{\mathbf{c}} = \mathbf{S}\mathbf{c}$, where $\mathbf{c}$ is available from (6.60); the gravitational forces become $\tilde{\mathbf{g}} = \mathbf{S}\mathbf{g}$, where $\mathbf{g}$ is available from (6.60); the elastic forces become $\tilde{\mathbf{f}}_E = \mathbf{S}\mathbf{f}_E$, where $\mathbf{f}_E$ is available from (6.44); and the input matrix becomes $\tilde{\mathbf{G}} = \mathbf{S}\mathbf{G}$, where $\mathbf{G}$ is available from (6.61).

Then, we obtain

$$\ddot{\tilde{\mathbf{q}}} = \mathbf{W} \begin{bmatrix} \begin{bmatrix} \mathbf{v}u_t \\ 0 \end{bmatrix} + \begin{bmatrix} \mathbf{0}_{2\times1} \\ k_e(\theta_{0m} - \theta_{01}) \end{bmatrix} - \begin{bmatrix} \bar{\beta}\dot{\theta}_{01}^2 \\ 0 \end{bmatrix} + \begin{bmatrix} 0 \\ m_s g \\ -g_4(\theta_{01}) \end{bmatrix} \\ \hline \begin{bmatrix} d_{G_x} u_t + u_r - \tau \\ \tau \end{bmatrix} + \begin{bmatrix} 0 \\ k_e(\theta_{01} - \theta_{0m}) \end{bmatrix} \end{bmatrix}, \tag{A.42}$$

where $\mathbf{v}$ is as in (A.31a). Notice that $\mathbf{W} = \tilde{\mathbf{M}}^{-1}$, where

$$\mathbf{W} = \left[ \begin{array}{c|c} \mathbf{W}_{11} \in \mathbb{R}^{3\times3} & \mathbf{0}_{3\times2} \\ \hline \mathbf{0}_{2\times3} & \mathbf{W}_{22} \in \mathbb{R}^{2\times2} \end{array} \right], \mathbf{W}_{11} = \left[ \begin{array}{c|c} \mathbf{W}_{11^1} \in \mathbb{R}^{2\times2} & \mathbf{W}_{11^{21}}^T \\ \hline \mathbf{W}_{11^{21}} \in \mathbb{R}^{1\times2} & W_{11^2} \in \mathbb{R} \end{array} \right] \in \mathbb{R}^{3\times3}, \quad (A.43)$$

and $\mathbf{W}_{22} = \mathrm{diag}([W_{22^1}, W_{22^2}]) \in \mathbb{R}^{2\times2}$. Now notice that $\ddot{\mathbf{p}}_m$ is available from the first two equations, $\ddot{\theta}_{01}$ from the third, $\ddot{\theta}_0$ from fourth, and $\ddot{\theta}_{0m}$ from the last equation of (A.42). By differentiating $\ddot{\mathbf{p}}_m$ twice w.r.t. time, and utilizing $\ddot{\theta}_0$ and $\ddot{\theta}_{0m}$ from (A.42) we obtain

$$\ddddot{\mathbf{p}}_m = \mathbf{W}_{11^1}\mathbf{v}\ddot{u}_t + \mathbf{W}_{11^1}\bar{\mathbf{v}}W_{22^1}u_t u_r - \mathbf{W}_{11^1}\bar{\mathbf{v}}W_{22^1}u_t\tau + \\ + \mathbf{W}_{11^{21}}^T W_{22^2}k_e\tau + \mathbf{h}_1(\tilde{\mathbf{q}}_r, \dot{\tilde{\mathbf{q}}}_r, u_t, \dot{u}_t), \quad (A.44)$$

where $\bar{\mathbf{v}}$ is as in (A.16), and $\tilde{\mathbf{q}}_r = \mathbf{S}\mathbf{q}_r$, with $\mathbf{q}_r = [\theta_0 \ \theta_{01} \ \theta_{0m}]^T \in \mathbb{R}^3$. Furthermore, by differentiating $\ddot{\theta}_{01}$ twice w.r.t. time, and utilizing $\ddot{\theta}_0$ and $\ddot{\theta}_{0m}$ from (A.42) we get

$$\ddddot{\theta}_{01} = \mathbf{W}_{11^{21}}\mathbf{v}\ddot{u}_t + \mathbf{W}_{11^{21}}\bar{\mathbf{v}}W_{22^1}u_t u_r - \mathbf{W}_{11^{21}}\bar{\mathbf{v}}W_{22^1}u_t\tau + \\ + W_{11^2}W_{22^2}k_e\tau + h_2(\tilde{\mathbf{q}}_r, \dot{\tilde{\mathbf{q}}}_r, u_t, \dot{u}_t). \quad (A.45)$$

Then using (A.44) and (A.45) we can write

$$\begin{bmatrix} \ddddot{\mathbf{p}}_m \\ \ddddot{\theta}_{01} \end{bmatrix} = \mathbf{h}(\mathbf{q}_r, \dot{\mathbf{q}}_r, u_t, \dot{u}_t) + \bar{\mathbf{G}}\bar{\mathbf{u}}, \quad (A.46)$$

where $\mathbf{h} = [\mathbf{h}_1^T \ h_2]^T \in \mathbb{R}^{3\times1}$ and

$$\bar{\mathbf{G}}\bar{\mathbf{u}} = \underbrace{\left[ \begin{array}{c|c|c} \mathbf{W}_{11^1}\mathbf{v} & \mathbf{W}_{11^1}\bar{\mathbf{v}}W_{22^1}u_t & \bar{\mathbf{G}}_{13} \\ \hline \mathbf{W}_{11^{21}}\mathbf{v} & \mathbf{W}_{11^{21}}\bar{\mathbf{v}}W_{22^1}u_t & \bar{\mathbf{G}}_{23} \end{array} \right]}_{\bar{\mathbf{G}}\in\mathbb{R}^{3\times3}} \underbrace{\begin{bmatrix} \ddot{u}_t \\ u_r \\ \tau \end{bmatrix}}_{\bar{\mathbf{u}}}, \quad (A.47)$$

where

$$\begin{aligned} \bar{\mathbf{G}}_{13} &= \mathbf{W}_{11^{21}}^T W_{22^2}k_e - \mathbf{W}_{11^1}\bar{\mathbf{v}}W_{22^1}u_t \\ \bar{\mathbf{G}}_{23} &= W_{11^2}W_{22^2}k_e - \mathbf{W}_{11^{21}}\bar{\mathbf{v}}W_{22^1}u_t \end{aligned} \quad (A.48)$$

and $\bar{\mathbf{G}}$ is the new input matrix, whose determinant is

$$|\bar{\mathbf{G}}| = -\frac{u_t k_e}{J_0 J_m m_s\left(J_1 m_s + m_1(m_0 + m_m)\|\mathbf{d}_1\|_2^2\right)}, \quad (A.49)$$

which is always invertible as long as $u_t \neq 0$ and $k_e \neq 0$ (if the elasticity is linear). This proofs that $[\mathbf{p}_m^T \ \theta_{01}]^T$ is exact linearizing output for Case EC. From Fact 1 it is a flat output as well. $\qquad\square$

# A.9 Proof of Proposition 9

Consider the CoM position of the overall system

$$\mathbf{p}_c = \frac{1}{m_s}\left(m_0\mathbf{p}_0 + \sum_{j=1}^{m}\left(\sum_{i=1}^{n^j}(m_{ij}\mathbf{p}_{ij} + m_{m_{ij}}\mathbf{p}_{m_{ij}})\right)\right). \tag{A.50}$$

Considering the computations given in Appendix B.1, the implicit dependency of $\mathbf{p}_c$ can be given as $\mathbf{p}_c = \mathbf{p}_c(\mathbf{p}_0, \mathbf{q}_r) = \mathbf{p}_c(\mathbf{y})$. Notice also the fact that

$$\ddot{\mathbf{p}}_c = \begin{bmatrix} -\sin(\theta_0) \\ -\cos(\theta_0) \end{bmatrix} u_t + \begin{bmatrix} 0 \\ m_s g \end{bmatrix}. \tag{A.51}$$

Defining the vector $\mathbf{w} = \mathbf{w}(\mathbf{y}, \dot{\mathbf{y}}, \ddot{\mathbf{y}}) = \ddot{\mathbf{p}}_c - [0\ g]^T = [w_x\ w_z]^T \in \mathbb{R}^2$, which is a function of flat outputs. It is clear that $\mathbf{w} = -\frac{u_t}{m_s}[\sin(\theta_0)\ \cos(\theta_0)]^T$. Therefore $\theta_0 = \text{atan2}(-w_x, -w_z)$ and $u_t = m_s\|\mathbf{w}\|$. Furthermore, differentiating $\theta_0(w_x, w_z)$ we obtain $\dot{\theta}_0(w_x, w_z, \dot{w}_x, \dot{w}_z)$ and $\ddot{\theta}_0(w_x, w_z, \dot{w}_x, \dot{w}_z, \ddot{w}_x, \ddot{w}_z)$, which are all functions of the derivatives of $\mathbf{p}_c$ from the second up to the fourth order.

Now considering the rotational dynamics of the last link of each manipulator, we can retrieve the $n^\mu$-th motor torque as

$$\tau_{n^\mu} = \mathbf{m}_{0n^\mu}^T(\theta_{0n^\mu})\ddot{\mathbf{p}}_0 + \sum_{l=1}^{n^\mu-1} m_{ln^\mu}(\theta_{0l^\mu}, \theta_{0n^\mu})\ddot{\theta}_{0l^\mu} + $$
$$+ \mathbb{J}_{n^\mu}\ddot{\theta}_{0n^\mu} + c_{r_{n^\mu}}(\mathbf{q}_{r^\mu}, \dot{\mathbf{q}}_{r^\mu}) + g_{r_{n^\mu}}(\theta_{0n^\mu}), \tag{A.52}$$

where $c_{r_{n^\mu}}$ and $g_{r_n^\mu}$ are the $n^\mu$-th elements of vectors $\mathbf{c}_r$ and $\mathbf{g}_r$, which are corresponding to the Coriolis and gravitational forces acting on the center of the $n$-th link of the $\mu$-th manipulator, respectively (See Appendix B.2 for the details). Hence, $\tau_{n^\mu}$ can be represented solely as a function of the flat outputs $\mathbf{y}$ and of its time derivatives $\dot{\mathbf{y}}$, $\ddot{\mathbf{y}}$.

Starting from the last joint, we can recursively compute all the joint torques of the $\mu$-th arm as functions of the flat outputs up to their final derivative. This means we can write the control torque of the $\nu$-th joint of the $\mu$-th manipulator in form of $\tau_{\nu^\mu} = \tau_{\nu^\mu}(\mathbf{y}, \dot{\mathbf{y}}, \ddot{\mathbf{y}})$, and $\nu^\mu = \{1, 2, \cdots, n^\mu\}$, using

$$\tau_{\nu^\mu} = \tau_{\nu^\mu+1} + \mathbf{m}_{0\nu^\mu}^T(\theta_{0\nu^\mu})\ddot{\mathbf{p}}_0 + c_{r_{\nu^\mu}}(\mathbf{q}_{r^\mu}, \dot{\mathbf{q}}_{r^\mu}) + \mathbb{J}_{\nu^\mu}\ddot{\theta}_{0\nu^\mu} + $$
$$+ g_{r_{\nu^\mu}}(\theta_{0\nu^\mu}) + \sum_{l=1, l\neq\nu^\mu}^{n^\mu} m_{l\nu^\mu}(\theta_{0l^\mu}, \theta_{0\nu^\mu})\ddot{\theta}_{0l^\mu}, \quad (A.53)$$

where it is clear that for $\nu^\mu = n^\mu$, $\tau_{\nu^\mu+1} = 0$, because the $(n^\mu + 1)$-th motor does not exist. In this way, one can compute all the input torques of all the manipulators, until the very first ones, as functions of sole flat outputs and their finite numbers of derivatives. Hence, $\tau_{1^\mu}$ will also be the sole function of the flat outputs as well. Then we can utilize $\tau_{1^\mu}$ in the third equation of the system dynamics to compute $u_r$:

$$u_r = J_0\ddot{\theta}_0 + \sum_{j=1}^{m}\tau_{1^j} - d_{G_x}u_t. \tag{A.54}$$

Notice that $\theta_0$ and $u_t$ have been computed above as functions of the flat output and a finite number of its derivatives only. Hence this holds for $u_r$ too. Since $\ddot{\theta}_0$ is a function of $\ddddot{\mathbf{y}}$, then so is $u_r$, implying the relative degree of the system is four times the dimension of $\mathbf{y}$, i.e. $r = 4(2 + n) = 8 + 4n$. This concludes the proof. □

# A.10 Proof of Proposition 10

The proof is analogous to that of Prop. 9. Knowing the fact that the CoM position of the overall system, given in (A.50) and its dynamics as in (A.51) are sole functions of the flat outputs and derivatives, we can write again $\theta_0 = \text{atan2}(-w_x, -w_z)$ and $u_t = m_s||\mathbf{w}||$. Furthermore, differentiating $\theta_0(w_x, w_z)$ we obtain $\dot{\theta}_0(w_x, w_z, \dot{w}_x, \dot{w}_z)$ and $\ddot{\theta}_0(w_x, w_z, \dot{w}_x, \dot{w}_z, \ddot{w}_x, \ddot{w}_z)$, which are all functions of the derivatives of $\mathbf{p}_c$ from the second up to the fourth order.

Assume that all elastic joints are linear. Consider the $\mu$-th manipulator, and let's focus on the torque input of its last joint, i.e., $\tau_{n^\mu}$. If this last joint is rigid, then its expression is identical to (A.52), while $\theta_{m_{0n^\mu}}$ is clearly undefined. If instead this last, i.e., $n^\mu$-th joint is elastic, then we first need to compute $\theta_{m_{0n^\mu}}$. This can be written from its link-side dynamics as

$$\theta_{m_{0n^\mu}} = \theta_{0n^\mu} + \frac{1}{k_{e_{n^\mu}}}\left(\mathbf{m}_{0n^\mu}^T(\theta_{0n^\mu})\ddot{\mathbf{p}}_0 + \mathbf{g}_{r_{n^\mu}} + \mathbf{c}_{r_{n^\mu}}(\mathbf{q}_{r^\mu}, \dot{\mathbf{q}}_{r^\mu}) + \right.$$
$$\left. + \sum_{l=1}^{n^\mu - 1} m_{ln^\mu}(\theta_{0l^\mu}, \theta_{0n^\mu})\ddot{\theta}_{0l^\mu} + (\mathbb{J}_{n^\mu} - J_{m_{n^\mu}})\ddot{\theta}_{0n^\mu}\right). \quad \text{(A.55)}$$

Notice the similarity between this and (A.52). Hence, $\theta_{m_{0n^\mu}}$ is represented as a function of flat outputs and derivatives. Then, $\tau_{n^\mu}$ is available from the last equation of the system dynamics of the $\mu$-th manipulator:

$$\tau_{n^\mu} = J_{n^\mu}\ddot{\theta}_{m_{0n^\mu}} + k_{e_{n^\mu}}\theta_{m_{0n^\mu}} - k_{e_{n^\mu}}\theta_{0n^\mu}, \quad \text{(A.56)}$$

where $\theta_{m_{0n^\mu}}$ and its derivatives are available from (A.55). Hence, we see that $\tau_{n^\mu}$ can always be represented as a function of flat outputs and derivatives, together with $\theta_{0n^\mu}$ even when the $n^\mu$-th joint (last joint of the $\mu$-th manipulator) is elastic.

Now, let's focus on the generic link number $\nu^\mu < n^\mu$, again for the $\mu$-th manipulator for now. We can proceed recursively from top to bottom, assuming we have already computed $\tau_{\nu^\mu+1}$. If the $\nu^\mu$-th link is rigid, then its expression is identical to (A.53) and $\theta_{m_{0\nu^\mu}}$ is not defined. If it is elastic, we first need to compute $\theta_{m_{0\nu^\mu}}$. This can be done with:

$$\theta_{m_{0\nu^\mu}} = \theta_{0\nu^\mu} + \frac{1}{k_{e_{\nu^\mu}}}\left(\tau_{\nu^\mu+1} + \mathbf{m}_{0\nu^\mu}^T(\theta_{0\nu^\mu})\ddot{\mathbf{p}}_0 + \sum_{l=1, l\neq\nu^\mu}^{n^\mu} m_{l\nu^\mu}(\theta_{0l^\mu}, \theta_{0\nu^\mu})\ddot{\theta}_{0l^\mu} + \right.$$
$$\left. + (\mathbb{J}_{\nu^\mu} - J_{m_{\nu^\mu}})\ddot{\theta}_{0\nu^\mu} + \mathbf{g}_{r_{\nu^\mu}}(\theta_{0\nu^\mu}) + \mathbf{c}_{r_{\nu^\mu}}(\dot{\mathbf{q}}_{r^\mu}, \mathbf{q}_{r^\mu})\right). \quad \text{(A.57)}$$

Again, notice the similarity with (A.53). We observe that (A.57) can also be employed for $\nu^\mu = n^\mu$, simply setting the non-existing $\tau_{n^\mu+1}$ equal to zero. Then, $\tau_{\nu^\mu}$ can be easily computed from

$$\tau_{\nu^\mu} = J_{\nu^\mu}\ddot{\theta}_{m_{0\nu^\mu}} + k_{e_{\nu^\mu}}\theta_{m_{0\nu^\mu}} - k_{e_{\nu^\mu}}\theta_{0\nu^\mu}, \quad \text{(A.58)}$$

which can be directly employed also for $\nu^\mu = n^\mu$.

Until this point, we showed that all control torques of the $\mu$-th robotic arm, regardless of their connection type (rigid or elastic), can be represented as sole functions of flat outputs and their derivatives. This means that the equations above are valid for the each robotic arm. Now, finally from the third equation of the system dynamics we retrieve

$$u_r = J_0 \ddot{\theta}_0 + \sum_{j=1}^{m} \tau_{1^j} - d_{G_x} u_t, \tag{A.59}$$

in which $\tau_{1\mu}$ is utilized from either (A.53) or (A.58), depending on the type of the actuation. Moreover $\ddot{\theta}_0$ and $u_t$ are available from previous computations. Hence, PVTOL torque is also represented using only the flat outputs. □

From Remark 13, it is possible to compute the relative degree of the overall system. Recalling that $k^\mu$ is the number of elastic joints in link $\mu$, and defining $\tilde{k}^\mu = \max(1, k^\mu)$, then

$$r = 4 + 4 \max_\mu \tilde{k}^\mu + \sum_{\mu=1}^{m} (2 + 2\tilde{k}^\mu) n^\mu \tag{A.60}$$

where it can be seen a quadratic dependence on the number of elastic joints. The term $\max_\mu \tilde{k}^\mu$ returns the value $\tilde{k}^\mu$ for the manipulator arm with the highest number of elastic joint. For a better understanding, let us give the following examples:

**Example A.10.1.** *Protocentric Aerial Manipulator (PAM) with $m$ number of manipulator arms, each having only rigid actuators. Notice that this actually corresponds to Case R. Since there are no compliant actuators, $k = k^\mu = \max_\mu \tilde{k}^\mu = 0$, and so $\tilde{k}^\mu = 1$. Then (A.60) becomes $r = 4 + 4 + 4n = 8 + 4n$, which is a perfect match to Appendix. A.9.*

**Example A.10.2.** *PAM with $m$ number of manipulator arms, each having some rigid actuators and each having only one compliant actuator. This means that $k^\mu = 1$. Then $\tilde{k}^\mu = 1$, and it means $r = 4 + 4 + 4n = 8 + 4n$. This means that if each manipulator has only one elastic joint, then the total relative degree is the same with the case if all joints were rigid (this result is in line with that of Yüksel et al. (2016b)).*

**Example A.10.3.** *PAM with 2 number of manipulator arms with mixed rigid-/elastic-joints. Let's say for the first arm it is $n^1 = 5, k^1 = 4$ and for the second one it is $n^2 = 7, k^2 = 3$. This is a highly complicated PAM, with two arms in total 12 actuators and links, and 7 compliant joints. Then we see that $\max_\mu \tilde{k}^\mu = 4$. Hence the total relative degree of the system is $r = 4 + 4 * 4 + (2 + 2 * 4) * 5 + (2 + 2 * 3) * 7 = 126$.*

# Appendix B

# Technical Computations

## B.1 Computation of the Inertia Matrix for (6.3.1)

Let us start with the positions and the orientations of the elements of the robotic system, which consist of a PVTOL equipped with $m$ fully actuated manipulators; the $\mu$-th manipulator has $n^\mu$ DoF. Then the absolute orientation of the $\nu^\mu$-th joint will be $\theta_{0\nu^\mu} = \theta_0 + \sum_{i=1}^{\nu^\mu} \theta_{i^\mu}$. For example, the absolute orientation of the second link of the third manipulator will be $\theta_{02^3} = \theta_0 + \theta_{1^3} + \theta_{2^3}$. Moreover, the rotation matrix corresponding to $\theta_{0\nu^\mu}$ will be $\mathbf{R}(\theta_{0\nu^\mu}) = \mathbf{R}_0(\theta_0) \prod_{i=1}^{\nu^\mu} \mathbf{R}_{i^\mu}(\theta_{i^\mu})$. Then we can write the following distance vectors

$$\mathbf{p}_G = \mathbf{p}_0 + \mathbf{R}_0 \mathbf{d}_G$$

$$\mathbf{p}_{n^\mu} = \mathbf{p}_0 + \mathbf{R}_{01^\mu} \bar{\mathbf{d}}_{1^\mu} + \cdots + \mathbf{R}_{0(n^\mu-1)} \bar{\mathbf{d}}_{(n^\mu-1)} + \mathbf{R}_{0n^\mu} \mathbf{d}_{n^\mu}$$

$$= \mathbf{p}_0 + \underbrace{\sum_{i=1}^{n^\mu-1} \mathbf{R}_{0i^\mu} \bar{\mathbf{d}}_{i^\mu}}_{:=0, \ \ if \ n^\mu=1} + \mathbf{R}_{0n^\mu} \mathbf{d}_{n^\mu}$$

$$\mathbf{p}_{e^\mu} = \mathbf{p}_0 + \sum_{i=1}^{n^\mu} \mathbf{R}_{0i^\mu} \bar{\mathbf{d}}_{i^\mu},$$

where $\bar{\mathbf{d}}_i = \mathbf{d}_i + \tilde{\mathbf{d}}_i$, $i = \{1, 2, ..., n^\mu\}$, and for the motors,

$$\mathbf{p}_{m_{n^\mu}} = \mathbf{p}_0 + \mathbf{R}_{01^\mu} \bar{\mathbf{d}}_{1^\mu} + \mathbf{R}_{02^\mu} \bar{\mathbf{d}}_{2^\mu} + \cdots + \mathbf{R}_{0(n^\mu-1)} \bar{\mathbf{d}}_{(n^\mu-1)}$$

$$= \mathbf{p}_0 + \underbrace{\sum_{i=1}^{n^\mu-1} \mathbf{R}_{0i^\mu} \bar{\mathbf{d}}_{i^\mu}}_{:=0, \ \ if \ n^\mu=1}.$$

Notice that this is due to the Assumption A2. The following gives the translational velocities for the $\nu^\mu$-the link and motor, and the end-effector:

$$\dot{\mathbf{p}}_{\nu^\mu} = \dot{\mathbf{p}}_0 + \underbrace{\sum_{i=1}^{\nu^\mu-1} \bar{\mathbf{R}}_{0i^\mu} \bar{\mathbf{d}}_{i^\mu} \dot{\theta}_{0i^\mu}}_{:=0, \ \ if \ \nu^\mu=1} + \bar{\mathbf{R}}_{0\nu^\mu} \mathbf{d}_{\nu^\mu} \dot{\theta}_{0\nu^\mu}$$

$$\dot{\mathbf{p}}_{m_{\nu^\mu}} = \dot{\mathbf{p}}_0 + \underbrace{\sum_{i=1}^{\nu^\mu-1} \bar{\mathbf{R}}_{0i^\mu} \bar{\mathbf{d}}_{i^\mu} \dot{\theta}_{0i^\mu}}_{:=0, \ \ if \ \nu^\mu=1}$$

$$\dot{\mathbf{p}}_{e^\mu} = \dot{\mathbf{p}}_0 + \sum_{i=1}^{n^\mu} \bar{\mathbf{R}}_{0i^\mu} \bar{\mathbf{d}}_{i^\mu} \dot{\theta}_{0i^\mu},$$

where from the definition $\nu^\mu = \{1, 2, \cdots, i, \cdots, n^\mu\}$, and $\bar{\mathbf{R}}_* = \frac{\partial \mathbf{R}_*}{\partial \theta_*}$.

Now we can write the energy of the system. We start by considering rigid manipulators. The kinetic energy is

$$K = \underbrace{\frac{1}{2}m_0\dot{\mathbf{p}}_0^T\dot{\mathbf{p}}_0 + \frac{1}{2}J_0\dot{\theta}_0^2}_{pvtolbase} + \underbrace{K_m}_{manipulators}, \tag{B.1}$$

where

$$K_m = \sum_{j=1}^{m}\left(\frac{1}{2}\sum_{i=1}^{nj}\left(m_{ij}\dot{\mathbf{p}}_{ij}^T\dot{\mathbf{p}}_{ij} + m_{m_{ij}}\dot{\mathbf{p}}_{m_{ij}}^T\dot{\mathbf{p}}_{m_{ij}} + (J_{ij} + J_{m_{ij}})\dot{\theta}_{0ij}^2\right)\right).$$

The potential energy is

$$V = -gm_0\mathbf{p}_0.\mathbf{e}_2 - g\sum_{j=1}^{m}\left(\sum_{i=1}^{nj}\left[m_{ij}\mathbf{p}_{ij} + m_{m_{ij}}\mathbf{p}_{m_{ij}}\right]\right).\mathbf{e}_2, \tag{B.2}$$

where $\mathbf{e}_2 = [0\ 1]^T$. Let us now write the well known Lagrange equation

$$\frac{d}{dt}\left(\frac{\partial L}{\partial\dot{\mathbf{q}}}\right) - \frac{\partial L}{\partial\mathbf{q}} = \mathbf{f} = \mathbf{G}\mathbf{u}, \tag{B.3}$$

where $L = K - V$, computed above. It is clear that using $K = \frac{1}{2}\dot{\mathbf{q}}^T\mathbf{M}(\mathbf{q})\dot{\mathbf{q}}$ we can find the inertia matrix as

$$\mathbf{M} = \begin{bmatrix} \mathbf{M}_p & * \\ \mathbf{M}_{pr} & \mathbf{M}_r \end{bmatrix} = \mathbf{M}^T \in \mathbb{R}^{(3+n)\times(3+n)}, \tag{B.4}$$

where

$$\mathbf{M}_p = \begin{bmatrix} m_s & * & * \\ 0 & m_s & * \\ 0 & 0 & J_0 \end{bmatrix} = \mathbf{M}_p^T \in \mathbb{R}^{3\times3}$$

is the pvtol-side inertia matrix; the sum $m_s$ of all masses is given by

$$m_s = m_0 + \sum_{j=1}^{m}\left(\sum_{i=1}^{nj}m_{ij} + m_{m_{ij}}\right).$$

The arm-side inertia matrix is

$$\mathbf{M}_r = \begin{bmatrix} \mathbf{M}_{r^1} & * & \cdots & * \\ 0 & \mathbf{M}_{r^2} & \cdots & * \\ \vdots & \vdots & \ddots & \vdots \\ 0 & 0 & \cdots & \mathbf{M}_{r^m} \end{bmatrix} = \mathbf{M}_r^T \in \mathbb{R}^{n\times n},$$

where for the $\mu$-th manipulator it is

$$\mathbf{M}_{r^\mu} = \begin{bmatrix} \mathbb{J}_{1^\mu} & * & \cdots & * \\ m_{12^\mu}(\theta_{01^\mu}, \theta_{02^\mu}) & \mathbb{J}_{2^\mu} & \cdots & * \\ \vdots & \vdots & \ddots & * \\ m_{1n^\mu}(\theta_{01^\mu}, \theta_{0n^\mu}) & m_{2n^\mu}(\theta_{02^\mu}, \theta_{0n^\mu}) & \cdots & \mathbb{J}_{n^\mu} \end{bmatrix} = \mathbf{M}_{r^\mu}^T \in \mathbb{R}^{n^\mu\times n^\mu}.$$

For the $\nu$-th joint of the $\mu$-th manipulator it is

$$\mathbb{J}_{\nu^\mu} = J_{\nu^\mu} + J_{m_{\nu^\mu}} + m_{\nu^\mu}\boldsymbol{\alpha}_{\nu^\mu}^T\boldsymbol{\alpha}_{\nu^\mu} + \underbrace{\sum_{i=\nu^\mu+1}^{n^\mu}(m_i + m_{m_i})\bar{\boldsymbol{\alpha}}_{\nu^\mu}^T\bar{\boldsymbol{\alpha}}_{\nu^\mu}}_{:=0, \quad if \ \nu=n^\mu}, \ \nu^\mu = \{1, ..., n^\mu\},$$

and for the coupling between the $\nu$-th and $\xi$-th joints of the $\mu$-th manipulator it is

$$m_{\nu\xi^\mu}(\theta_{0\nu^\mu}, \theta_{0\xi^\mu}) = m_{\xi^\mu}\bar{\boldsymbol{\alpha}}_{\nu^\mu}^T\boldsymbol{\alpha}_{\xi^\mu} + \underbrace{\sum_{i=\xi^\mu+1}^{n^\mu}(m_i + m_{m_i})\bar{\boldsymbol{\alpha}}_{\nu^\mu}^T\bar{\boldsymbol{\alpha}}_{\xi^\mu}}_{:=0, \quad if \ \xi^\mu=n^\mu},$$

$$\nu^\mu = \{1, ..., n^\mu - 1\},$$
$$\xi^\mu = \{1, ..., n^\mu\}, \ \nu^\mu < \xi^\mu$$
$$\boldsymbol{\alpha}_{\nu^\mu}(\theta_{0\nu^\mu}) = \bar{\mathbf{R}}(\theta_{0\nu^\mu})\mathbf{d}_{\nu^\mu}$$
$$\bar{\boldsymbol{\alpha}}_{\nu^\mu}(\theta_{0\nu^\mu}) = \bar{\mathbf{R}}(\theta_{0\nu^\mu})\bar{\mathbf{d}}_{\nu^\mu}.$$

Notice that $\mathbb{J}_{\nu^\mu}$ is state independent. The coupling term between the PVTOL and the arm side inertia is given with the following,

$$\mathbf{M}_{pr} = \begin{bmatrix} \mathbf{M}_{pr^1}^T & \mathbf{M}_{pr^2}^T & \cdots & \mathbf{M}_{pr^m}^T \end{bmatrix}^T \in \mathbb{R}^{n\times 3},$$

where for the $\mu$-th manipulator it is

$$\mathbf{M}_{pr^\mu} = \begin{bmatrix} \mathbf{m}_{01^\mu}^T(\theta_{01^\mu}) & 0 \\ \mathbf{m}_{02^\mu}^T(\theta_{02^\mu}) & 0 \\ \mathbf{m}_{03^\mu}^T(\theta_{03^\mu}) & 0 \\ \vdots & \vdots \\ \mathbf{m}_{0n^\mu}^T(\theta_{0n^\mu}) & 0 \end{bmatrix} \in \mathbb{R}^{n^\mu\times 3}.$$

For the $\nu$-th joint of the $\mu$-th manipulator:

$$\mathbf{m}_{0\nu^\mu}(\theta_{0\nu^\mu}) = m_{\nu^\mu}\boldsymbol{\alpha}_{\nu^\mu} + \underbrace{\sum_{i=\nu^\mu+1}^{n^\mu}(m_i + m_{m_i})\bar{\boldsymbol{\alpha}}_{\nu^\mu}}_{:=0, \quad if \ \nu^\mu=n^\mu} \in \mathbb{R}^{2\times 1}.$$

This completes the computation of the generalized inertia matrix.

# B.2 Computation of the Gravity and Coriolis Forces for (6.80) and (6.81)

Following Appendix B.1, since $L = K - V$, we can write

$$\frac{d}{dt}\left(\frac{\partial L}{\partial \dot{\mathbf{q}}}\right) - \frac{\partial L}{\partial \mathbf{q}} = \mathbf{M}\ddot{\mathbf{q}} + \underbrace{\dot{\mathbf{M}}\dot{\mathbf{q}} - \frac{\partial L}{\partial \mathbf{q}}}_{\mathbf{c+g}} = \mathbf{M}\ddot{\mathbf{q}} + \underbrace{\dot{\mathbf{M}}\dot{\mathbf{q}} - \frac{\partial K}{\partial \mathbf{q}}}_{\mathbf{c}} + \underbrace{\frac{\partial V}{\partial \mathbf{q}}}_{\mathbf{g}} = \mathbf{f}.$$

Hence from the fact that $\mathbf{g} = \frac{\partial V}{\partial \mathbf{q}}$ and using (B.2) we can find the gravitational forces as presented in (6.80).

The Coriolis/centrifugal forces are shown above as

$$\mathbf{c} = \dot{\mathbf{M}}\dot{\mathbf{q}} - \frac{\partial K}{\partial \mathbf{q}}. \tag{B.5}$$

For the first term of the right side of the equality, we have the following

$$\dot{\mathbf{M}} = \begin{bmatrix} \dot{\mathbf{M}}_p & * \\ \dot{\mathbf{M}}_{pr} & \dot{\mathbf{M}}_r \end{bmatrix} = \dot{\mathbf{M}}^T \in \mathbb{R}^{(3+n)\times(3+n)},$$

$$\dot{\mathbf{M}}_{pr} = \begin{bmatrix} \dot{\mathbf{M}}_{pr^1} \\ \dot{\mathbf{M}}_{pr^2} \\ \vdots \\ \dot{\mathbf{M}}_{pr^m} \end{bmatrix} \in \mathbb{R}^{n\times 3},$$

where

$$\dot{\mathbf{M}}_p = \mathbf{0}_3, \quad \dot{\mathbf{M}}_{pr^\mu} = \begin{bmatrix} \dot{m}_{01^\mu}(\theta_{01^\mu})^T & 0 \\ \dot{m}_{02^\mu}(\theta_{02^\mu})^T & 0 \\ \dot{m}_{03^\mu}(\theta_{03^\mu})^T & 0 \\ \vdots & \vdots \\ \dot{m}_{0n^\mu}(\theta_{0n^\mu})^T & 0 \end{bmatrix} \in \mathbb{R}^{n^\mu \times 3},$$

and

$$\dot{\mathbf{M}}_r = \begin{bmatrix} \dot{\mathbf{M}}_{r^1} & * & \cdots & * \\ 0 & \dot{\mathbf{M}}_{r^2} & \cdots & * \\ \vdots & \vdots & \ddots & \vdots \\ 0 & 0 & \cdots & \dot{\mathbf{M}}_{r^m} \end{bmatrix} = \dot{\mathbf{M}}_r^T \in \mathbb{R}^{n\times n},$$

where

$$\dot{\mathbf{M}}_{r^\mu} = \begin{bmatrix} 0 & * & \cdots & * \\ \dot{m}_{12^\mu}(\theta_{01^\mu}, \theta_{02^\mu}) & 0 & \cdots & * \\ \vdots & \vdots & \ddots & * \\ \dot{m}_{1n^\mu}(\theta_{01^\mu}, \theta_{0n^\mu}) & \dot{m}_{2n^\mu}(\theta_{02^\mu}, \theta_{0n^\mu}) & \cdots & 0 \end{bmatrix} \tag{B.6}$$

$$= \dot{\mathbf{M}}_{r^\mu}^T \in \mathbb{R}^{n^\mu \times n^\mu}.$$

For the second term of the equality in (B.5), we put the kinetic energy in the following form

$$K = \frac{1}{2}\dot{\mathbf{q}}^T \mathbf{M}\dot{\mathbf{q}} = K_0 + K_1 + K_2, \tag{B.7}$$

where

$$K_0 = \frac{1}{2} m_s \dot{\mathbf{p}}_0^T \dot{\mathbf{p}}_0 + \frac{1}{2} J_0 \dot{\theta}_0^2 + \sum_{j=1}^{m} \sum_{i=1}^{n^j} \frac{1}{2} \mathbb{J}_{ij} \dot{\theta}_{0ij}^2$$

$$K_1 = \sum_{j=1}^{m} \dot{\mathbf{p}}_0^T \sum_{i=1}^{n^j} \mathbf{m}_{0ij} \dot{\theta}_{0ij} \tag{B.8}$$

$$K_2 = \sum_{j=1}^{m} \sum_{l=1}^{n^j-1} \left( \sum_{i=l+1}^{n^j} m_{lij} \dot{\theta}_{0lj} \dot{\theta}_{0ij} \right).$$

Now remember that $\frac{\partial K}{\partial \mathbf{q}} = \frac{\partial K_0}{\partial \mathbf{q}} + \frac{\partial K_1}{\partial \mathbf{q}} + \frac{\partial K_2}{\partial \mathbf{q}}$. It is clear that $\frac{\partial K_0}{\partial \mathbf{q}} = \mathbf{0}_{(n+3) \times 1}$. Moreover notice the following equality

$$\frac{\partial K_1}{\partial \mathbf{q}} = \begin{bmatrix} 0 \\ 0 \\ 0 \\ \dot{\mathbf{p}}_0^T \dot{\mathbf{m}}_{011} \\ \vdots \\ \dot{\mathbf{p}}_0^T \dot{\mathbf{m}}_{0n^1} \\ \vdots \\ \dot{\mathbf{p}}_0^T \dot{\mathbf{m}}_{01^m} \\ \vdots \\ \dot{\mathbf{p}}_0^T \dot{\mathbf{m}}_{0n^m} \end{bmatrix} = \begin{bmatrix} \dot{\mathbf{M}}_p \\ \dot{\mathbf{M}}_{pr} \end{bmatrix} \dot{\mathbf{q}}_p. \tag{B.9}$$

Recalling that

$$\dot{\mathbf{M}}\dot{\mathbf{q}} = \begin{bmatrix} \dot{\mathbf{M}}_p \\ \dot{\mathbf{M}}_{pr} \end{bmatrix} \dot{\mathbf{q}}_p + \begin{bmatrix} \dot{\mathbf{M}}_{pr}^T \\ \dot{\mathbf{M}}_r \end{bmatrix} \dot{\mathbf{q}}_r, \tag{B.10}$$

we have

$$\mathbf{c} = \dot{\mathbf{M}}\dot{\mathbf{q}} - \frac{\partial K}{\partial \mathbf{q}}$$

$$= \begin{bmatrix} \dot{\mathbf{M}}_{pr}^T \\ \dot{\mathbf{M}}_r \end{bmatrix} \dot{\mathbf{q}}_r - \frac{\partial K_2}{\partial \mathbf{q}} = \begin{bmatrix} \sum_{j=1}^{m} \sum_{i=1}^{n^j} \bar{\mathbf{m}}_{0ij} \dot{\theta}_{0ij}^2 \\ 0 \\ \mathbf{c}_r(\mathbf{q}_r, \dot{\mathbf{q}}_r) \end{bmatrix} \in \mathbb{R}^{(3+n) \times 1}, \tag{B.11}$$

where $\bar{\mathbf{m}}_{0ij} = \frac{\partial \mathbf{m}_{0ij}}{\partial \theta_{0ij}} \in \mathbb{R}^{2 \times 1}$; noticing that $\frac{\partial \bar{\mathbf{R}}_*}{\partial \theta_*} = -\mathbf{R}_*$ :

$$\bar{\mathbf{m}}_{0\nu^\mu}(\theta_{0\nu^\mu}) = m_{\nu^\mu}\tilde{\boldsymbol{\alpha}}_{\nu^\mu} + \underbrace{\sum_{i=\nu^\mu+1}^{n^\mu} (m_i + m_{m_i})\bar{\tilde{\boldsymbol{\alpha}}}_{\nu^\mu}}_{:=0,\quad if\ \nu^\mu=n^\mu} \in \mathbb{R}^{2 \times 1},$$

$$\nu^\mu = \{1, ..., n^\mu\}$$
$$\tilde{\boldsymbol{\alpha}}_{\nu^\mu}(\theta_{0\nu^\mu}) = -\mathbf{R}(\theta_{0\nu^\mu})\mathbf{d}_{\nu^\mu}$$
$$\bar{\tilde{\boldsymbol{\alpha}}}_{\nu^\mu}(\theta_{0\nu^\mu}) = -\mathbf{R}(\theta_{0\nu^\mu})\bar{\mathbf{d}}_{\nu^\mu},$$

and $\mathbf{c}_r(\mathbf{q}_r, \dot{\mathbf{q}}_r) \in \mathbb{R}^n$ is the arm side Coriolis forces in the form of

$$\mathbf{c}_r(\mathbf{q}_r, \dot{\mathbf{q}}_r) = [\mathbf{c}_{r1}^T(\mathbf{q}_{r1}\ \dot{\mathbf{q}}_{r1}) \ \cdots \ , \mathbf{c}_{rm}^T(\mathbf{q}_{rm}\ \dot{\mathbf{q}}_{rm})]^T \in \mathbb{R}^n.$$

Now, from (B.8) we can write $K_2 = \sum_{j=1}^m K_{2j}$ and for the $\mu$-th manipulator it is $K_{2\mu} = \frac{1}{2}\dot{\mathbf{q}}_{r\mu}^T \mathbf{B}_{r\mu} \dot{\mathbf{q}}_{r\mu}$, where

$$\begin{aligned}
\mathbf{B}_{r\mu} &= \mathbf{M}_{r\mu} - \text{diag}\{\mathbb{J}_{1\mu}, \mathbb{J}_{2\mu}, \cdots, \mathbb{J}_{n\mu}\} \\
&= \begin{pmatrix}
0 & * & \cdots & * \\
m_{12\mu}(\theta_{01\mu}, \theta_{02\mu}) & 0 & \cdots & * \\
\vdots & \vdots & \ddots & * \\
m_{1n\mu}(\theta_{01\mu}, \theta_{0n\mu}) & m_{2n\mu}(\theta_{02\mu}, \theta_{0n\mu}) & \cdots & 0
\end{pmatrix} \\
&= \mathbf{B}_{r\mu}^T \in \mathbb{R}^{n^\mu \times n^\mu}.
\end{aligned}$$

Then we can write for the $\mu$-th component of $\mathbf{c}_r$

$$\begin{aligned}
\mathbf{c}_{r\mu}(\mathbf{q}_{r\mu}, \dot{\mathbf{q}}_{r\mu}) &= \dot{\mathbf{M}}_{r\mu}\dot{\mathbf{q}}_{r\mu} - \frac{\partial K_{2\mu}}{\partial \mathbf{q}_{r\mu}} \\
&= \dot{\mathbf{M}}_{r\mu}\dot{\mathbf{q}}_{r\mu} - \frac{1}{2}\frac{\partial\left(\dot{\mathbf{q}}_{r\mu}^T \mathbf{B}_{r\mu} \dot{\mathbf{q}}_{r\mu}\right)}{\partial \mathbf{q}_{r\mu}},
\end{aligned} \tag{B.12}$$

where $\dot{\mathbf{M}}_{r\mu}$ is available from (B.6). Then we computed the followings

$$\dot{\mathbf{M}}_{r\mu}\dot{\mathbf{q}}_{r\mu} = \begin{bmatrix}
0 + \sum_{i=2}^{n^\mu} \dot{m}_{1i\mu}\dot{\theta}_{0i\mu} \\
\dot{m}_{12\mu}\dot{\theta}_{01\mu} + \sum_{i=3}^{n^\mu} \dot{m}_{2i\mu}\dot{\theta}_{0i\mu} \\
\dot{m}_{13\mu}\dot{\theta}_{01\mu} + \dot{m}_{23\mu}\dot{\theta}_{02\mu} + \sum_{i=4}^{n^\mu} \dot{m}_{3i\mu}\dot{\theta}_{0i\mu} \\
\vdots \\
\sum_{i=1}^{n^\mu-1} \dot{m}_{in\mu}\dot{\theta}_{0i\mu} + 0
\end{bmatrix}, \tag{B.13}$$

and

$$\frac{\partial K_{2^\mu}}{\partial \mathbf{q}_{r^\mu}} = \begin{bmatrix} 0 + \sum_{i=2}^{n^\mu} {}^1\bar{m}_{1i^\mu}\dot\theta_{0i^\mu}\dot\theta_{01^\mu} \\ \left( {}^2\bar{m}_{12^\mu}\dot\theta_{01^\mu} + \sum_{i=3}^{n^\mu} {}^2\bar{m}_{2i^\mu}\dot\theta_{0i^\mu} \right)\dot\theta_{02^\mu} \\ \left( {}^3\bar{m}_{13^\mu}\dot\theta_{01^\mu} + {}^3\bar{m}_{23^\mu}\dot\theta_{02^\mu} + \sum_{i=4}^{n^\mu} {}^3\bar{m}_{3i^\mu}\dot\theta_{0i^\mu} \right)\dot\theta_{03^\mu} \\ \vdots \\ \sum_{i=1}^{n^\mu-1} {}^{n^\mu}\bar{m}_{in^\mu}\dot\theta_{0i^\mu}\dot\theta_{0n^\mu} + 0 \end{bmatrix}, \tag{B.14}$$

where ${}^k\bar{m}_{kl} = \frac{\partial m_{kl}(\theta_{0k},\theta_{0l})}{\partial \theta_{0k}}$ (note that here $k = \nu^\mu$ and $l = \xi^\mu$), and

$$\begin{aligned} {}^k\bar{m}_{kl}(\theta_{0k},\theta_{0l}) &= m_l\bar{\bar{\boldsymbol\alpha}}_k^T\boldsymbol\alpha_l + \underbrace{\sum_{i=l+1}^{n}(m_i+m_{m_i})\bar{\bar{\boldsymbol\alpha}}_k^T\tilde{\boldsymbol\alpha}_l}_{:=0,\ \ if\ l=n} \qquad \text{if } k < l \\ {}^k\bar{m}_{kl}(\theta_{0k},\theta_{0l}) &= m_k\bar{\boldsymbol\alpha}_l^T\tilde{\boldsymbol\alpha}_k + \underbrace{\sum_{i=k+1}^{n}(m_i+m_{m_i})\bar{\boldsymbol\alpha}_l^T\bar{\bar{\boldsymbol\alpha}}_k}_{:=0,\ \ if\ k=n} \qquad \text{if } k > l \end{aligned} \tag{B.15}$$

$$\begin{aligned} \boldsymbol\alpha_k(\theta_{0k}) &= \bar{\mathbf{R}}(\theta_{0k})\mathbf{d}_k \\ \tilde{\boldsymbol\alpha}_k(\theta_{0k}) &= -\mathbf{R}(\theta_{0k})\mathbf{d}_k \\ \bar{\boldsymbol\alpha}_k(\theta_{0k}) &= \bar{\mathbf{R}}(\theta_{0k})\bar{\mathbf{d}}_k \\ \bar{\bar{\boldsymbol\alpha}}_k(\theta_{0k}) &= -\mathbf{R}(\theta_{0k})\bar{\mathbf{d}}_k, \end{aligned}$$

with $\bar{\mathbf{R}}_* = \frac{\partial \mathbf{R}_*}{\partial \theta_*}$ and we have used $\frac{\partial \bar{\mathbf{R}}_*}{\partial \theta_*} = -\mathbf{R}_*$. Now utilizing (B.13) and (B.14) in (B.12), one can write the $\nu$-th element of $\mathbf{c}_{r^\mu}$ as

$$c_{r_{\nu^\mu}} = \sum_{i=1}^{\nu^\mu-1}\left( \dot{m}_{i\nu^\mu} - {}^{\nu^\mu}\bar{m}_{i\nu^\mu}\dot\theta_{0\nu^\mu} \right)\dot\theta_{0i} + \sum_{i=\nu^\mu+1}^{n}\left( \dot{m}_{\nu^\mu i} - {}^{\nu^\mu}\bar{m}_{\nu^\mu i}\dot\theta_{0\nu^\mu} \right)\dot\theta_{0i},$$

which is equivalent to

$$\begin{aligned} c_{r_{\nu^\mu}} &= \sum_{i=1...n^\nu,\ i\neq\nu^\mu}\left( \dot{m}_{i\nu^\mu} - {}^{\nu^\mu}\bar{m}_{i\nu^\mu}\dot\theta_{0\nu^\mu} \right)\dot\theta_{0i} \\ &= \sum_{i=1...n^\mu,\ i\neq\nu^\mu} {}^i\bar{m}_{i\nu^\mu}\dot\theta_{0i}^2. \end{aligned} \tag{B.16}$$

This actually means that, due to the A.3 in Section 6.3, there are no Coriolis forces appearing from the motion of one arm to another. Moreover, since we choose the absolute orientations as the generalized coordinates, $c_{r_{\nu^\mu}}$ contains only the centrifugal terms.

## B.3 Computation of the Control Input Matrix for (6.83)

Following Appendix B.2, the generalized body-fixed forces are written in the form of $\mathbf{f} = \mathbf{Gu}$, where the corresponding control input matrix is

$$\mathbf{G} = \begin{bmatrix} \mathbf{G}_p & \mathbf{0}_{2\times n} \\ \mathbf{G}_d & \mathbf{G}_r \end{bmatrix} \in \mathbb{R}^{(n+3)\times(n+2)}, \mathbf{G}_p = \begin{bmatrix} -\sin(\theta_0) & 0 \\ -\cos(\theta_0) & 0 \end{bmatrix}$$

$$\mathbf{G}_d = \begin{bmatrix} d_{G_x} & 1 \\ \mathbf{0}_{n\times 1} & \mathbf{0}_{n\times 1} \end{bmatrix} \in \mathbb{R}^{(n+1)\times 2}, \mathbf{G}_r = \begin{bmatrix} \mathbf{G}_{rp} \\ \mathbf{G}_{rr} \end{bmatrix} \in \mathbb{R}^{(n+1)\times n},$$

where

$$\mathbf{G}_{rp} = \begin{bmatrix} -1 & \mathbf{0}_{1\times(n^1-1)} & \cdots & -1 & \mathbf{0}_{1\times(n^m-1)} \end{bmatrix} \in \mathbb{R}^{1\times n},$$

$$\mathbf{G}_{rr} = \begin{bmatrix} \mathbf{G}_{rr^1} & \cdots & * \\ \vdots & \ddots & \vdots \\ \mathbf{0} & \cdots & \mathbf{G}_{rr^m} \end{bmatrix} \in \mathbb{R}^{n\times n},$$

$$\mathbf{G}_{rr^\mu} = \begin{bmatrix} 1 & -1 & 0 & \cdots & 0 & 0 \\ 0 & 1 & -1 & \cdots & 0 & 0 \\ \vdots & \vdots & \vdots & \ddots & \vdots & \vdots \\ 0 & 0 & 0 & \cdots & 1 & -1 \\ 0 & 0 & 0 & \cdots & 0 & 1 \end{bmatrix} \in \mathbb{R}^{n^\mu \times n^\mu}.$$

# Appendix C

# Nonlinear Systems, Stability and Control

This chapter is dedicated to clarify some frequently used concepts of nonlinear systems, their *stability, stabilizability* and control. Specifically, we hope that this chapter helps to fix the ideas on the nonlinear control theory we used within this thesis (especially passivity-based control), and make a bridge to the existing literature. We *stress* the fact that everything presented in this chapter already exists in the literature. It is our desire to remind the connection from the *stability* of the nonlinear system to the benefits of using *passivity-based controllers*.

For this part of the thesis, we greatly benefited from Sepulchre et al. (1997), Isidori (1995) and Khalil (2001); and the graduate level course in *Nonlinear Control* taught by Prof. Dr.-Ing Frank Allgöwer, Dr.-Ing. Rainer Blind and M.Sc. Jan Maximilian Montenbruck in the summer semester of 2015 at the *Institute for Systems Theory and Automatic Control* in the University of Stuttgart[1].

Our plan, in the following, is to briefly give the important concepts of nonlinear systems theory in sense of stability and control, and *pinpoint* their locations in the aforementioned books properly (e.g. in the following, Definition XXX in [pYYY] means the *Definition* XXX in *page* YYY.). It is in fact a repetition of most of the Definitions, Theorems, and Lemmas especially of Khalil (2001) but in a connected and hopefully simplified way. We would like to note that the information given in this section extends the limits of this thesis, yet it *superficially* presents the some important concepts by avoiding further details. We hope that this section brings convenience to the reader, as it does to the author for partially organizing and summarizing the vast knowledge available in the literature.

Note that for preserving the similarity with the literature; no boldfaced characters are used for vectors and matrices as it was for the rest of the thesis.

---

[1]http://www.ist.uni-stuttgart.de/index.en.html

# C.1 Known Concepts for the Nonlinear Systems Stability

In this Section we aim to give the basic stability concepts of the nonlinear systems. Let us first give the following nonlinear system equations

$$\dot{x} = f(x), \quad y = h(x), \quad x(0) = x_0 \tag{C.1}$$

$$\dot{x} = f(t, x), \quad f(t, 0) = 0, \quad x(0) = x_0 \tag{C.2}$$

$$\dot{x} = f(x, u), \quad y = h(x, u), \quad x(0) = x_0 \tag{C.3}$$

$$\dot{x} = f(x) + g(x)u, \quad y = h(x, u), \quad x(0) = x_0 \tag{C.4}$$

$$\dot{x} = f(t, x, u), \quad y = h(t, x, u), \quad x(0) = x_0 \tag{C.5}$$

where $f : D \to \mathbb{R}^n$ is a locally Lipschitz map from a domain $D \subset \mathbb{R}^n$ into $\mathbb{R}^n$. The first equation is known as a *autonomous* or *time invariant* system; the second one is a *nonautonomous* (time varying) system, where the origin is an equilibrium point because $f(t, 0) = 0, \forall t \geq 0$; the third one is a *forced time invariant* system; and the fourth one is a *control-affine system*. The last system is a *forced nonautonomous* system.

**For systems with No Input**

Let us give a definition of the *stability* of an autonomous system:

---

**Definition C.1.1** (Stability-Definition 4.1 in [p112] of Khalil (2001)). *The origin, $x(0)$ of* (C.1) *is an equilibrium point, and it is*

*(i)* **stable**, *if* $\forall \epsilon > 0, \exists \delta(\epsilon) > 0$ *s.t.* $||x(0)|| < \delta \iff ||x(t)|| < \epsilon, \forall t \geq 0$.

*(ii)* **unstable** *if not stable.*

*(iii)* **asymptotically stable** *if stable and* $||x(0)|| < \delta \Rightarrow \lim_{t \to \infty} = 0$.

---

Notice that the stability of a non-autonomous system in form of (C.2) is available in Definition 4.4 in [p149] of Khalil (2001), where $\delta$ might depend not only on $\epsilon$ but also $t_0$. For brevity, in the following and in the general of the thesis by stability of a system we mean its stability in its equilibrium.

Now, notice the following theorem for Lyapunov stability:

---

**Theorem C.1.1** (Lyapunov Stability-Theorem 4.1 in [p114] of Khalil (2001)). *Let* $x = 0$ *be an equilibrium point for* (C.1) *and* $D \subset \mathbb{R}^n$ *be a domain containing* $x = 0$. *Let* $V : D \to \mathbb{R}$ *be a continuously differentiable function. s.t.* $V(0) = 0$ *and* $V(x) > 0 \in D \setminus \{0\}$. *Then* $x = 0$ *is*

(i) **stable,** *if* $\dot{V} \leq 0 \in D$.

(ii) **asymptotically stable,** *if stable and* $\dot{V} < 0 \in D \setminus \{0\}$.

(iii) **globally asymptotically stable,** *if it is asymptotically stable and* $V(x)$ *is radially unbounded, i.e.* $||x|| \to \infty \Rightarrow V(x) \to \infty$. *Also see Theorem 4.2 in p124 of Khalil (2001).*

(iv) **exponential stable,** *if* $||x(t)|| \leq k||x(0)||e^{-\lambda k}, \forall t \geq 0, \ k \geq 1, \lambda > 0,$ $\forall ||x(0)|| < c, \quad c > 0$. *Also notice that exponential stability implies asymptotic stability.*

Notice the similarities between Def. C.1.1 and Theorem C.1.1; the former defines the stability of a nonlinear system by introducing a *bound* on its state using some *gains*, and the latter does the same for an *energetic* function of the system. Different stability concepts connect to each other, since all introduce some kind of a *gain condition* for rendering the nonlinear system to a *well-behaving* one. In the following we will discuss this a bit more, but a general comparison between different stability concepts are given in Section C.2.

Let us note that the Lyapunov stability for linear systems is explained in Theorems 4.6 in [p136] and 4.7 in [p139] of Khalil (2001). The Lyapunov theorem for stability given in Theorem C.1.1 can be extended to *global exponential stability* using Theorem 4.10 in [p154] of Khalil (2001). Furthermore, the stability properties of a system can tell us great deal about the existence of a *Lyapunov candidate*. These *converse theorems* help to seek for a proper Lyapunov function for the further steps. For example, the converse theorem for *global exponential stability* is given in Theorem 4.14 in [p163] of Khalil (2001). Also for *global asymptotic stability* using converse theorem is shown in Theorems 4.16 and 4.17 in [p167] of Khalil (2001).

While the above definitions and theorems of stability are for the autonomous systems, for non-autonomous systems a way to redefine the concepts of stability or asymptotic stability is needed, so that they hold uniformly in the initial time $t_0$. For this, *comparison functions* will be used.

---

**Definition C.1.2** (Comparison Functions-Definitions 4.2 and 4.3 in [p144] of Khalil (2001)). *Let* $\mathcal{K}$, $\mathcal{K}_\infty$ *and* $\mathcal{KL}$ *be special classes. Then a continuous function* $\alpha(r) : \mathbb{R} \to \mathbb{R}$ *has the following properties;*

- $\alpha(r) \in \mathcal{K}$, *if* $\frac{\partial \alpha(r)}{\partial r} > 0$ *(it is strictly increasing)*,

- $\alpha(r) \in \mathcal{K}_\infty$, *if* $\alpha(r) \in \mathcal{K}$ *and* $\lim_{r \to \infty} \alpha(r) = \infty$.

*Moreover, a continuous function* $\beta(s, r)$ *is said to belong class* $\mathcal{KL}$, *i.e.* $\beta(s, r) \in \mathcal{KL}$, *if*

- $\frac{\partial \beta(r,s)}{\partial r} > 0$, $\quad \frac{\partial \beta(r,s)}{\partial s} < 0$ *and*

- $\lim_{s \to \infty} \beta(r, s) = 0$

---

Notice the similarity between $V$, $\dot{V}$ of Theorem C.1.1 and $\alpha$, $\beta$ of Def. C.1.2, respectively. Further properties of $\mathcal{K}$, $\mathcal{K}_\infty$ and $\mathcal{KL}$ can be found in Lemmas 4.2 and 4.3 in [p145] of Khalil (2001). For the stability definitions using comparison functions, we refer the reader to the Lemma 4.5 in [p150] of Khalil (2001). The uniform stability using comparison functions for non-autonomous systems is given in Lemma 4.5 in [p150] of Khalil (2001), where also uniform asymptotic stability, and global uniform asymptotic stability conditions are shown. Furthermore using comparison functions together with Lyapunov theorem, the uniform asymptotic stability, and global uniform asymptotic stability are proven in Theorems 4.8 in [p151] of Khalil (2001) and 4.9 in [p252] of Khalil (2001), respectively.

**For systems with Inputs**

So far, we have talked about the stability of systems with no inputs (e.g. (C.1) and (C.2)), where the only *degree of freedom* given to us for achieving the stability was observing the characteristics of the system, $f(x)$, and varying the initial condition of the system, $x_0$. When the nonlinear-system has inputs, e.g. (C.3), (C.4), this brings another freedom for achieving the stability. In the presence of a bounded input (with no output considered), we will talk about *Input-to-State Stability* (ISS). When an output is also involved, then *Input-to-Output (I/O) Stability* will be in our focus.

The definition of ISS is given as Definition 4.7 in [p175] of Khalil (2001), where the sufficient condition for ISS is given in Theorem 4.19 in [p176] of Khalil (2001). Also from the Lemma 4.6 in [p176] of Khalil (2001), we can say that ISS implies the existence of a Lyapunov function, and vice versa.

A non-linear system with inputs make us think, how this control input can be used for *stabilization* of a system, which brings the following question: under which condition a system is *stabilizable*? The *stabilizability* of a system is discussed using *Control Lyapunov Function* (CLF) in the following:

---

**Theorem C.1.2** (Control Lyapunov Function (CLF)-Theorem 9.4.1 in [p449] of Isidori (1995)). *Consider the system in* (C.4) *(with no output considered), and $f$ and $g$ are smooth vector fields and $f(0) = 0$. Then, $\exists$ almost smooth $u = \alpha(x)$, which globally asymptotically stabilizes the equilibrium $x = 0$ of* (C.4) $\iff$ $\exists V(x) > 0$ *which is proper smooth and*

*(i) $L_g V(x) = 0$ implies $L_f V(x) < 0, \forall x \neq 0$,*

*(ii) $\forall \epsilon > 0, \exists \gamma > 0$, s.t. if $x_0 \neq 0$ satisfies $\|x\| < \gamma$, then there is some $u$ with $|u| < \epsilon$ s.t. $L_f V(x) + L_g V(x) u < 0$.*

*Then $V$ is called* Control Lyapunov Function *(CLF).*

---

where $L_f V(x) = \frac{\partial V(x)}{\partial x} f(x)$ and $L_g V(x) = \frac{\partial V(x)}{\partial x} g(x)$. Notice that CLF is actually a Lyapunov function for closed-loop systems. The first condition means that if $L_g V(x) = 0$, then the control input $u$ cannot be used to *stabilize* the system, because it has no influence. This implies that the unforced system dynamics should be stable by itself, which is exactly what $L_f V(x) < 0, \forall x \neq 0$ means. The second condition implies that, if $L_g V \neq 0$, then $u$ can be used to dominate the term on the left-side of the inequality for stabilizing the system, using a *small $u$* in price of a *small $x$* (small control property).

**For systems with Inputs and Outputs**

Let us now consider the non-linear systems not only with inputs $u$ but also the outputs $y$. One way to realize this is to have an input-output model of the system with no state equation representing the system's internal structure. In this thesis we are more interested in those which are together with their state-space models. Without giving the details, with $\mathcal{L}_p$-stability, it is possible to investigate how an output $y \in \mathcal{L}_p$ behaves with an input $u \in \mathcal{L}_p$, without knowing the state-space model of the system. Motivating cases could be the systems where modeling is hard or almost impossible (biology or social sciences). There, $\mathcal{L}_p$ refers to a space, where $p$ stands for the type of $p$-norm used to define that space. The $\mathcal{L}_p$ stability is explained in the Definition 5.1 in [p197] of Khalil (2001), using a comparison function $\alpha \in \mathcal{K}$ as the gain of the control input. Notice that if this gain is a nonnegative constant, then it is called *finite-gain* $\mathcal{L}_p$ stability. For systems with a state-space model, $\mathcal{L}_p$ stability and *finite-gain* $\mathcal{L}_p$ stability are given in Theorem 5.1 in [p202] of Khalil (2001).

In case of $p = 2$, the $\mathcal{L}_2$ stability has a special sense in the system analysis, since it is natural to work with square integrable signals, e.g. finite energy. For the systems with state-space model, Theorem 5.5 in [p211] of Khalil (2001) shows the $\mathcal{L}_2$ stability condition. Input-Output (I/O) stability connects itself to the concept of *passivity* through $\mathcal{L}_2$ stability, as shown in Lemma 6.5 in [p242] of Khalil (2001). Passivity is a special case of *dissipativity* (see Def. 1 and Willems (1972)), which implies that the increase of the system energy can only be done by external means (see Def. 2). This *energy preserving/bounding* property is strongly related to the *well-behaveness* of the system output, as discussed in Chapter 4 of this thesis. Chapter 6 of Khalil (2001) is another reference for the readers interested in passivity. In this thesis we benefited from the concept of passivity in Chapter 4, for a quadrotor aerial robot when it is physically interacting with its environment. Especially using IDA-PBC, there we showed how to assign a desired physical property to such system and achieve aerial physical interaction using a passivity-based controller.

## Summary

Notice that *stability* is in some sense describing the *well-behaveness* of a system, with a relation between what we are varying in this system and what we observe. In fact, stability implies a *gain condition* between what we are varying and what we are observing. If a gain exist, then we say that the system is *well-behaved*. Hence with stability, we talk about existence of a gain. For Lyapunov stability, we vary the initial condition $x_0$ and observe the state $||x||$. For ISS, we do the same by also varying the control input $u$. When there is a possibility of finding such a gain by imposing a desired control input, we think of *stabilizability* of a system, as in CLF. For the systems with inputs and outputs, we talk about the well-behaveness of the system from $u$ to $y$. Already by thinking of stability as the well-behaveness of a system under the condition of existence of a gain (as an element of $\mathcal{K}$ or $\mathcal{KL}$ class depending on the gain), we can see that different stability concepts are not so much of different from each other. To further strengthen the relationship between the concepts given above, we introduce the following section.

As mentioned in Sec.1.4.2, more methods with stabilizing properties exist in the literature, e.g. *Sliding Mode Control* (see Chapter 14 of Khalil (2001)) and *Back-stepping/-forwarding Control* (see Sepulchre et al. (1997) and Krstic et al. (1995)), which we did not study within this thesis.

## C.2 Connections Between Different Stability Concepts

In Section C.1 we briefly reported the different stability concepts of different non-linear systems. While stability implies existence of a *gain* for the *well-behaveness* of a system, different conditions apply for different system models, e.g. systems with no inputs, with inputs, and with inputs and outputs.

In this section, as a supplement to Sec. C.1, we will try to report the relationships between different stability concepts. Notice that the information given here, as in Sec. C.1, extends the limits of this thesis, and our goal is to keep it as brief as possible.

---

**Definition C.2.1.** *A system is said to be $\varnothing$ G.E.S, if for $u = 0$ its equilibrium is globally exponential stable (G.E.S). Same notation applies for exponential stable (E.S), asymptotic stable (A.S) and globally asymptotic stable (G.A.S).*

---

### Exponential Stability and Hurwitz

Using the Corollary 4.3 in [p166] of Khalil (2001), we say that the origin, $x_0 = x(0)$, as the equilibrium of (C.1) is *exponentially stable* $\iff A = [\frac{\partial f}{\partial x}](x_0)$ is Hurwitz.

### Exponential Stability and ISS

From the Lemma 4.6 in [p176] of Khalil (2001), the system $f(t, x, u)$ is ISS if:

*(i)* $f(t, x, u)$ is continuously differentiable and globally Lipschitz in $(x, u)$, uniformly in $t$,

*(ii)* and the system is ∅G.E.S, i.e. the unforced system $f(t, x, 0)$ has a *globally exponentially stable* equilibrium at the origin $x = 0$.

For the conditions on ∅G.E.S, see the Theorem 4.10 in [p154] of Khalil (2001) and Theorem 4.14 in [p162] of Khalil (2001).

**Asymptotic Stability and ISS**

From the Theorem 4.19 in [p176] of Khalil (2001), the system $\dot{x} = f(t, x, u)$ is ISS if:

*(i)* $f(t, x, u)$ is ∅G.A.S, i.e. the unforced system $f(t, x, 0)$ is *globally asymptotic stable* at the equilibrium $x = 0$.

*(ii)* and $\frac{\partial V}{\partial x} f(x, u) \leq -W(x)$, $\forall ||x|| \geq \rho(||u||) > 0$ with $\rho \in \mathcal{K}$, $W > 0$.

For understanding ∅G.A.S, one can see the Theorems 4.16 and 4.17 in [p167] of Khalil (2001). Notice also the Lemma 4.7 in [p180] of Khalil (2001), that a cascade connection of one *globally uniformly asymptotic stable system* (G.U.A.S) and one ISS system is also G.U.A.S.

**Exponential Stability and $\mathcal{L}_p$ Stability**

From the Theorem 5.1 in [p202] of Khalil (2001); if the system in (C.5) is ∅G.E.S (recall Theorem 4.14 in [p162] of Khalil (2001)), then it is *finite-gain $\mathcal{L}_p$ stable* with the following conditions on $f$ and $h$:

$$||f(t, x, u) - f(t, x, 0)||_p \leq L||u||_p$$
$$||h(t, x, u)||_p \leq \eta_1 ||x_1||_p + \eta_2 ||u||_p$$

with $L, \eta_1, \eta_2 \in \mathbb{R}_{\geq 0}$, and $|| * ||_p$ indicating the *p-norm* of $*$. If the system is ∅E.S, then it is *small-signal finite-gain $\mathcal{L}_p$ stable*.

**Uniform Asymptotic Stability and $\mathcal{L}_\infty$ Stability**

From the Theorem 5.2 in [p206] of Khalil (2001); if the system in (C.5) is ∅G.U.A.S (recall Theorem 4.9 in [p152] of Khalil (2001)), then it is also *small-signal $\mathcal{L}_\infty$ stable*, if $f$ and $h$ satisfy:

$$||f(t, x, u) - f(t, x, 0)|| \leq \alpha_1 ||u||$$
$$||h(t, x, u)|| \leq \alpha_2 ||x_1|| + \alpha_3 ||u|| + \eta$$

where $\alpha_i \in \mathcal{K}$ and $\eta \in \mathbb{R}_{\geq 0}$. Also see Corollary 5.3 in [p208] of Khalil (2001).

**ISS and $\mathcal{L}_\infty$ Stability**

The Theorem 5.3 in [p215] of Khalil (2001) clears that; if the system in (C.5) is ISS, then it is also $\mathcal{L}_\infty$ stable if $h$ satisfies: $||h(t, x, u)|| \leq \alpha_1(||x||) + \alpha_2(||u||) + \eta$ for $\alpha_1, \alpha_2 \in \mathcal{K}$ and $\eta \in \mathbb{R}_{\geq 0}$.

## Dissipativity and CLF

Consider the system in (C.3). Say $H(x)$ is a proper storage function, and $\rho(u, y)$ is a supply rate as in Def. 1 and Def. 2. If $\forall x \neq 0, \exists u \in \mathbb{R}^m$, s.t. $L_f H(x) = \nabla H(x).f(x, u) \leq \rho(u, h(x)) < 0$, then $H(x)$ is CLF.

## Dissipativity and ISS

The system in (C.3) is ISS, if and only if when it is dissipative w.r.t. the supply rate $s\rho(t) = -\alpha_1(||x||) + \alpha_2(||u||)$ with $\alpha_1, \alpha_2 \in \mathcal{K}$.

## Asymptotic Stability and $\mathcal{L}_2$ Stability

Without the details, we refer the reader to Lemmas 5.1 and 5.2 in [p215] of Khalil (2001).

## $\mathcal{L}_p$ Stability of the Feedback Interconnected Systems

Considering two feedback interconnected systems, as in Fig. 5.1 in [p218] of Khalil (2001), and each being finite-gain $\mathcal{L}_p$ stable. Then from the Theorem 5.6 in [p218] of Khalil (2001), the feedback connection is also finite-gain $\mathcal{L}_p$ stable if $\gamma_1 \gamma_2 < 1$, where $\gamma_1$ and $\gamma_2$ are the *non-negative* gains of the system input as shown in Def. 5.1 in [p197] of Khalil (2001), for first and the second system respectively.

## Passivity and Feedback Interconnection

As stated in Theorem 6.1 in [p247] of Khalil (2001), the feedback interconnection of two passive systems is passive. For the feedback connection of the dissipative systems and their stability, see Theorem 2.30 in [p50] of Sepulchre et al. (1997).

## Passivity and $\mathcal{L}_2$ Stability

Following the Lemma 6.4 in [p242] of Khalil (2001), if the output of (C.3) is strictly passive with $u^T y \geq \dot{V} + \delta y^T y$, for $V$ is the Lyapunov candidate and $\delta > 0$, then it is finite-gain $\mathcal{L}_2$ stable and its $\mathcal{L}_2$ gain is $\gamma \leq 1/\delta$.

## Passivity and Lyapunov Stability

If the system in (C.3) is passive with a storage function $H(x) > 0$, then its origin $\dot{x} = f(x, 0)$ is stable (see Lemma 6.6 in [p242] of Khalil (2001)).

## Passivity and Asymptotic Stability

Let us first make the following definition:

---

**Definition C.2.2.** *A system in form of* (C.3) *is* Zero State Detectable *(ZSD) for* $u = 0$ *(unforced system) if no solution of this system can stay identically outside of* $S \subset \mathbb{R}^n$, *where* $S$ *is the largest positively invariant set contained in* $\{x \in \mathbb{R}^n : h(x, 0) = 0)\}$. *This makes the origin of the system* $x = 0$ *asymptotic stable conditionally to* $S$. *If* $S = \{0\}$, *then the system is* Zero State Observable *(ZSO).*

---

From the Lemma 6.7 in [p243] of Khalil (2001), the origin of the system in (C.3) is ∅A.S if,

- *strictly passive*, or

- *output strictly passive* (see Def. 6.1 in [p231] of Khalil (2001)) and *Zero State Observable* (ZSO) (see Def. C.2.2).

If the storage function is radially unbounded, then it is ∅G.A.S.

**Passivity and Asymptotic Stability via Output Feedback: Damping Injection**

From the Theorem 14.4 in [p604] of Khalil (2001), if the system in (C.3) is

*(i)* passive with a radially unbounded positive definite storage function and

*(ii)* ZSO (see Def. C.2.2),

then the origin $x = 0$ can be rendered G.A.S by the output feedback $u = -\xi(y)$, with $\xi(.)$ is locally Lipschitz, $\xi(0) = 0$, and $y^T \xi(y) > 0, \forall y \neq 0$.

**Concluding Remark**

In Appendix C, we tried to summarize important nonlinear system theory tools available in the literature for the stability, stabilizability and stabilization of the nonlinear systems. We also informally remind the fact that those concepts are not entirely different from each other, in fact, they mean most of the time the same thing with some variances. However let us note, that the knowledge summarized here can be conceptualize as a tiny little snow flake, standing on top of a great iceberg. What author knows is way less than what author seems to know, and this is way less than what is to know. There is more to understand for the author, and much more to start discovering. See Fig. C.1 for a sketch of what we are talking about.

Hence, we recommend further readings for better understanding of the subject, e.g. Byrnes and Isidori (1991), Isidori (2013) and of course Isidori (1995) and Khalil (2001).

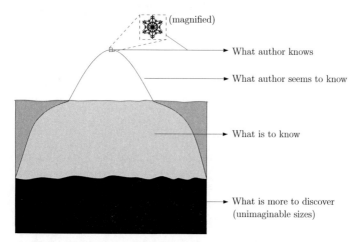

Figure C.1: A comparative sketch illustrating what author knows about the subject studied in Appendix C, and its similarity to a snow flake on top of an iceberg. Snow flake is magnified for the convenience of the reader. From snow flake to the dark colored part is depicting the knowledge author wishes to gather over the years. Dark colored part illustrates the great sizes of volumes, which are yet in shadows and to be discovered.

# Bibliography

D. Abeywardena, S. Kodagoda, G. Dissanayake, and R. Munasinghe. Improved state estimation in quadrotor MAVs: A novel drift-free velocity estimator. *IEEE Robotics & Automation Magazine*, 20(4):32–39, 2013.

J. Acosta, M. Sanchez, and A. Ollero. Robust control of underactuated aerial manipulators via ida-pbc. In *2014 IEEE Conf. on Decision and Control*, pages 673–678, LA, California, USA, December 2014.

AeRoArms. EU Collab. Project ICT-644271. `www.aeroarms-project.eu`, 2015-2019.

AEROworks. EU Collab. Project ICT-644128. `www.aeroworks2020.eu`.

ARCAS. EU Collab. Project ICT-287617. `www.arcas-project.eu`, 2011-2015.

F. Augugliaro and R. D'Andrea. Admittance control for physical human-quadrocopter interaction. In *12th European Control Conference*, pages 1805–1810, Zurich, Switzerland, Jul. 2013.

K.-J. Bathe. *Finite Element Procedures*. Prentice Hall, 1 edition, 1996. ISBN 978-0979004957.

S. Bellens, J. De Schutter, and H. Bruyninckx. A hybrid pose/wrench control framework for quadrotor helicopters. In *2012 IEEE Int. Conf. on Robotics and Automation*, pages 2269–2274, St.Paul, MN, May 2012.

S. Bouabdallah, M. Becker, and R. Siegwart. Autonomous miniature flying robots: Coming soon! *IEEE Robotics & Automation Magazine*, 13(3):88–98, 2007.

J. B. Brandt and M. S. Selig. Propeller performance data at low reynolds numbers. In *AIAA Aerospace Sciences Meeting*, Orlando, FL, Jan. 2011.

D. J. Braun, H. Matthew, and S. Vijayakumar. Optimal variable stiffness control: formulation and application to explosive movement tasks. *Autonomous Robots*, 33(3):237–253, 2012.

D. J. Braun, F. Petit, and F. Huber. Robots driven by compliant actuators: Optimal control under actuation constraints. *IEEE Trans. on Robotics*, 29(5):1085–1101, 2013.

D. Brescianini and R. D'Andea. Design, modeling and control of an omni-directional aerial vehicle. In *2016 IEEE Int. Conf. on Robotics and Automation*, pages 3261–3266, Stockholm, Sweden, May 2016.

D. Brescianini and R. D'Andrea. Design, modeling and control of an omni-directional aerial vehicle. In *2016 IEEE Int. Conf. on Robotics and Automation*, pages 3261–3266, Stockholm, Sweden, May 2016.

L. J. Bridgeman and J. R. Forbes. The exterior conic sector lemma, 2015.

C. I. Byrnes and A. Isidori. Passivity, feedback equivalence, and the global stabilization of minimum phase nonlinear systems. *IEEE Trans. on Automatic Control*, 36(11):1228–1240, 1991.

G. Cai, J. Dias, and L. Seneviratne. A survey of small-scale unmanned aerial vehicles: Recent advances nd future development trends. *Unmanned Systems*, 2(2):1–25, 2014.

H. Campion and W. Chung. Wheeled robots. In B. Siciliano and O. Khatib, editors, *Springer Handbook of Robotics*, pages 391–410. Springer, 2008.

R. Cano, C. Perez, F. Pruano, A. Ollero, and G. Heredia. Mechanical Design of a 6-DOF Aerial Manipulator for assembling bar structures using UAVs. In *2nd IFAC Work. on Research, Education and Development of Unmanned Aerial Systems*, 2013.

W. Chen, D. J. Ballance, P. J. Gawthrop, and J. O'Reilly. A nonlinear disturbance observer for robotic manipulators. *IEEE Trans. on Industrial Electronics*, 47(4):932–938, 2000.

M. Cutler, N. K. Ure, B. Michini, and J. P. How. Comparison of fixed and variable pitch actuators for agile quadrotors. In *AIAA Conf. on Guidance, Navigation and Control*, Portland, OR, Aug. 2011.

A. De Luca. Decoupling and feedback linearization of robots with mixed rigid/elastic joints. In *1996 IEEE Int. Conf. on Robotics and Automation*, pages 816–821, Minneapolis, USA, Apr. 1996.

A. De Luca. Decoupling and feedback linearization of robots with mixed rigid/elastic joints. *International Journal of Robust and Nonlinear Control*, 8(11):965–977, 1998. ISSN 1099-1239. doi: 10.1002/(SICI)1099-1239(199809)8:11<965::AID-RNC371>3. 0.CO;2-4. URL http://dx.doi.org/10.1002/(SICI)1099-1239(199809)8:11<965:: AID-RNC371>3.0.CO;2-4.

A. De Luca and W. Book. Robots with flexible elements. In B. Siciliano and O. Khatib, editors, *Springer Handbook of Robotics*, pages 287–319. Springer, 2008.

A. De Luca and G. Oriolo. Trajectory planning and control for planar robots with passive last joint. *The International Journal of Robotics Research*, 21(5-6):575–590, 2002.

V. Duggal, M. Sukhwani, K. Bipin, G. S. Reddy, and K. M. Krishna. Plantation monitoring and yield estimation using autonomous quadcopter for precision agriculture. In *2016 IEEE Int. Conf. on Robotics and Automation*, pages 5121–5127, May 2016.

D. Falanga, E. Mueggler, M. Faessler, and D. Scaramuzza. Aggressive quadrotor flight through narrow gaps with onboard sensing and computing. In *2017 IEEE Int. Conf. on Robotics and Automation*, May, submitted.

I. Fantoni and R. Lozano. *Non-linear Control for Underactuated Mechanical Systems*. Springer, 2002. ISBN 978-1447101772.

M. Fliess, J. Levine, and P. Rouchon. Flatness and defect of nonlinear systems: Introductory theory and examples. *International Journal of Control*, 61:1327–1361, 1995.

J. R. Forbes. Extensions of input-output stability theory and the control of aerospace systems. Doctoral Thesis, University of Toronto, https://tspace.library.utoronto.ca/handle/1807/31751, 2012.

A. Franchi, C. Masone, V. Grabe, M. Ryll, H. H. Bülthoff, and P. Robuffo Giordano. Modeling and control of UAV bearing-formations with bilateral high-level steering. *The International Journal of Robotics Research, Special Issue on 3D Exploration, Mapping, and Surveillance*, 31(12):1504–1525, 2012a.

A. Franchi, C. Secchi, M. Ryll, H. H. Bülthoff, and P. Robuffo Giordano. Shared control: Balancing autonomy and human assistance with a group of quadrotor UAVs. *IEEE Robotics & Automation Magazine, Special Issue on Aerial Robotics and the Quadrotor Platform*, 19(3):57–68, 2012b.

M. Fumagalli, S. Ivaldi, M. Randazzo, L. Natale, G. Metta, G. Sandini, and F. Nori. Force feedback exploiting tactile and proximal force/torque sensing. *Autonomous Robots*, 33: 381–398, 2012a.

M. Fumagalli, R. Naldi, A. Macchelli, R. Carloni, S. Stramigioli, and L. Marconi. Modeling and control of a flying robot for contact inspection. In *2012 IEEE/RSJ Int. Conf. on Intelligent Robots and Systems*, pages 3532–3537, Vilamoura, Portugal, Oct 2012b.

G. Garimella and M. Kobilarov. Towards model-predictive control for aerial pick-and-place. In *2015 IEEE/RSJ Int. Conf. on Intelligent Robots and Systems*, pages 4692–4697, May 2015.

G. Gioioso, M. Ryll, D. Prattichizzo, H. H. Bülthoff, and A. Franchi. Turning a near-hovering controlled quadrotor into a 3D force effector. In *2014 IEEE Int. Conf. on Robotics and Automation*, pages 6278–6284, Hong Kong, China, May. 2014.

F. Gomez-Estern and A. J. van der Schaft. Physical damping in IDA-PBC controlled underactuated mechanical systems. *European Journal of Control*, 10(5):451–468, 2004.

V. Grabe, M. Riedel, H. H. Bülthoff, P. Robuffo Giordano, and A. Franchi. The TeleKyb framework for a modular and extendible ROS-based quadrotor control. In *6th European Conference on Mobile Robots*, pages 19–25, Barcelona, Spain, Sep. 2013.

C. Ha, D. J. Lee, and H.-. N. Nguyen. Mechanics and control of quadrotors for tool operation. *Automatica*, 61:289–301, 2015.

S. Haddadin and K. Krieger. On impact decoupling properties of elastic robots and time optimal velocity maximization on joint level. In *2012 IEEE/RSJ Int. Conf. on Intelligent Robots and Systems*, pages 5089–5096, Vilamoura, Portugal, October 2012.

R. V. Ham, T. G. Sugar, B. Vanderborght, K. W. Hollander, and D. Lefeber. Compliant actuator designs. *IEEE Robotics & Automation Magazine*, 16(3):81–94, 2009.

J. Hauser, S. Sastry, and G. Meyer. Nonlinear control design for slightly non-minimum phase systems: Application to v/stol aircraft. *Automatica*, 28(4):665–679, 1992.

M. Hofer, M. Muehlebach, and R. D'Andrea. Application of an approximate model predictive control scheme on an unmanned aerial vehicle. In *2016 IEEE Int. Conf. on Robotics and Automation*, pages 2952–2957, May 2016.

B. Houska, H. Ferreau, and M. Diehl. ACADO Toolkit – An Open Source Framework for Automatic Control and Dynamic Optimization. *Optimal Control Applications and Methods*, 32(3):298–312, 2011.

ICARUS. EU 7th Framework Programme 285417. http://www.fp7-icarus.eu/, 2011-2015.

A. Isidori. *Nonlinear Control Systems, 3rd edition*. Springer, 1995. ISBN 3540199160.

A. Isidori. The zero dynamics of a nonlinear system: From the origin to the latest progress of a ling successful story. *European Journal of Control*, 19:369–378, 2013.

A. Isidori, L. Marconi, and A. Serrani. *Robust Autonomous Guidance*. Springer, 2003. ISBN 978-1447100119.

*QBMove-VSA. Variable Stiffness Actuator.* http://www.qbrobotics.com/.

A. E. Jimenez-Cano, J. Martin, G. Heredia, A. Ollero, and R. Cano. *Control of an aerial robot with multi-link arm for assembly tasks. In* 2013 IEEE Int. Conf. on Robotics and Automation, *pages 4916–4921, May 2013.*

S. Kajita and B. Espiau. Legged robots. In B. Siciliano and O. Khatib, editors, Springer Handbook of Robotics, *pages 361–387. Springer, 2008.*

H. K. Khalil. Nonlinear Systems. *Prentice Hall, 3rd edition, 2001. ISBN 978-0130673893.*

S. Kim, S. Choi, and H. J. Kim. Aerial manipulation using a quadrotor with a two dof robotic arm. In 2013 IEEE/RSJ Int. Conf. on Intelligent Robots and Systems, *pages 4990–4995, Tokyo, Japan, November 2013.*

K. Kondak, K. Krieger, A. Albu-Schäffer, M. Schwarzbach, M. Laiacker, I. Maza, A. Rodriguez-Castano, and A. Ollero. *Closed-loop behavior of an autonomous helicopter equipped with a robotic arm for aerial manipulation tasks.* International Journal of Advanced Robotic Systems, *10:1–9, 2013.*

K. Kondak, F. Hubert, M. Schwarzbach, M. Laiacker, D. Sommer, M. Bejar, and A. Ollero. *Aerial manipulation robot composed of an autonomous helicopter and a 7 degrees of freedom industrial manipulator. In* 2014 IEEE Int. Conf. on Robotics and Automation, *pages 2108–2112, Hong Kong, China, May 2014.*

T. J. Koo and S. Sastry. Differential flatness based fill authority helicopter control design. In 1999 IEEE Conf. on Decision and Control, *pages 1982–1987, December 1999.*

C. Korpela, M. Orsag, and P. Oh. Towards valve turning using a dual-arm aerial manipulator. In 2014 IEEE/RSJ Int. Conf. on Intelligent Robots and Systems, *pages 3411–3416, Sep 2014.*

M. Krstic, I. Kanellakopoulos, and P. V. Kokotović. Nonlinear and Adaptive Control Design. *John Wiley, 1995. ISBN 978-0471127321.*

D. J. Lee and C. Ha. *Mechanics and control of quadrotors for tool operation. In 2012 ASME Dynamic Systems and Control Conference, Fort Lauderdale, FL, Oct. 2012.*

D. J. Lee, A. Franchi, H. I. Son, H. H. Bülthoff, and P. Robuffo Giordano. *Semi-autonomous haptic teleoperation control architecture of multiple unmanned aerial vehicles.* IEEE/ASME Trans. on Mechatronics, Focused Section on Aerospace Mechatronics, *18 (4):1334–1345, 2013.*

T. Lee, M. Leoky, and N. H. McClamroch. *Geometric tracking control of a quadrotor UAV on SE(3). In 49th IEEE Conf. on Decision and Control, pages 5420–5425, Atlanta, GA, Dec. 2010.*

Y. Liu, J. M. Montenbruck, P. Stegagno, F. Allgöwer, and A. Zell. *A robust nonlinear controller for nontrivial quadrotor maneuvers: Approach and verification. In 2013 IEEE/RSJ Int. Conf. on Intelligent Robots and Systems, pages 5410–5416, Hamburg, Germany, Sep. 2015.*

L. Ljung. System Identification: Theory for the User. *Prentice Hall, 1986. ISBN 978-0136566953.*

G. Loianno, C. Brunner, G. McGrath, and V. Kumar. *Estimation, control, and planning for aggressive flight with a small quadrotor with a single camera and imu. In 2017 IEEE Int. Conf. on Robotics and Automation, May, submitted.*

S. Lupashin and R. D'Andrea. *Stabilization of a flying vehicle on a taut tether using inertial sensing. In 2013 IEEE/RSJ Int. Conf. on Intelligent Robots and Systems, pages 2432–2438, Tokyo, Japan, Nov 2013.*

S. Lupashin, A. Schöllig, M. Sherback, and R. D'Andrea. *A simple learning strategy for high-speed quadrocopter multi-flips. In 2010 IEEE Int. Conf. on Robotics and Automation, pages 1642–1648, Anchorage, AK, May 2010.*

R. Mahony and V. Kumar. *Aerial robotics and the quadrotor.* IEEE Robotics & Automation Magazine, *19(3):19, 2012.*

R. Mahony, V. Kumar, and P. Corke. *Multirotor Aerial Vehicles: Modeling, Estimation, and Control of Quadrotor.* IEEE Robotics & Automation Magazine, *19(3):20–32, 2012.*

R. Mahony, R. W. Beard, and V. Kumar. Modeling and Control of Aerial Robots, *pages 1307–1334. Springer, 2016. ISBN 978-3-319-32552-1. doi: 10.1007/978-3-319-32552-1_52. URL* http://dx.doi.org/10.1007/978-3-319-32552-1_52.

L. Marconi, R. Naldi, and L. Gentili. *Modeling and control of a flying robot interacting with the environment.* Automatica, *47(12):2571–2583, 2011.*

P. Martin, R. M. Murray, and P. Rouchon. *Flat systems, equivalence and trajectory generation. In 2003 CDS Technical Report, 2003.*

C. D. McKinnon and A. P. Schoellig. *Unscented external force and torque estimation for quadrotors.* In arXiv, Aug 2016. URL `https://arxiv.org/abs/1603.02772`.

D. Mellinger, N. Michael, and V. Kumar. *Trajectory generation and control for precise aggressive maneuvers with quadrotors.* In 12th Int. Symp. on Experimental Robotics, Delhi, India, Dec. 2010.

A. Y. Mersha, R. Carloni, and S. Stramigioli. *Port-based modeling and control of underactuated aerial vehicles.* In 2011 IEEE Int. Conf. on Robotics and Automation, *pages 14–19, Shanghai, China, May 2011.*

V. Mistler, A. Benallegue, and N. K. M'Sirdi. *Exact linearization and noninteracting control of a 4 rotors helicopter via dynamic feedback.* In 10th IEEE Int. Symp. on Robots and Human Interactive Communications, *pages 586–593, Bordeaux, Paris, France, Sep. 2001.*

M. W. Mueller and R. D'Andrea. *Relaxed hover solutions for multicopters: Application to algorithmic redundancy and novel vehicles.* The International Journal of Robotics Research, *35(8):873–889, 2016.*

M. W. Mueller, S. Luphasin, and R. D'Andrea. *Quadrocopter ball juggling.* In 2011 IEEE/RSJ Int. Conf. on Intelligent Robots and Systems, *pages 5113–5120, Sep 2011.*

R. M. Murray, Z. Li, and S. S. Sastry. A mathematical introduction to robotic manipulation. *CRC, 1994. ISBN 0849379814.*

R. M. Murray, M. Rathinam, and W. Sluis. *Differential flatness of mechanical control systems: A catalog of prototype systems.* In ASME Int. Mechanical Eng. Congress and Exposition, *San Francisco, CA, Nov. 1995.*

G. Muscio, F. Pierri, M. A. Trujillo, E. Cataldi, G. Giglio, G. Antonelli, F. Caccavale, A. Viguria, S. Chiaverini, and A. Ollero. *Experiments on coordinated motion of aerial robotic manipulators.* In 2016 IEEE Int. Conf. on Robotics and Automation, *pages 1224–1229, Stockholm, Sweden, May 2016.*

*myCopter Project. EU 7th Framework Programme 266470.* `http://www.mycopter.eu/`, *Jan 2011- Dec 2014.*

R. Naldi. *Prototyping, modeling and control of a class of vtol aerial robots.* Doctorate Thesis, University of Bologna, 2008.

R. Naldi, L. Gentili, L. Marconi, and A. Sala. *Design and experimental validation of a nonlinear control law for a ducted-fan miniature aerial vehicle.* Control Engineering Practice, *18(7):747–760, 2010.*

H. Nguyen and D. Lee. *Hybrid force/motion control and internal dynamics of quadrotors for tool operation.* In 2013 IEEE/RSJ Int. Conf. on Intelligent Robots and Systems, *pages 3458–3464, Tokyo, Japan, Nov. 2013.*

H.-N. Nguyen, S. Park, and D. J. Lee. *Aerial tool operation system using quadrotors as rotating thrust generators.* In 2015 IEEE/RSJ Int. Conf. on Intelligent Robots and Systems, *pages 1285–1291, Hamburg, Germany, Oct. 2015.*

A. Nikoobin and R. Haghighi. *Lyapunov-based nonlinear disturbance observer for serial n-link robot manipulators.* Journal of Intelligent & Robotics Systems, *55(2-3):135–153, 2009.*

K. Ogata. Modern Control Engineering. *Prentice Hall, 5 edition, 2010. ISBN 978-0136156734.*

H. Olsson, K. J. Astrom, M. Gafvert, C. C. D. Wit, and P. Lischinsky. *Friction models and friction compensation.* European Journal of Control, *4(3):176–195, 1998.*

R. Ortega, A. van der Schaft, B. Maschke, and G. Escobar. *Interconnection and damping assignment passivity-based control of port-controlled Hamiltonian systems.* Automatica, *38(4):585–596, 2002.*

M. C. Ozparpucu and S. Haddadin. *Optimal control for maximizing link velocity of visco-elastic joints. In* 2013 IEEE/RSJ Int. Conf. on Intelligent Robots and Systems, *pages 3035–3042, Tokyo, Japan, November 2013.*

G. D. Padfield. HELICOPTER FLIGHT DYNAMICS, 2nd Edition. *Blackwell Publishing, 2007. ISBN 978-1405118170.*

C. Papachristos, K. Alexis, and A. Tzes. *Dual–authority thrust–vectoring of a tri–tiltrotor employing model predictive control.* Journal of Intelligent & Robotics Systems, *81(3): 471–504, 2015.*

M. Quigley, K. Conley, B. Gerkey, J. Faust, T. Foote, J. Leibs, R. Wheeler, and A. Y. Ng. *ROS: an open-source Robot Operating System. In* Workshop on Open Source Software in Robotics at the 2009 IEEE Int. Conf. on Robotics and Automation, *Kobe, Japan, May 2009.*

S. Rajappa, M. Ryll, H. H. Bülthoff, and A. Franchi. *Modeling, control and design optimization for a fully-actuated hexarotor aerial vehicle with tilted propellers. In* 2015 IEEE Int. Conf. on Robotics and Automation, *pages 4006–4013, Seattle, WA, May 2015.*

R. K. Remple and M. B. Tischler. Aircraft and Rotorcraft System Identification, 2nd Edition. *AIAA Education, 2012. ISBN 978-1600868207.*

B. Ren, S. Ge, C. Chen, C.-H. Fua, and T. Lee. Modeling, Control and Coordination of Helicopter Systems. *Springer, 2012. ISBN 978-1461415633.*

F. Ruggiero, J. Cacace, H. Sadeghian, and V. Lippiello. *Impedance control of VToL UAVs with a momentum-based external generalized forces estimator. In* 2014 IEEE Int. Conf. on Robotics and Automation, *pages 2093–2099, Hong Kong, China, Apr. 2014.*

F. Ruggiero, M. A. Trujillo, R. Cano, H. Ascorbe, A. Viguria, C. Perez, V. Lippiello, A. Ollero, and B. Siciliano. *A multilayer control for multirotor uavs equipped with a servo robot arm. In* 2016 IEEE Int. Conf. on Robotics and Automation, *pages 4014–4020, May 2016.*

M. Ryll, H. H. Bülthoff, and P. Robuffo Giordano. *A novel overactuated quadrotor unmanned aerial vehicle: modeling, control, and experimental validation.* IEEE Trans. on Control Systems Technology, *23(2):540–556, 2015.*

M. Ryll, D. Bicego, and A. Franchi. *Modeling and control of FAST-Hex: a fully-actuated by synchronized-tilting hexarotor*. In 2016 IEEE/RSJ Int. Conf. on Intelligent Robots and Systems, *pages 1689–1694, Daejeon, South Korea, Oct. 2016.*

F. Schiano, J. Alonso-Mora, K. Rudin, P. Beardsley, and R. B. Siciliano. *Towards estimation and correction of wind effects on a quadrotor uav*. In IMAV 2014: International Micro Air Vehicle Conference and Competition, *Delft, Netherlands, August 2014.*

G. S. Schmidt, C. Ebenbauer, and F. Allgöwer. *Output regulation for control systems on se(n): A separation principle based approach*. IEEE Trans. on Automatic Control, *59 (11):3057–3062, 2014.*

C. Secchi, S. Stramigioli, and C. Fantuzzi. Control of Interactive Robotic Interfaces: a port-Hamiltonian Approach. *Tracts in Advanced Robotics. Springer, 2007. ISBN 978-3540497127.*

R. Sepulchre, M. Jankovic, and P. Kokotovic. Constructive Nonlinear Control. *Communications and Control Engineering Series. Springer, 1997. ISBN 3540761276.*

SHERPA. *EU 7th Framework Programme 600958.* `http://www.sherpa-project.eu/`, *2013- 2017.*

B. Siciliano and O. Khatib. Handbook of Robotics. *Springer, 2008. ISBN 9783540382195.*

B. Siciliano, L. Sciavicco, L. Villani, and G. Oriolo. Robotics: Modelling, Planning and Control. *Springer, 2009. ISBN 978-1-84628-641-4.*

M. W. Spong. *Underactuated mechanical systems*. In B. Siciliano and K. P. Valavanis, editors, Control Problems in Robotics and Automation, *pages 135–150. Springer, 1998.*

K. Sreenath and V. Kumar. *Dynamics, control and planning for cooperative manipulation of payloads suspended by cables from multiple quadrotor robots*. In Robotics: Science and Systems, *Berlin, Germany, June 2013.*

J. Thomas, J. Polin, K. Sreenath, and V. Kumar. *Avian-Inspired Grasping for Quadrotor Micro UAVs*. In 2013 ASME Int. Design Engineering Technical Conf. and Computers and Information in Engineering Conf., *Portland, OR, Aug. 2013.*

M. Tognon, B. Yüksel, G. Buondonno, and A. Franchi. *Dynamic decentralized control for protocentric aerial manipulators*. In 2017 IEEE Int. Conf. on Robotics and Automation, *Singapore, May 2017.*

T. Tomic and S. Haddadin. *A unified framework for external wrench estimation, interaction control and collision reflexes for flying robots*. In 2014 IEEE/RSJ Int. Conf. on Intelligent Robots and Systems, *pages 4197–4204, Sep 2014.*

Z. Wang, P. Goldsmith, and J. Gu. *Regulation control of underactuated mechanical systems based on a new matching equation of port-controlled Hamiltonian systems*. In 2009 IEEE Int. Conf. on Robotics and Automation, *pages 992–997, Kobe, Japan, May 2009.*

J. C. Willems. *Dissipative dynamical systems part I: General theory*. Archive for Rational Mechanics and Analysis, *45(5):321–351, 1972.*

H. Yang and D. J. Lee. *Dynamics and control of quadrotor with robotic manipulator*. In 2014 IEEE Int. Conf. on Robotics and Automation, *Hong Kong, China, May. 2014*.

Y. Yu and X. Ding. *A quadrotor test bench for six degree of freedom flight*. Journal of Intelligent & Robotics Systems, *68(3):323–338, 2012*.

B. Yüksel and A. Franchi. *Pvtol aerial manipulators with a rigid or an elastic joint: Analysis, control, and comparison*. submitted to, *(hal-01388462), Oct 2016*. URL `https://hal.archives-ouvertes.fr/hal-hal-01388462`.

B. Yüksel, C. Secchi, H. H. Bülthoff, and A. Franchi. *Reshaping the physical properties of a quadrotor through IDA-PBC and its application to aerial physical interaction*. In 2014 IEEE Int. Conf. on Robotics and Automation, *pages 6258–6265, Hong Kong, China, May. 2014a*.

B. Yüksel, C. Secchi, H. H. Bülthoff, and A. Franchi. *A nonlinear force observer for quadrotors and application to physical interactive tasks*. In 2014 IEEE/ASME Int. Conf. on Advanced Intelligent Mechatronics, *pages 433–440, Besançon, France, Jul. 2014b*.

B. Yüksel, S. Mahboubi, C. Secchi, H. H. Bülthoff, and A. Franchi. *Design, identification and experimental testing of a light-weight flexible-joint arm for aerial physical interaction*. In 2015 IEEE Int. Conf. on Robotics and Automation, *pages 870–876, Seattle, WA, May 2015*.

B. Yüksel, G. Buondonno, and A. Franchi. *Differential flatness and control of protocentric aerial manipulators with any number of arms and mixed rigid-/elastic-joints*. In 2016 IEEE/RSJ Int. Conf. on Intelligent Robots and Systems, *pages 561–566, Daejeon, South Korea, Oct. 2016a*.

B. Yüksel, N. Staub, and A. Franchi. *Aerial robots with rigid/elastic-joint arms: Single-joint controllability study and preliminary experiments*. In 2016 IEEE/RSJ Int. Conf. on Intelligent Robots and Systems, *pages 1667–1672, Daejeon, South Korea, Oct. 2016b*.

B. Yüksel, C. Secchi, H. H. Bülthoff, and A. Franchi. *Aerial physical interaction via ida-pbc*. submitted, *2017*.

G. Zames. *On the input-output stability of time-varying nonlinear feedback systems part i: Conditions derived using concepts of loop gain, coicity, and positivity*. IEEE Trans. on Automatic Control, *11(2):228–238, 1966*.

X. Zhan. *Extremal eigenvalues of real symmetric matrices with entries in an interval*. SIAM Journal on Matrix Analysis and Applications, *27(3):851–860, 2006*.

C. B. Zilles and J. K. Salisbury. *A constraint-based god-object method for haptic display*. 1995 IEEE/RSJ Int. Conf. on Intelligent Robots and Systems, *3(Pittsburgh, PA):146–151, 1995*.